22nd International Conference on Building Materials, Products and Technologies (ICBMPT 2019)

AA006713

IOP Conference Series: Materials Science and Engineering
Volume 583

Lednice, Czech Republic
4 – 6 June 2019

ISBN: 978-1-5108-9448-8
ISSN: 1757-8981

Printed from e-media with permission by:

Curran Associates, Inc.
57 Morehouse Lane
Red Hook, NY 12571

Some format issues inherent in the e-media version may also appear in this print version.

Institute of Physics
Dirac House, Temple Back
Bristol BS1 6BE UK

Phone: 44 1 17 929 7481
Fax: 44 1 17 920 0979

techtracking@iop.org

Additional copies of this publication are available from:

Curran Associates, Inc.
57 Morehouse Lane
Red Hook, NY 12571 USA
Phone: 845-758-0400
Fax: 845-758-2633
Email: curran@proceedings.com
Web: www.proceedings.com

Table of contents

Volume 583

International Conference Building Materials, Products and Technologies

4–6 June 2019, Lednice, Czech Republic

Accepted papers received: 3 July 2019
Published online: 14 August 2019

Preface

Papers

Inorganic binders

IOP Conf. Series: Materials Science and Engineering **583** (2019) 011001 doi:10.1088/1757-899X/583/1/011001

Preface

The 22nd International Conference ICBMPT 2019 – International Conference on Building Materials, Products and Technologies organized by the Research Institute for Building Materials was held in Lednice, Czech Republic, on 4—6 June 2019. The main topics of ICBMPT 2019 were: Inorganic Binders, Secondary raw materials, Composite materials and Technology and test methods. The aim of the conference was to bring researchers from universities and industrial enterprises together to share their experience and to create an international forum for the exchange of both scientific and practical knowledge in the field of building materials.

36 papers were accepted after strict peer-reviewing by expert reviewers, we would like to thank all the participants for their valuable contribution to the conference and these proceedings. We would like to express our gratitude to the reviewers for their time, valuable recommendations and constructive evaluation, in order to guarantee high quality of all the papers. The organizing committee is very proud that more than 90 participants from several European countries were attracted. Due to this fact, the spectrum of the contributions reflected many aspects of the state of the art in this fast-developing field of applications. We hope that the conference made a significant contribution to the knowledge in this scientific field.

IOP Conf. Series: Materials Science and Engineering **583** (2019) 011002 doi:10.1088/1757-899X/583/1/011002

Peer review statement

All papers published in this volume of *IOP Conference Series: Materials Science and Engineering* have been peer reviewed through processes administered by the proceedings Editors. Reviews were conducted by expert referees to the professional and scientific standards expected of a proceedings journal published by IOP Publishing.

Possibilities of ammonium ions fixation into mixed binders

J Hajzler[1], R Novotny[1], T Opravil[1] and P Ptacek[1]

[1] Brno University of Technology, Faculty of Chemistry, Purkyňova 464/118, Brno Cz-612 00, Czech Republic

E-mail: xchajzlerj@fch.vutbr.cz

Abstract. The purpose of this study is to analyze the usability of ammonium ions fixation in mixed binders with the addition of high temperature fly ash. The experiment used high-temperature fly ash from operations in which the secondary flue gas denitrification methods have been implemented, specifically by the method of the selective non-catalytic reduction, known as DeNOx. The fixation of ammonium ions was performed by the addition of sodium hexanitrocobaltate(III). The study examines the influence of addition of cobalt(III) complex compound on the release of ammonia into the environment during the mixing the mixed binder. The released ammonia was quantitatively captured into the acid and subsequently titrated. The effect of cobalt(III) complex compound on the rate of hydration and mechanical properties of the mixed binders was also studied. The influence on the environment in terms of leachability was observed.

1. Introduction

Combustion of coal or other materials in thermal power plants generates, in addition to solids parts, gaseous substances, including carbon oxides, sulphur and nitrogen oxides. Nitrogen oxides (referred to as NO_x) are formed by the oxidation of nitrogen in fuels with air oxygen and are counted among the most dangerous pollutants. NO_x contained in flue gases consists of NO (90%) and NO_2 (5%). They both are formed during combustion. NO (90–95%), which is initially formed colourless, odourless and then is transformed into a reddish-brown, acrid smelling NO_2. The residual nitrogen oxides are represented to a small extent (5%) and are often generated by the interaction of NO with NO_2 [1, 2, 3].

Emissions are air pollutants and therefore emission limits are set (ELV–emission limit values). There are ELVs to protect human health against the negative effects of NO_x. The following table 1 states the NO_x emission limits in mg/Nm^3 for each category of combustion equipment accepted by the European Commission in 2017. These limits will come into force in the Member States at the turn of 2021/2022 [4].

Table 1. BAT-associated emission levels (BAT-AELs) for NOx emission to air from combustion of coal and/or lignite [4].

Combustion plant total rated thermal input (MW)	BAT-AEL (mg/Nm^3) Yearly average	
	New plant	Existing plant
< 100	100–150	100–270
100–300	50–100	100–180
≥ 300	50–85	85–150

To achieve these emission values, it is necessary to use the best available techniques (BAT). These techniques may be divided into two large groups. Techniques based on suppression of forming NO_x are called primary measures and secondary techniques, which consist of elimination of the formed NO_x. Secondary techniques contain selective reduction, namely selective catalytic reduction (SCR) and selective non-catalytic reduction (SNCR) [5].

The SNCR method passes without the presence of a catalyst at higher temperatures of 950–1100°C depending on the reducing agent used. The removal efficiency of the method is usually around 50%. The SCR method takes place in the presence of the catalyst at temperatures of 170–510°C depending on the catalyst used. SCR efficiency is up to 90%. Ammonia water, urea and cyanuric acid are used as reducing agents for selective measures [6, 7].

Both methods (SCR and SNCR) are effective and chemically well-researched. Due to the secondary measures, in the flue gay is NOx reduces to 30 mg/m^3, it fulfils the strictest ELV [2, 8].

High-temperature fly ash is used as an admixture for concrete according to ČSN EN 206-1. Addition of the fly ash brings a number of benefits. The ash can optimize the grain curve, improves processability, increases the resistance of concrete in a chemically aggressive environment, reduces reversible contraction of concrete, etc. Secondary raw material partly substitutes cement, it reduces primary raw material consumption and reduces environmental impacts from the point of view of the production of greenhouse gases in cement production [9].

Another way to reduce the environmental impact of Portland cement production is the use of cement kiln by-pass dust (CKD). By-pass cement kiln dust is a fine-grained alkaline particulate material similar in appearance to Portland cement collected in the air-pollution control devices during the production of cement clinker. This material contains high concentrations of heavy metals and therefore should be disposed as a hazardous waste. Its chemical composition is not suitable for returning back into feedstock and, therefore, it has to be discharged. Nevertheless, the economic pressure caused by CKD disposal tends to its re-addition to the final cement product. Addition of CKD is about 3 wt. % into cement [10, 11, 12].

There are differences in chemical and mineralogical composition between the classic high-temperature fly ash and fluidized fly ash. Fluidized fly ash contains up to 20 wt. % SO_3, up to 15 wt. % CaO and an annealing loss up to 15 wt. %. For these reasons this fly ash cannot be evaluated according to ČSN EN 450, because it does not fulfil most of the prescribed technical criteria and cannot be used for the production of concrete according to ČSN EN 206-1 [9, 13].

After introduction of the SNCR process, high-temperature fly ash contains ammonium ions most often in the form of ammonium sulphate and ammonium hydrogen sulphate. The ammonia content is about 40 ppm (40 mg in 1 kg fly ash), while reducing NO_x below 200 m/Nm^3. The pH increases by mixing the mixture of fly ash and cement with water. Ammonia is released very easily and quickly in an alkaline environment. High concentration of ammonia in the air has a negative impact on human health and industrial equipment. The release of ammonia is the reason why this fly ash is not used for the production of concrete [14, 15].

There are developing methods of ammonium ions elimination in mixtures containing by-products. These by-products are from the process of NO_x conversion to nitrogen and water. One of these methods is elimination using tannin, which builds an ammonium ion into its structure. An insoluble ammonium salt is formed and ammonia is no longer released into the environment. Another method is found on the reaction of ammonium ions with sodium hexanitrocobaltate, when a conversion happens and an insoluble stable compound is formed [5, 7, 16].

In this article was researched the possibility of fixing ammonium ions in mixed binders where high-temperature fly ash was used after the $DeNO_x$ process. Elimination was performed using sodium hexanitrocobaltate added as an aqueous solution.

2. Experimental

2.1. Methods

The manual spectrometric method ČSN ISO 7150-1 was used to determine the ammonium ions in the tested fly ash [17]. The possibility of elimination was performed on the principle of capturing of released ammonia into hydrochloric acid. An airtight container was filled by the fly ash and cement. Subsequently, it was rinsed with nitrogen and the outlet from the container was led into hydrochloric acid. The water with dissolved complex was added into the loose mixture. The experiment lasted for one hour. The acid was subsequently titrated. Mechanical properties, compressive strength and flexural strength were measured on the complex strength test of building materials equipment DESTTEST 3310 (Betonsystem). Flexural strength and compressive strength were measured on each test specimen. The size of the individual prisms was 40 × 40 × 160 mm. The testing specimens were kept in a humid environment. Mechanical tests were performed after 1, 2, 4, 7, and 28 days. The effect of cobalt complex compound on the hydration process was studied on an isothermal calorimeter (TAM Air). Aqueous extracts were prepared from mixtures after mechanical tests after 1, 2, 4, 7 and 28 days. Ten times larger volume of the distilled water was added to the weighted amount. Samples were placed in a head-heel shaker for 24 hours. Thereafter, measurements were made and cobalt values were determined. The method was performed by means of inductively coupled plasma emission spectroscopy (ICP – OES). In this work the ICP–OES method was performed using the ULITMA 2 (HORIBA Scientific).

2.2. Materials

In the experiment, high-temperature fly ash was used after the SNCR process from the Počerady coal-fired power plant. The main crystalline phases are formed by mullite and quartz. Furthermore, hematite, magnetite and anatase are represented. The residue is an amorphous. The phase composition of CKD was determined with X-ray diffraction (XRD). CKD mainly consisted of KCl, K_2SO_4, CaO, K_2O and C_2S. Cement 42.5 R (plant Mokrá, CZ) was also used. Its specific surface area was 2.438 m^2/g measured at high-speed surface area analyzer NOVA 2200e (Quantachrome INSTRUMENTS), and the chemical composition is given in table 2. For the production of testing specimen was used coarse, medium and fine. Sodium hexanitrocobaltate has been prepared at the Faculty of Chemistry [12, 18].

Table 2. Chemical composition of cement (% by mass).

Component	SiO_2	Al_2O_3	Fe_2O_3	CaO	MgO	SO_3	Na_2O	K_2O
Cement	19.1	4.2	3.5	63.4	1.0	2.6	0.1	0.9

2.3. Sample composition

Five mixtures were prepared. The composition of the individual mixtures is shown in table 3. Individual mixtures differ only in the addition of sodium hexanitrocobaltate to water. The water to binder (w/b) ratio was selected to be 0.5. The water coefficient is the ratio of water to binder. Because the cement is replaced by fly ash, the water coefficient is calculated to the sum of the added fly ash and cement. The ratio of coarse, medium and fine sand was 1:1:1. The ratio of sand to binder was 3:1.

Table 3. The composition of mixtures for mechanical properties.

Sample	Fly ash [g]	Cement [g]	Sand [g] F/M/C	Water [g]	Sodium hexanitrocobaltate(III) [mg]
A	225	225	450/450/450	225	0
B	225	225	450/450/450	225	37.5
C	225	225	450/450/450	225	75
D	225	225	450/450/450	225	150
E	225	225	450/450/450	225	500

The composition of the individual mixtures in which was researched the influence of the addition of sodium hexanitrocobaltate on releasing of gaseous ammonia is shown in table 4. The mixture was formed from fly ash and cement in a ratio of 1:1. The cobalt complex was dissolved in water and subsequently added to the mixture. The water to binder ratio was 0.21. The low coefficient was selected to humidification the mixture and subsequently to initiate hydration and to produce alkaline pH. Gaseous ammonia was then released. If the water coefficient is higher, the gaseous ammonia could be captured into the surplus of water in the mix

Table 4. The composition of mixtures for titration.

Sample	Fly ash [g]	Cement [g]	Water [g]	Sodium hexanitrocobaltate(III) [mg]
F	600	600	252	0
G	600	600	252	100
H	600	600	252	200
I	600	600	252	400
J	600	600	252	1 333

3. Result and discussion

By manual spectrometric method, it was determined that high temperature fly ash after the SNCR process from the Počerady coal-fired power plant contained 36.11 ±0.42 ppm of ammonia.

The ammonia released by mixing the particular mixtures, was quantitatively captured into the acid and subsequently titrated. From the results of consumptions of individual titrations does not ensue the dependence. Differences in individual consumptions are within the error rate of the method.

The time dependence on flexural and compressive strengths on the formed mixtures measured on testing beams after 1, 2, 4, 7 and 28 days is shown in figure 1 and figure 2.

Mixture B with a cobalt complex content of 37.5 mg has the highest flexural strength values. The maximum value was 4.67 MPa. With the increasing content of the added complex, the strengths are not significantly influenced. The most significant decrease was observed in the 28-day strength of the mixture D. Overall, it is possible to state that the differences among the measurements of the individual mixtures are due to statistical deviation.

A similar trend can be seen in figure 2, The highest strength values were observed again in mixture B, where the highest value reached 23.86 MPa. The mechanical resistance regularly increases with increasing time in all mixtures. There was no significant decrease in any value compared to the values without content of sodium hexanitrocobaltate.

In general, it is possible to state that the flexural and compressive strengths increase with increasing time both in the mixture without the addition of sodium hexanitrocobaltate and with its gradual addition.

Therefore, it was concluded that the addition of sodium hexanitrocobaltate in the selected concentrations to the mixing water does not influence the resultant mechanical properties of the tested specimen.

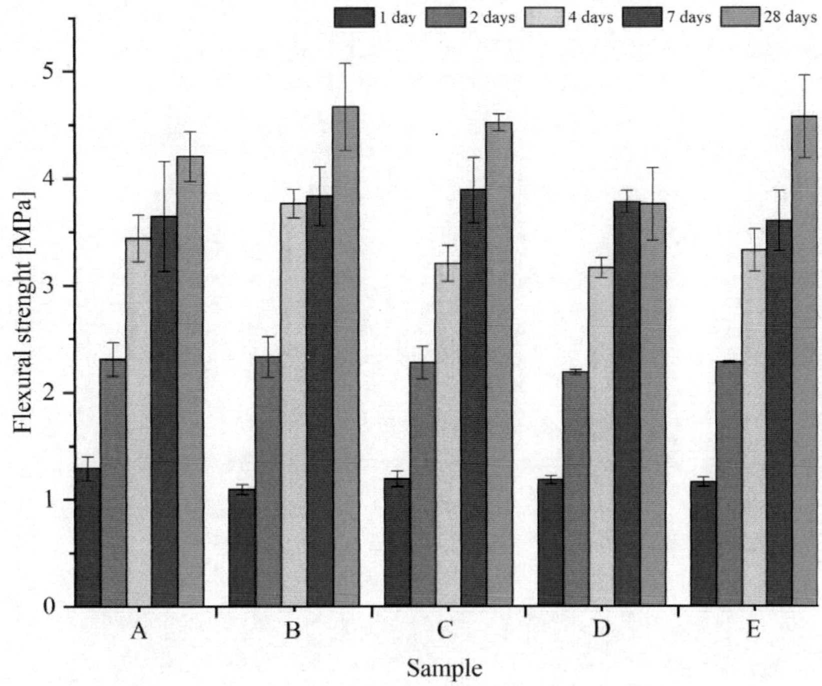

Figure 1. Flexural strength.

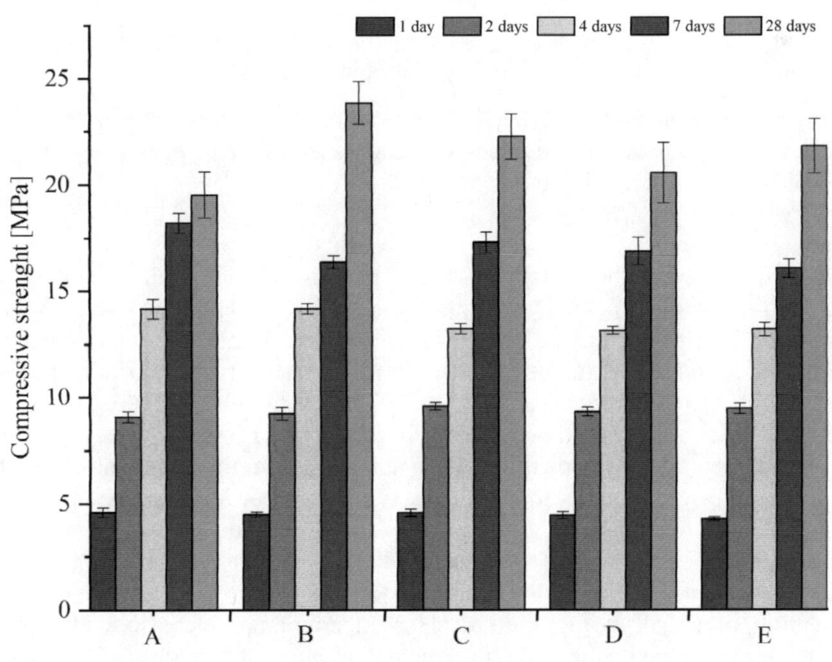

Figure 2. Compressive strength.

The calorimetry curve, shown in figure 3, shows the hydration process of the prepared mixtures. The standard hydration process of cement can be observed. Samples were composed of fly ash and cement in a ratio of 1:1. The hydration process is the same as hydration process of reference without the addition of other compounds. High-temperature fly ash has only a dilution effect here because the heat flow acquires half the value [19]. A powder complex was subsequently added to the dry mixture. From the calorimetry curves is evident that the hydration rate is not influenced by the addition of cobalt complex.

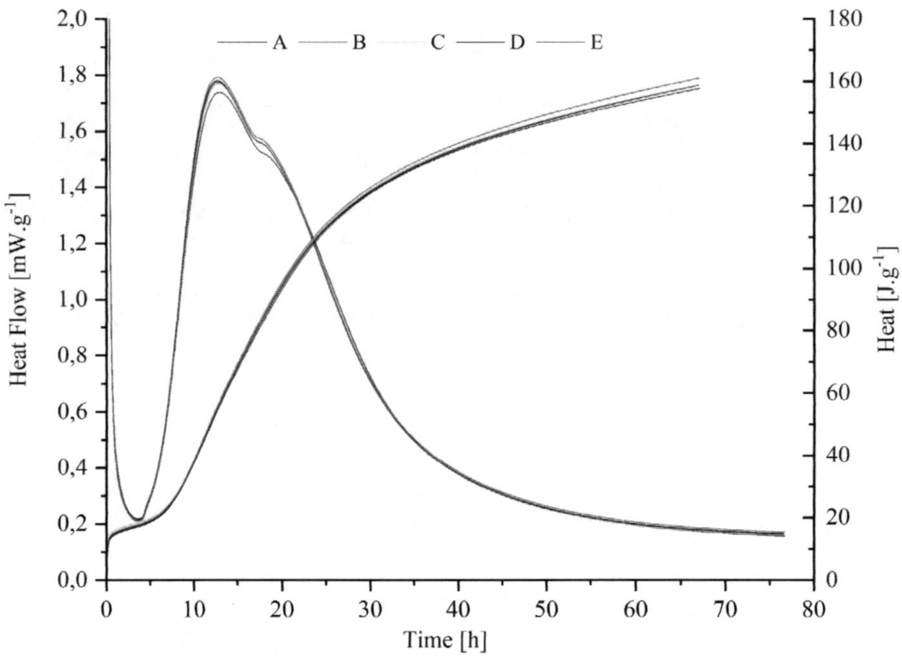

Figure 3. Calorimetry curve.

Aqueous extracts were analysed by the ICP–OES method. The devices detection limit is 35 $mg \cdot l^{-1}$. All analysed samples are below the devices detection limit. Cobalt is not leached into water from the cobalt complex added to the mixing water.

4. Conclusion

This work was focused on the possibilities of fixation of ammonium ions in mixed binders. An ammonium ion elimination method was tested by using sodium hexanitrocobaltate, which was added to the mixing water. Testing specimens were formed from mixtures to testing of mechanical properties and the samples were subjected to measurements on the isothermal calorimeter and their aqueous extracts were subjected to measurements on ICP–OES to determining the cobalt.

From the measured results of the mechanical properties of the formed specimens ensue that the addition of sodium hexanitrocobaltate does not influence the resultant strengths. Differences in flexural and compressive strengths among the individual mixtures are within the statistical deviation. Calorimetry curves also confirmed that the addition of cobalt compound does not influence the hydration process. Added cobalt compound is not released from the testing specimens, which is proved by leachability measurements. Mixing the mixtures and the capturing of ammonia into the acid didn't confirm the assumption that the amount of released ammonia decrease with increasing addition of the complex. This is probably due to the insufficient addition of the complex. The mixture contained not only ammonium ions from high-temperature fly ash, but also potassium ions from the addition of bypass dust in cement.

Acknowledgement
This work was supported by specific research FCH-S-18-5194 and with the financial support by the project: GA19-16646S "The elimination of the negative impact of zinc in Portland cement by accelerating concrete admixtures", with financial support from the Czech Science Foundation.

References

[1] Gómez-García M A, Pitchon V and Kiennemann A 2005 Pollution by nitrogen oxides: an approach to NOx abatement by using sorbing catalytic materials *Environment International. Elsevier Ltd.* **31**(3) pp 445–467
[2] Gohlke O 2010 A new process for NO_x reduction in combustion systems for the generation of energy from waste *Waste Management* **30**(7) pp 1348–1354
[3] Shi-long F, Song Q and Yao Q 2015 Study on the catalysis of CaCO 3 in the SNCR deNOx process for cement kilns *Chemical Engineering Journal* **262** pp 9–17
[4] Prováděcí rozhodnutí komise (EU) 2017/1442: kterým se stanoví závěry o nejlepších dostupných technikách (BAT) podle směrnice Evropského parlamentu a Rady 2010/75/EU pro velká spalovací zařízení 2017 (Brusel) L212
[5] Hajzler J 2016 *Možnosti eliminace čpavkového skluzu v technologických vodách elektráren* (Brno)
[6] Van Caneghem J, De Greef J, Block Ch and Vandecasteele C 2016 NOx reduction in waste incinerators by selective catalytic reduction (SCR) instead of selective non catalytic reduction (SNCR) compared from a life cycle perspective: a case study *Journal of Cleaner Production Elsevier Ltd* **112**(5) pp 4452–4460
[7] Hajzler J, Opravil T, Pořízka J and Ptáček P 2018 Possibilities of elimination of ammonia slip from technological water in power plants *IOP Conference Series: Materials Science and Engineering. IOP Publishing* **379** pp 1–7
[8] De Greef J, Villani K, Goethals J, Van Belle H, Van Caneghem J and Vandecasteele C 2013 Optimising energy recovery and use of chemicals, resources and materials in modern waste-to-energy plants *Waste Management. Elsevier Ltd* **33**(11) pp 2416–2424
[9] Popílek a jeho použití do betonu: Vyhovující nové betonářské normě ČSN EN 206-1 (Svaz výrobců cementu ČR) p 4
[10] Kalina L, Bílek V, Kiripolský T, Novotný R and Másilko J 2018 Cement Kiln By-Pass Dust: An Effective Alkaline Activator for Pozzolanic Materials *Materials* **11**(9)
[11] Eckert J O and Guo Q 1998 Heavy metals in cement and cement kiln dust from kilns co-fired with hazardous waste-derived fuel: application of EPA leaching and acid-digestion procedures *Journal of Hazardous Materials* **59**(1) pp 55–93
[12] Abdel-Ghani N T, El-Sayed H A and El-Habak A A 2018 Utilization of by-pass cement kiln dust and air-cooled blast-furnace steel slag in the production of some "green" cement products *HBRC Journal* **14**(3) pp 408–414
[13] ČSN EN 450-1:2013 *Popílek do betonu: Část 1: Definice, specifikace a kritéria shody* (Praha)
[14] Beranová D, Opravil T, Ptáček P and Snop R 2018 Release of ammonia from conventional power plant fly ash after the introduction of SNCR process *IOP Conference Series: Materials Science and Engineering* **379** p 6
[15] Elektrárna Počerady 2019 (Praha: Skupina ČEZ: Výroba elektřiny)
[16] A method of stabilizing residual ammonia in a mixture containing energy by-products using tannin 2017
[17] ČSN ISO 7150-1:1994 *Jakost vod stanovení amonných iontů: Manuální spektrometrická metoda* (Praha)
[18] Bílek V, Pařízek L and Kalina L 2015 Effect of the by-pass cement-kiln dust and fluidized-bed-combustion fly ash on the properties of fine-grained slag-based composited *Materiali In Tehnologije* **49**(4) pp 549–552
[19] Šiler P, Kolářová I, Novotný R, Másilko J, Pořízka J, Bednárek J and Švec J 2018 Application

of isothermal and isoperibolic calorimetry to assess the effect of zinc on cement hydration *Journal of Thermal Analysis and Calorimetry. Cham: Springer International Publishing* **133**(1) pp 27–40

IOP Conf. Series: Materials Science and Engineering **583** (2019) 012002 doi:10.1088/1757-899X/583/1/012002

Effect of geopolymer composition on the mechanical and rheological properties

L Halasta[1], P Rovnanik[1] and O Mikhailova[1]

[1] Faculty of Civil Engineering, Brno University of Technology, Veveří 95, Brno 602 00, Czech Republic

E-mail: lukas.halasta@seznam.cz

Abstract. Coating systems, not only in the building industry, have historically their irreplaceable role. Modern times, however, take on increasing responsibility for both the environment, the versatility of product use, and the reduction of production costs. The development of the new multifunctional geopolymer coating, therefore, clearly covers all of these global aspects. In this paper, the mechanical and rheological properties of various geopolymer pastes were investigated. The geopolymer mixtures were composed of metakaolin and various dosage of water glass solution with three different SiO_2/Na_2O ratios. The best mechanical properties were achieved for the mixture with content 90% of water glass and SiO_2/Na_2O ratio of 1.4. The rheological measurements showed high thixotropy of all mixtures and very low yield stress values.

1. Introduction

Protective coatings play an irreplaceable role in protection of a very wide range of construction and building materials. They provide the substrates with the required properties and protect them in particular against their degradation due to weather and operating influences. The role of them is growing as the range of protected substrates (especially steel, zinc and aluminium alloys, wood, concrete, masonry and mineral materials) is constantly expanding. A new option for the protect of these substrates can be materials with higher utility properties than conventional protective coatings, namely geopolymers [1]. In addition to environmental, durability and applicability advantages over conventional coating systems, geopolymers also exhibit high temperature resistance of up to 800 degrees Celsius [2] and are thus destined to be used on structures to be protected against high temperatures.

Geopolymers are materials composed predominantly of amorphous substances with minimal occurrence of crystalline phase. At higher temperatures, the amorphous structure is microcrystalline [3]. For the production of geopolymers, aluminosilicate materials such as metakaolin, blast furnace slag or fly ash are commonly used and these are activated with a suitable alkaline activator. The most commonly used reagent to start the geopolymerization reaction is water glass. Water glass is a colloidal solution of alkaline silicates produced by dissolving alkaline silica glass (mostly sodium) in water. Soluble silicates are essentially a combination of alkali metal oxide of Na_2O, SiO_2 and water. The base formula is $x SiO_2 \cdot Na_2O \cdot z H_2O$, where x is the degree of polymerization and z is the number of water molecules. Sodium silicate is defined by the molar ratio of SiO_2/Na_2O, also referred to as the silicate modulus (M_s) [4].

The pH of the alkali silicate solution is dependent on the SiO_2/Na_2O molar ratio, but also on the concentration of alkalis. The pH of the solution is between 10.9 and 13.5; value is a very important characteristic determining the stability of high modulus silicate solutions, i.e. their tendency to gelation or coagulation. The stability of alkaline silicate solutions increases with increasing pH [4].

The optimum silicate modulus is variable depending on the raw material and aging conditions. In the case of insufficient alkaline activation of the aluminosilicate by the activator, a lower modulus is preferred. Otherwise, a higher silicate modulus is preferred [5], especially because of easier and more friendly handling of the fresh geopolymer mixture.

Sodium water glass was chosen as an alkaline activator. The most commonly used and most available alkaline activators are sodium activators mentioned above ($Na_2O{\cdot}nSiO_2$ silicates, NaOH hydroxides). Some potassium activators, such as potassium hydroxide KOH or $K_2O{\cdot}nSiO_2$ potassium hydroxide (the properties of sodium and potassium compounds are very similar) were used in laboratory studies, however their use is greatly limited by their availability and cost. For the purposes of this work, sodium water glass with different silicate modulus 1.2; 1.4; 1.6 was used.

The metakaolin (Mefisto L05) was used as a basic component. Metakaolin is aluminosilicate material produced by the burning of kaolin, kaolin clay in the temperature range 600–900°C. The basic component of kaolin is the kaolinite mineral ($Al_2Si_2O_5(OH)_4$) resulting from the weathering of feldspars. Metakaolin can be used as an active admixture for concrete to improve physical and mechanical properties. It increases strength, improves water absorption and reduces efflorescence [6].
table 1 shows the chemical and physical properties of the metakaolin.

Table 1. Chemical and physical properties of Mefisto L05 [7].

Component	Typical value (%)	Guaranteed value
Al_2O_3 content	40.1	min. 38.0
SiO_2 content	54.1	max. 57.0
K_2O content	0.8	max. 0.9
Fe_2O_3 content	1.1	max. 1.80
TiO_2 content	1.8	max. 2.00
MgO content	0.18	max. 0.4
CaO content	0.13	max. 0.2
Loss of ignition (%)	2.2	0.3 - 0.4
Humidity (%)	0.5	-
Specific surface (m^2/g)	12.69	-
Pozzolan activity (°C)	4.3	min. 4.0
Degree of whiteness (%)	60	-

The most important properties of the coating compositions are their viscosity, which is closely related to the flow properties. Rheological properties affect how the mixture will behave, how it will be transferred to the substrate and spread over its surface. We are interested not only in technology, but also in energy. To measure the rheological properties, the rotary rheometer with concentric cylinders' geometry was used. This geometry is optimal for the measurement of building binders because it provides almost no free surface of the measured material and hence, it prevents the mixing water from evaporation which would otherwise influence the measurement. The basic rheological parameters, plastic viscosity and yield stress, were determined from the decreasing branches of the measured flow curves using a Bingham model. The thixotropic parameter was determined as a surface area between increasing and decreasing branches of the flow curve. The flowing test were performed again only on the mixtures with 90% content of alkaline activator and different silicate modulus.

The fundamentals of the research are focussed on the formulation of a universal coating system that will be applicable to all building materials, non-flammable, UV stable, breathable for water vapour, resistant to weathering and frost, resistant to the effects of dilute acids and alkalis, and is insoluble in water.

The benefit of this work will be a new, durable painting universal system, its advantages will be simple applicability, independence from petroleum products in its production, and ecological production compared to conventional coating systems.

2. Material and sample preparation

Geopolymer mixtures were prepared from metakaolin and sodium water glass was used as an alkaline activator. The metakaolin was supplied by ČLUŽ company Czech Republic. The activating solution was prepared by mixing 1000 g of commercial sodium water glass with an SiO_2/Na_2O ratio of 1.6 and 30 g of NaOH p.a, in order to adjust the SiO_2/Na_2O value to 1.4, and mixing 1000 g of commercial sodium water glass with an SiO_2/Na_2O ratio of 1.6 and 70 g of NaOH, in order to adjust the SiO_2/Na_2O value to 1.2. The chemical composition of water glasses is given in table 2.

Table 2. Chemical composition of water glass (wt. %).

Component	Water glass 1.6	Water glass 1.4	Water glass 1.2
SiO_2	26.43	25.66	24.70
Na_2O	16.61	18.38	20.59
H_2O	56.96	55.96	54.71

The L05 material granulometry can be represented as a percentage of the cumulative representation where the particle size of the d_{50} is about 3 μm and with the d_{90} is around 10 μm. The particle size distribution of aluminosilicate precursor (metakaolin) was determined by laser granulometry measurement and the distribution curves are presented in figure. 1. The d_{50} and d_{90} value indicates that 50% and 90% all grains are smaller than a given value.

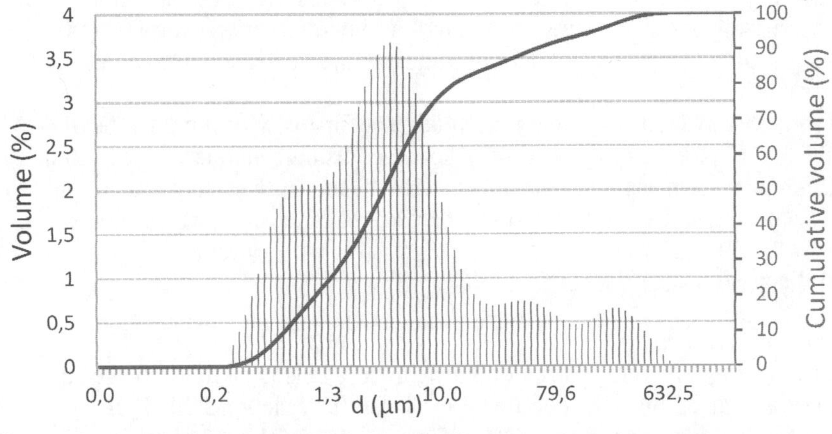

Figure 1. Particle size distributions of metakaolin L05 determined by laser granulometry measurement.

Rheological properties were determined for the pastes, which were composed of 40 g of metakaolin and 70%, 80%, 90% a 100% of water glass of a different silicate modulus by weight of metakaolin. The water coefficient was determined as the amount of water in which two spherical perpendicular measurements were taken on average 100 mm ±10% when the individual mixtures were spilled onto the substrate. Water content of all mixtures is showed in table 3.

The specimens prepared for the mechanical testing were cast in steel moulds (40 x 40 x 160 mm) and sealed with PE foil. After 48 h were demoulded and stored in plastic bags under laboratory condition. The mixtures were made and stored at ambient temperature 22 ±2°C and relative humidity 60% ±10%. After 28 days of curring were the speciments under mechanical tests (flexural strength and compressive strength).

Table 3. Water content of individual samples.

Sample	Water glass (silicate modulus)	Content of water glass from metakaolin (%)	Water content (ml)
1	1.6	70	5.5
2		80	3.5
3		90	3.0
4		100	1.5
5	1.4	70	5.0
6		80	3.5
7		90	3.0
8		100	1.5
9	1.2	70	6.0
10		80	4.0
11		90	2.0
12		100	2.0

3. Experimental and methods

Rheological measurements of the pastes were performed by means of a Discovery HR-1 (TA Instruments) hybrid rheometer and TRIOS 4.0.2.30774 software was used for data evaluation. The measurements were performed in a Peltier Concentric Cylinder system with a DIN rotor at 25°C. The standard gap for the DIN cylinder system (5.917 mm) was employed. The rheological measurements were started after a period of rest of 60 s. Paste properties were estimated from the flow curves determined for increasing and decreasing values within the shear stress range from 0.1 to 100 s^{-1}. Yield stress values and plastic viscosities were determined from the decreasing branch of the flow curve using the Bingham model [8].

Mechanical properties were tested using prismatic specimens with dimensions of 40 x 40 x 160 mm at the age of 28 days. Flexural strengths were determined using a standard three point bending test and compressive strengths were measured on the far edge of each of the two residual pieces obtained from the flexural test according to the EN 196-1 standard. For each mixture a set of three specimens was used.

4. Results and discussion

4.1. Mercury intrusion porosimetry

Pore size was carried out by mercury intrusion porosimetry. This is method for the study of the structure of porous substances, based on the phenomenon of capillary depression. Pore size distribution was determined only for the mixtures with 90% content of water glass which appeared to be the most promising activator dosage from the viewpoint of the mechanical properties. The results of the porosimetry measurements are presented in figure 2 and figure 3, showing the comparison of cumulative pore volume vs. pore diameter and differential pore volume vs. pore diameter, respectively, for different silicate modulus of alkaline activator.

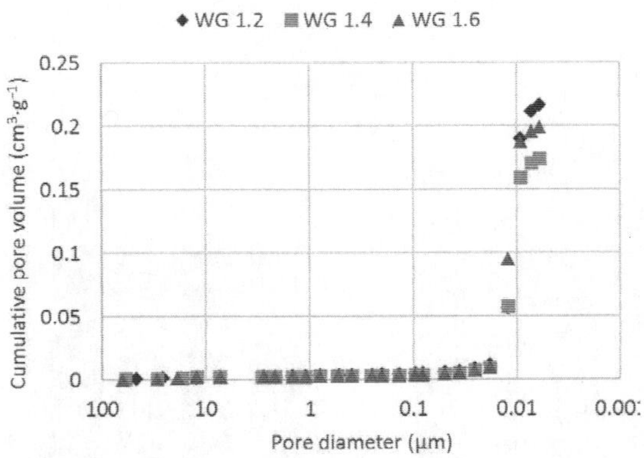

Figure 2. Cumulative intruded volume vs. pore diameter for the mixtures with 90% content of water glass having different silicate modulus.

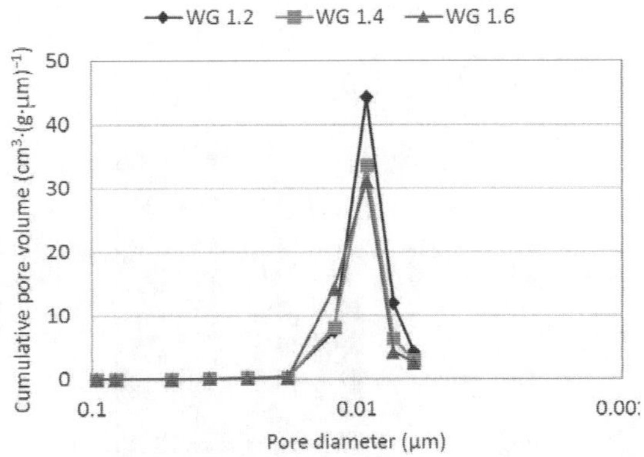

Figure 3. Differential pore volume vs. pore diameter for the mixtures with 90% content of water glass having different silicate modulus.

A cumulative pore distribution showed that the maximum of the pore volume is associated with small pores below 20 nm, which is common for metakaolin based geopolymers [9]. The lowest value of the total intruded volume was achieved for the mixture with M_s = 1.4 and the highest one for M_s = 1.2. These results clearly correspond to the trend observed for the compressive strength. The curves of the differential pore volume show that the dominant pore size in all three tested mixtures is ca. 10 nm. However, the number of pores is higher almost by one half when the activator with M_s = 1.2 was used.

4.2. Mechanical properties

The compressive strength of the geopolymer mixtures with various silicate modulus of water glass ratios are presented in figure 4. The tests were performed at samples geopolymer paste ages 28 days. The lowest 28 days compressive strength (47.5 MPa) was achieved for the mixture with 70% content of water glass and silicate modulus 1.6, the highest 28 days compressive strength (81.9 MPa) was achieved for the mixture with 90% content of water glass a silicate modulus 1.4. On the other hand, the lowest 28 days flexural strength (1.2 MPa) was achieved for the mixture with 80% content of water glass and silicate modulus 1.4, the highest 28 days flexural strength (5.7 MPa) was achieved for the mixture with 90% content of water glass a silicate modulus 1.2.

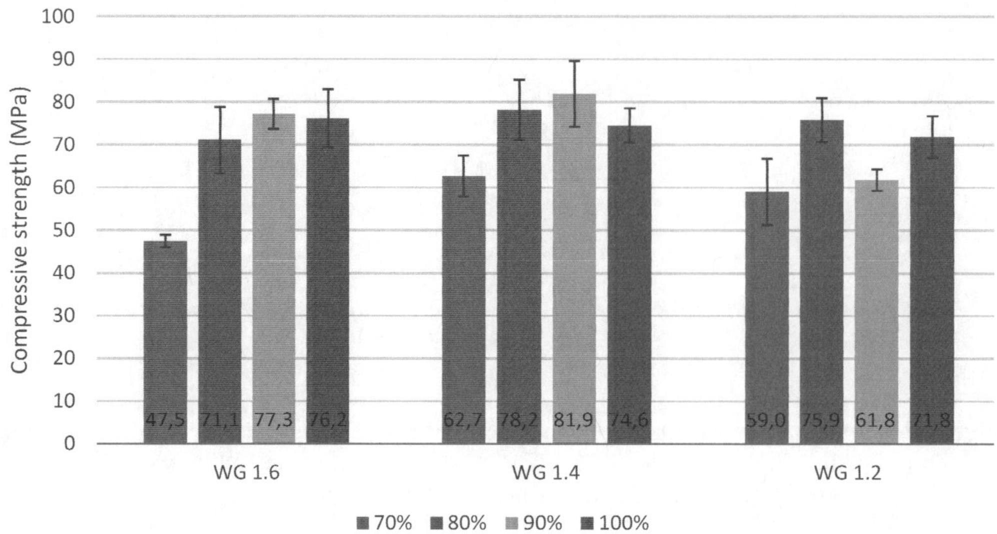

Figure 4. Compressive strengths of geopolymers with various ratio of water glasses.

It is difficult to find any trend for flexural strengths because the standard deviations of the results are very high. This is caused mainly by the absence of any aggregates that usually prevent the binder from propagation of microcracks. If any microcracks appears in the binder, e.g. due to drying shrinkage, it acts as a stress concentrator and the failure of the beam occurs at lower applied stress. This phenomenon can also explain the paradox that the mixtures with $M_s = 1.4$ show the lowest flexural strengths although the compressive strengths are the highest ones.

4.3. Rheological properties

The results of the assessed rheological parameters are summarized in table 4. The plastic viscosity increased with decreasing silicate modulus of water glass. This effect can be generally attributed to the alkaline activating solution because the decreasing silicate modulus is associate with higher alkaline content. Especially, sodium ions are extremely hydrophilic and adsorb a lot of water in the solvating layer. It causes a reduction of free rheological water and rise in viscosity. The same trend was also observed for the yield stress values but with respect to the flow curves presented in figure 5 and literature data [10], it is assumed that all tested mixtures show practically very low yield stress.

Table 4. Rheological parameters of samples with 90% content of water glass.

	Viscosity (Pa·s)	Yield stress (Pa)	Thixotropy ($Pa \cdot s^{-1}$)
WG 1.6	5.43	5.59	4588
WG 1.4	7.56	13.83	5404
WG 1.2	12.54	25.60	14654

For non-Newtonian fluids, the internal structure changes due to deformation, which is reflected in the macroscopic behaviour of the liquid. A mechanical decrease in viscosity over time with constant mechanical stress may occur and such fluid are called thixotropic. The thixotropic parameter of tested geopolymer mixtures considerably increased with lowering silicate modulus (table 4). While thixotropic parameter for higher modulus (1.6 and 1.4) are almost similar, the shear thinning behaviour of mixture with $M_s = 1.2$ was more pronounced as the value of the thixotropic parameter increased almost three times.

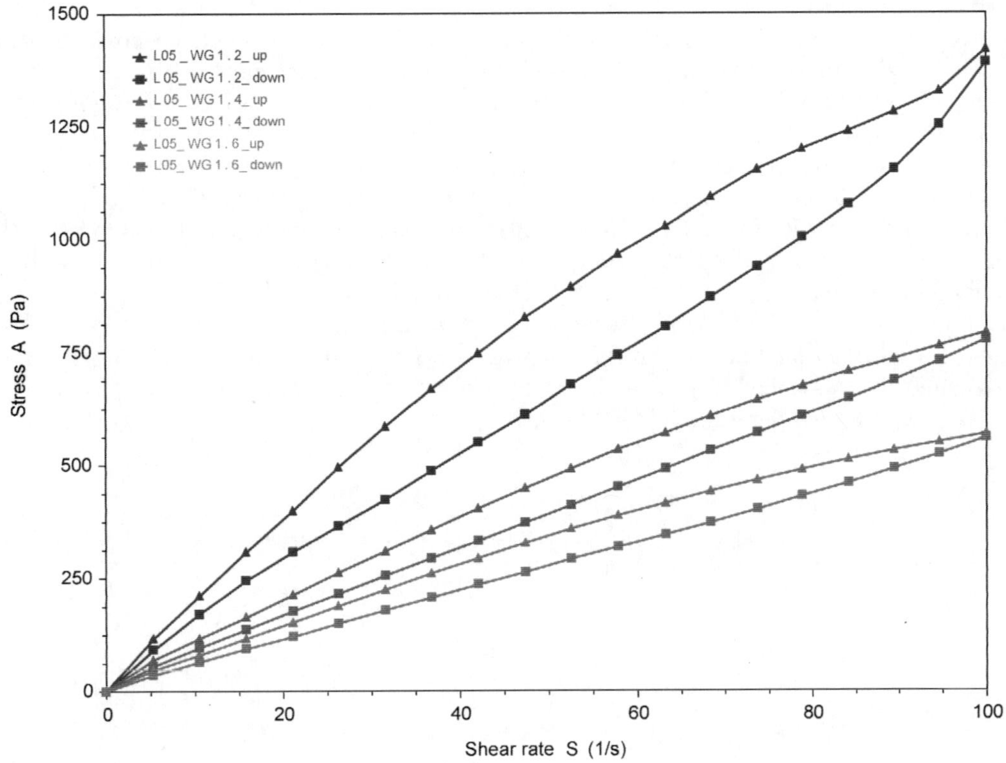

Figure 5. Flow curve of geopolymer with 90% of water glass having different silicate modulus.

5. Summary

Four different geopolymer mixtures with metakaolin and sodium water glass with various silicate modulus were tested for their mechanical and rheological properties.

This paper evaluates the mechanical and rheological properties of the geopolymer composition by applying sodium water glass as an alkaline activator with different silicate modulus – 1.6; 1.4 and 1.2. The best effect on mechanical properties was obtained for the samples with content 90% of water glass with respect to the weight of metakaolin. This trend manifested itself in both compressive strengths and flexural strengths.

For rheological measurements, all samples exhibited thixotropic behaviour which increased with lower silicate modulus of water glass. For all samples very low yield stress and relatively plastic viscosity was observed. However, the viscosity increased when the silicate modulus was lowered.

Acknowledgements
This outcome has been achieved with the financial support of the Ministry of Education, Youth and Sport of the Czech Republic under the "National Sustainability Programme I", project No. LO1408 (AdMaS UP).

Reference

[1] Herrmann F and Schiller M 2019 Zkoušení nátěrových hmot a ochranných povlaků (*Online:* www.sczl.cz/download/download/2006-vk-solan/04 synpo herrmann.pdf)
[2] Temuujin J, Rickard W, Lee M and Riessen A 2011 *Journal of Non-Crystalline Solid,* p 1399–1404
[3] Opravil T, Brandštetr J, Havlica J and Frank V 2008 *Metakaolin (*Brno) pp 80–86
[4] Davidovits J 2008 *Geopolymer - Chemistry and Applications* 2nd ed. p 592
[5] Bernal S, Gutierrez R, Provis J and Rose V 2010 *Cement and Concrete research* pp 898–907
[6] Xu H and Van Deventer J S J 2002 *Geopolymerisation of multiple minerals* pp 1131–1139
[7] Mefisto L05 (*Online:* http://www.cluz.cz/cz/mefisto-l05)
[8] E.C. Bingham 1922 *Fluidity and Plasticity* (New York)
[9] Rovnaník P 2010 *Construction and Building Materials* pp 1176–1183
[10] Rovnaník P, Vyšvařil M, Rovnaníková P, Grzeszczyk S and Janowska-Renkas E 2018 *Construction building material* pp 924–933

IOP Conf. Series: Materials Science and Engineering **583** (2019) 012003 doi:10.1088/1757-899X/583/1/012003

Possibilities of fly ash activation in alumina recovery process

M Marko[1], T Opravil[1], J Masilko[1] and J Porizka[1]

[1] Brno University of Technology, Faculty of Chemistry, Purkynova 468/118, 612 00 Brno, Czech Republic

Email: xcmarkom@fch.vut.cz

Abstract. This paper summarises various possibilities of high-temperature fly ash (HTA) and fluidised-bed filter ash (FFA) activation in metal and oxides recovery processes (esp. alumina recovery). The study deals with chemical and phase composition ash modification to improve leachability of aluminium compounds into sulphuric acid solution. Samples of HTA and FFA produced by power plant industry in the Czech Republic were subjected to diverse types of modification, such as sulphuric acid concentration, ash/acid ratio and sintering reactions using several types of solid agents. Sintering reaction were based on forming acid soluble calcium aluminates via donating the system with calcium oxide, - carbonate and - chloride. Phase and chemical composition were characterised by means of XRD and XRF analysis and the concentration of Al in acid solutions via ICP–OES. Appropriate modification of input material by high-temperature sintering reaction using calcium chloride combined with extraction into 10 wt. % sulphuric acid leads to dissolution up to 99% of aluminium from FFA and up to 94% form HTA. At least, irreplaceable dopant for cement production – gypsum, is formed as a by-product of leaching.

1. Introduction

The use of any energy source comprises some technological advantages and disadvantages, as well as environmental benefits and problems [1]. Worldwide, probably the most typical energy source is still coal. The coal firing power plants produce large amounts of an unburnable rest – fly ash (FA).

FA is generated from the various organic and inorganic components of burned coal at 700–900°C in case of fluidised bed burning and 1 000–1 300°C on a way of classical high–temperature combustion [2, 3]. The worldwide production of FA reaches about 500 mil. tonnes a year [4]. With This large amount of the secondary energy products needs to be stored. This process means deposition of FA mainly in waste ponds. These sludge lagoons could cause environmental contamination and other ecological problems in case of dissolution toxic substances like heavy metals. FA particles are very small (3–80 μm) and could be easily suspended in the air and have become also a major source of gas pollution [5].

Since 1950s many researches on FA application started. Since then the FA is utilised in cement and concrete production because of its pozzolanic properties, because FA react with lime and produce insoluble stable compounds. This category contains about 50% of the utilized FA. [6] Other applications include structural stabilization, waste stabilization, mining applications, agriculture, subbases and other. The Current FA utilization reaches up to 40% in the USA and 47% in Europe [2, 7].

FA could be also strategical source in the search of new material base for the future because many strategical compounds like alumina, titanium oxide, silica as well as some scarce and valuable elements like Li, Rb, Sr, Nb, Ta, Ge, Se, Mo, Ce, Sm, Eu and many other could be recovered from this coal combustion product.

2. Alumina recovery

The alumina recovery from FA was firstly extend in 1950s in Poland due to bauxite embargo during the Cold war [1]. Since then many procedures leading to alumina recovery have been accomplished from the direct acid leaching methods to modified sintering reactions connected with acid or alkali dissolution.

2.1. Direct acid leaching

Direct acid leaching methods (DAL) are based on dissolution of selected components caused by mineral acids, such as sulphuric, nitric, hydrochloric, hydrofluoric or a mixture. DAL could be affected by increasing reaction temperature, FA/acid ratio, acid concentration. Many researches were published on that topic. [8–10] Based on progress results, especially sulphuric acid leaching appears to be one of the most suitable agents for this process. The publications submit a proper FA/acid ratio at least 5:1, sulphuric acid concentration in a range of 30–65 wt. % (approx. 4–10 $mol \cdot l^{-1}$) and they also recommend heating the ash and agent mixture up to 300°C combined with permanent stirring. Higher acid concentration decreases leachability of Al because the incipient aluminium sulphate is not soluble in this medium and sticks on the surface of FA particles. [8]

2.1.1. Sintering reactions. The sintering processes are usually based on a high-temperature reaction between FA and sintering agent to form easily soluble and stable aluminium (or other) compounds [11]. Calcium or sodium reagents could be used as sintering agents. Sintering of aluminium rich FA comes out from the original Pedersen process used in crude iron and calcium aluminate slags production from a mixture of bauxite, iron ore, coke and limestone. [5, 11]

Calcium sinter processes

This process starts with a magnetic separation to remove all magnetic compounds of iron to improve the efficiency. Then FA is mixed with CaO and $CaCO_3$ in a certain FA/agent ratio (1:1 to 1:3) and then is continues with mixture sintering at 1 300–1 400°C. At this temperature, the mullite and quartz phases react together with the calcium agent to form mainly $12CaO \cdot 7Al_2O_3$ (mayenite) and $2CaO \cdot SiO_2$ as a by-product, according to the equations (1–5). Other elements, such as Ti and non-magnetic Fe form $CaO \cdot Fe_2O_3$ and $CaO \cdot TiO_2$ respectively, described in equations (6) and (7). [13]

$$CaCO_3 \rightarrow CaO + CO_2, \tag{1}$$

$$7(3Al_2O_3 \cdot 2SiO_2) + 64CaO \rightarrow 3(12CaO \cdot 7Al_2O_3) + 14(2CaO \cdot SiO_2), \tag{2}$$

$$3Al_2O_3 \cdot 2SiO_2 + 13CaO \rightarrow 3(3CaO \cdot Al_2O_3) + 2(2CaO \cdot SiO_2), \tag{3}$$

$$3Al_2O_3 \cdot 2SiO_2 + 5CaO \rightarrow 3(CaO \cdot Al_2O_3) + 2(2CaO \cdot SiO_2), \tag{4}$$

$$2CaO + SiO_2 \rightarrow 2CaO \cdot SiO_2, \tag{5}$$

$$CaO + Fe_2O_3 \rightarrow CaO \cdot Fe_2O_3, \tag{6}$$

$$CaO + TiO_2 \rightarrow CaO \cdot TiO_2. \tag{7}$$

The auto-disintegration process takes place during the cooling below 500°C and it is attributed to the phase modification of generated $2CaO \cdot SiO_2$. The cooled sinter can be leached by recyclable Na_2CO_3 solution or by sulphuric acid. According to the research papers, the efficiency of alumina recovery reaches up to 80% depending on a way of dissolution into soda solution or a sulphuric acid. [5, 11, 13]

Another way of mayenite formation in alumina recovery process, using calcium chloride appears to be a suitable reaction agent. This technique simplifies the sintering process by decreasing reaction temperature to about 1 000°C combined with a natural cool down under 600°C. The proper FA/$CaCl_2$ ratio is between 1:0.8 to 1:1.2, depending on FA type. These high-temperature reactions could be described in equations (8–10).

$$3Al_2O_3 \cdot 2SiO_2 + 6CaCl_2 + SiO_2 + 6H_2O \rightarrow 3(2CaO \cdot Al_2O_3 \cdot SiO_2) + 12HCl, \quad (8)$$

$$3SiO_2 + 3CaCl_2 + Fe_2O_3 + 3H_2O \rightarrow 3CaO \cdot Fe_2O_3 \cdot 3SiO_2 + 6HCl, \quad (9)$$

$$7(3Al_2O_3 \cdot 2SiO_2) + 36CaCl_2 + 36H_2O \rightarrow 3(12CaO \cdot 7Al_2O_3) + 14SiO_2 + 72HCl. \quad (10)$$

This method is also followed by sulphuric acid leaching to recover up to 97% of alumina. The course of acid leaching could be described in equations (11) and (12). [15]

$$3CaO \cdot Fe_2O_3 \cdot 3SiO_2 + 6H_2SO_4 \rightarrow 3CaSO_4 \downarrow + 3H_4SiO_4 \downarrow + Fe_2(SO_4)_3, \quad (11)$$

$$12CaO \cdot 7Al_2O_3 + 33H_2SO_4 \rightarrow 12CaSO_4 \downarrow + 7Al_2(SO_4)_3 + 33H_2O. \quad (12)$$

3. Materials and methods

3.1. Raw materials

One sample of HTA (type 1) and FFA (type 2) from the Czech Republic production were used to undergo a couple of sintering reactions described above (CaO, $CaCO_3$ and $CaCl_2$ agents). At first, samples of FA were dried in laboratory dryer at 105°C to constant weight. Then a fraction with particle size lower than 63 μm was sieved. Then the chemical and phase composition, specific area and loss on ignition were determined. The used CaO and $CaCO_3$ agents were commercial sources; the lime with a specification Cl 90Q (min. free lime content of 90%) and the calcite very finely grounded (max. impurity content of 5%). The $CaCl_2$ agent was used in a p. a. form.

3.2. Sintering reactions conditions and leaching

Calcium oxide and calcium carbonate sintering reactions were carried out in super-kanthal laboratory furnace using temperature protocol as follows – lab. temperature to 1 250°C with a heating rate of 10°C·min^{-1}, followed by constant temperature at 1 250°C for 60 min, the cooling rate of 5°C·min^{-1} to 600°C and then a natural cooling to laboratory temperature. Sintering with calcium chloride agent was done in laboratory muffle furnace heating to 1 000°C, other parameters of the temperature protocol were the same as described above. The amounts of sintering agents were chosen as follows: agent/FA 0.5, 1.0 and 1.5; in case of $CaCl_2$ mixtures containing 0.6, 0.8 chloride agent have been prepared extra. All these mixtures were homogenised in a sealed bottle on a rotary shaker.

Clinker was disintegrated in a mortar and characterised by *x*-ray fluorescence spectrometry XRF (Olympus Delta professional) and *x*-ray diffraction XRD (Panalytical Empyrean) after sintering reaction. The samples of modified FA were subjected to dissolution of Al content into the sulphuric acid solution with acid/FA ratio of 15. Concentration of Al in the acid solution was quantified by inductively coupled plasma optical emission spectroscopy ICP–OES (Horiba Scientific – Ultima 2). The extraction of Al into sulphuric acid was performed using laboratory magnetic stirrer with constant stirring speed of 250 rpm.

4. Results

4.1. Raw materials

Results of the chemical composition determined by XRF method are given in table 1. The XRD Rietveld analysis showed presence of quartz, cristobalite, magnetite and mullite phases in case of the high-temperature FA and quartz, anhydrite, lime, calcite, portlandite, hematite, anatase and muscovite phases in FFA. The density (*D*) of FA was determined using pycnometer method using isopropyl alcohol as the immersion liquid. The specific surface area was measured by Blaine. The free lime content in FA samples was qualified by saccharose titration method. All these results inclusive loss on drying (*LOD*) and loss on ignition (*LOI*) are summarized in table 2.

Table 1. Chemical properties of FA, part 1.

Sample	Content (wt. %)								
	Al_2O_3	Fe_2O_3	TiO_2	MgO	CaO	SiO_2	K_2O	Na_2O	SO_3
HTA (type 1)	29.50	7.45	1.83	0.32	2.94	54.04	2.69	0.24	0.76
FFA (type 2)	27.93	7.03	3.36	0.38	21.24	32.07	0.93	0.47	6.42

Table 2. Chemical properties of FA, part 2.

Sample	*D* ($kg \cdot m^{-3}$)	Blaine ($m^2 \cdot kg$)	Free CaO (%)	*LOD* (%)	*LOI* (%)
HTA (type 1)	2249	315.4	–	0.29	1.12
FFA (type 2)	2716	633.6	4.76	0.47	2.42

It is obvious that the FA could be quite rich mainly in aluminium, iron and even silicon as well. In some kinds of HTA the content of aluminium oxides reaches up to 35%. To compare with the purity of nowadays bauxite ore of about 45% [15], the content of alumina in FA is not that low. The SiO_2 content prevents the traditional way of aluminium production from the Bayer method.

4.2. Sintered materials

4.2.1. $CaO/CaCO_3$ sintering. Sintered mixtures of HTA and $CaO/CaCO_3$ were disintegrated in mortar to give a particle size less than 63 µm. All samples were subjected to determination of unreacted agent as the free lime content using saccharate titration method. These results given in table 3 show, that the addition of the sintering agent above 1.5 FA weight could be considered as non-effective.

Table 3. Unreacted amount of reaction agent determination.

Agent addition (agent/FA)	CaO		$CaCO_3$	
	Free CaO (%)	Unreacted amount (%)	Free CaO (%)	Unreacted amount (%)
0.5	7.87	23.6	0.96	4.8
1.0	25.65	51.3	8.81	26.4
1.5	40.74	77.9	23.39	54.6

Calcium carbonate addition also affects the porosity of the sintered material, due to the limestone decomposition and release of CO_2, resulting in a pore aggregate in the material, which leads to easier disintegration than the lime addition. To remove the unreacted amount of the sintering agent, the 1 wt. % HCl washing was used. The volume of the washing agent was increased by 30% of the equimolar amount of the free lime. The XRD analysis shows the presence mainly of $12CaO \cdot 7Al_2O_3$ and other calcium-aluminate phases.

To compare leachability of aluminium from sintered and raw (ref) FA, a single-stage extraction into 10 wt. % sulphuric acid at room temperature was tested in laboratory. Leachability of Al was monitored in time using ICP-OES. The results shown in figures 1–4 represent dissolved aluminium contentment (%) in time.

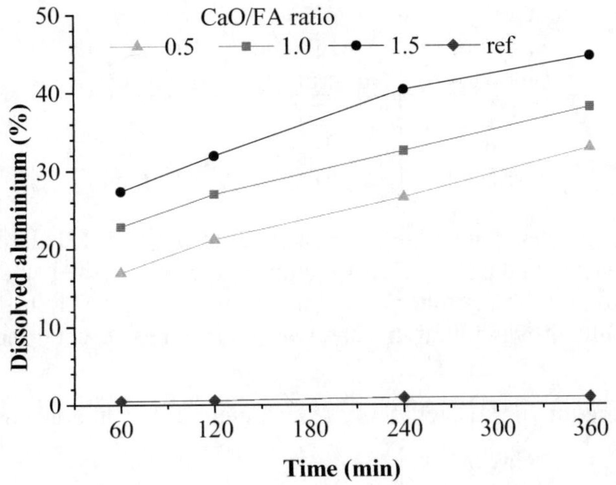

Figure 1. Leachability of aluminium with CaO addition into 10 wt. % H_2SO_4 medium; HTA.

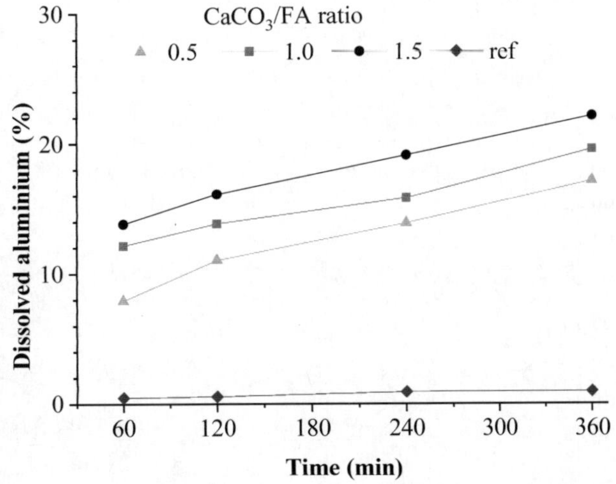

Figure 2. Leachability of aluminium with $CaCO_3$ addition into 10 wt. % H_2SO_4 medium; HTA.

According to the results, there has been a multiple increase of efficiency in Al dissolution than in case of raw FA under the same conditions. From the point of view of leachability, the most optimal is the addition of 1.5 CaO. Up to 44.7% of present Al could be dissolved. In terms of trends it is possible to talk about a similar course of reaction with limestone agent as in the case of CaO addition. From an economic point of view, however, the maximum addition method does not provide very satisfactory results since it obtains a relatively high percentage of unreacted proportion of the sintering. The possibility of the unreacted portion agent removing by washing using a weak solution of hydrochloric acid is easily feasible but prevents the regeneration of the unreacted sintering agent.

4.2.2. $CaCl_2$ sintering. Sintered mixtures of FA and chloride agent were cooled down at the room temperature. Subsequently, a part of each sample was washed using distilled water and the amount of unreacted chloride was determined by complexometric titration with chelate III. The results are given in table 4.

Table 4. Unreacted amount of reaction agent determination.

$CaCl_2$ addition (agent/FA)	HTA Unreacted amount (%)	FFA Unreacted amount (%)
0.5	12.5	17.8
1.0	24.3	35.3
1.5	51.9	58.6

From the results of the unreacted agent amount, the additions above 1.0 already might appear to be ineffective as it does not react with the FA. Selected samples of water washed, and non-washed materials were further extracted into 10 wt. % sulphuric acid to compare the leachability of Al in the previous method. Based on the results, it was found that the washed samples reached the leachability of Al by 30% higher (table 5).

Table 5. Comparation of Al leachability from water washed and non-washed samples.

$CaCl_2$ addition (agent/FA)	non-washed	washed
0.6	27.92	61.77
0.8	36.14	83.51
1.0	65.38	93.32

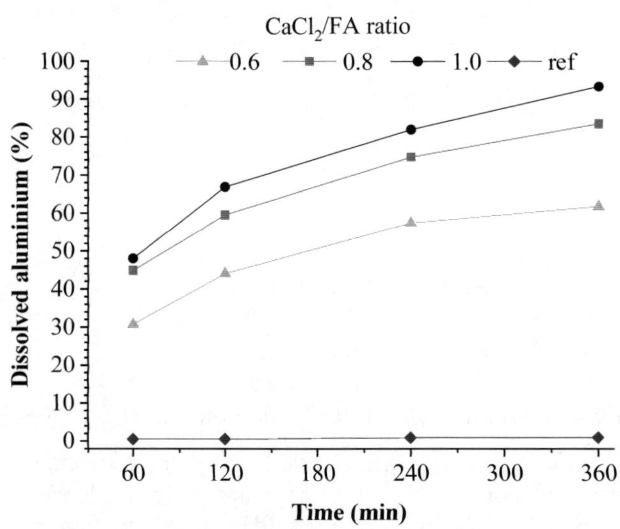

Figure 3. Leachability of aluminium with $CaCl_2$ addition into 10 wt. % H_2SO_4 medium; HTA, washed.

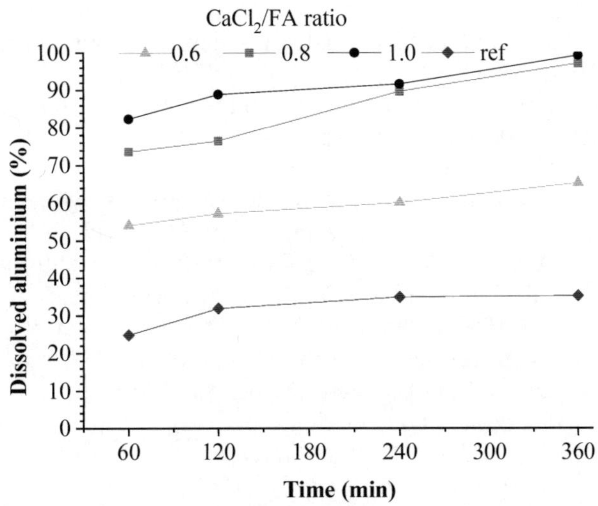

Figure 4. Leachability of aluminium with $CaCl_2$ addition into 10 wt. % H_2SO_4 medium; FFA, washed.

By comparing the results of leaching of HTA and FFA, this method could be referred to as universal and fully applicable for both types of FA. In terms of HTA, the optimum addition of $CaCl_2$ reaches 1.0, in case of FFA the sufficient addition is 0.8. As can be seen in the charts, it is possible to separate up to 99% of aluminium. The given method is even relatively selective for leaching Al, since dissolution max. 4% Ti and max. 27 %Fe. The chloride addition could also reach the leachability of Al two times higher compared to the addition of CaO. The course of the experiment was carried out in accordance to article [15] with the achievement of similar results. According to these data, the calcium chloride method could be probably used in other types of fly ash and other materials.

5. Conclusion

Within the scope of increasing the leachability of aluminium from HTA and FFA, some methods of chemical reaction were laboratory tested. FA phase composition modification by sintering with various melt-inducing agents forming of new, especially, calcium-aluminate phases have been evaluated. Modification of the FA composition using CaO in a high temperature reaction at 1 250°C gives a possibility to grow the efficiency of Al dissolution into 10 wt. % sulphuric acid up to 45%. In case of FA modified by addition of $CaCl_2$ and sintered at 1 000°C, it is possible to dissolute up to 99.3% Al. Based on the obtained results, sintering reaction with $CaCl_2$ could be labelled as a universal method for both types of FA (HTA and FFA). At the same time, this method seems to be also very selective and offers the possibility of unreacted sintering agent regeneration.

Acknowledgements
This outcome has been achieved with the financial support by the project of specific research FCH-S-18-5194.

References

[1] Izquierdo M and Querol X 2011 Leaching behaviour of elements from coal combustion fly ash: An overview *International Journal of Coal Geology* **94** pp 54–66
[2] Blissett R S and Rowson N A 2012 A review of the multi-component utilisation of coal fly ash *Fuel* **97** pp 1–23
[3] Scheetz B E and Earle R 1998 Utilization of fly ash *Current Opinion in Solid State & Material Science* **5** pp 510–20
[4] Ahmaruzzaman M 2009 A review on the utilization of fly ash *Progress in Energy and Combustion*

Science **36** pp 327–63

[5] Yao Z T, Xia M S, Sarker P K and Chen T 2013 A review of the alumina recovery from coal fly ash *Fuel* **120** pp 74–85

[6] Production and Use of Coal Combustion Products in the U.S. 2015 *American Coal Ash* Website

[7] Blanco F, Garcia M P and Ayala J 2004 Variation in fly ash properties with milling and acid leaching *Fuel* **84** pp 89–96

[8] LI Lai-shi *et al* 2011 Extraction of Alumina from Coal Fly Ash with Sulfuric Acid Leaching Method *The Chinese Journal of Process Engineering* **11** pp 254–58

[9] Matjie R H, Bunt J R and Van Heerden J H 2005 Extraction of alumina from coal fly ash generated from a selected low rank bituminous *South African coal Minerals Engineering* **18** pp 299–310

[10] Phillips C V and Wills K J 1982 Laboratory study of the extraction of Al_2O_3 of smelter grade from China clay micaceous residues by a nitric acid route *Hydrometallurgy* **9** pp 15–28

[11] Ding J *et al* 2016 Research and industrialization progress of recovering alumina from fly ash: A

[12] concise review *Waste Management* **60** pp 375–87

[13] Lin H Y, Wan L and Yang Y F 2012 Aluminium hydroxide ultrafine powder extracted from fly ash *Advanced Materials Research* pp 5212–515 1548–53

[14] Kim A G and Hesbach P 2009 Comparasion of fly ash leaching methods *Fuel* pp 926–37

[15] Sun Y *et al* 2017 Recovery of Alumina from Coal Fly Ash by $CaCl_2$ Calcination Followed by H_2SO_4 Leaching *Journal of Environmental & Analytical Toxicology*

IOP Conf. Series: Materials Science and Engineering **583** (2019) 012004 doi:10.1088/1757-899X/583/1/012004

The effect of burnt lime addition on hydration of Ultra-high performance cementitious composites

R Novotný[1], E Bartoníčková[1] and J Kotrla[1]

[1] Institute of Materials Chemistry, Brno University of Technology, Purkyňova 464/118, Brno 61200, Czech Republic

EmaiL: xcnovotny2@fch.vut.cz

Abstract. The paper deals with the burnt lime utilization as a source of calcium ions participated during hydration process of ultra-high performance cementitious composites. These type of ultra-high performance cementitious composites are based on the high content of silica fume that together with calcium ions form the binder phase. In the case of ordinary Portland cement, the hydration during induction period supposed to be too slow for dissolution of required amount of calcium ions necessary for the pozzolanic reaction of presented silica fume. The low addition of the lime should be a sufficient calcium source that allow the starting of the pozzolanic reaction before the acceleration period occurs. The calcium ions content from the beginning of the reaction was controlled by an addition of soft-burnt and dead-burnt lime. Its influence was studied in terms of hydration process by isothermal calorimetry. Based on the experimental data, the hydration reaction mechanism was proposed.

1. Introduction

Commonly used types of concretes could be typically characterized by their low homogeneity closely related to low tensile flexural strengths. The possible solution to solve this problem is to substitute coarse aggregates with fine sand, to increase content of Portland cement (OPC), addition of extremely fine reactive components (i.e. silica fume, precipitated silica, metakaolin, ground blast furnace slag) and to reduce the water to cement ratio by superplasticizer application [1]. These types of concretes are generally named as ultra-high performance cementitious composites (UHPC). High content of cement, reactive components and low water to cement ratio, typical for this type of composite system, cause volume changes that result in formation of micro-cracks due to plastic and autogenous shrinkage and self-exsiccation [2]. Due to the shrinkage phenomena, it is necessary to apply water curing during hydration and further hardening of prepared composites [2, 3].

High silica fume content, which in dispersion may behave as an aggregate causing less friction between cement particles, is also typical for UHPC [4]. Silica fume and calcium cations, coming from OPC hydration as a secondary product, react together and form binder phases which are beneficiary to mechanical properties of binder species [5]. Nevertheless, the velocity of calcium ions formation for pozzolanic reaction is directly connected with the cement hydration rate [6]. During the first hours of hydration, the induction period occurs and the hydration process is slowed down. Induction period can be prolonged due to the higher content of superplasticizers [7]. High bulk density also reduces the ionic transport, therefore the binders with high content of silica fumes against OPC demonstrate slower strength development [8].

Paper deals with the possibility of calcium oxide usage as a source of calcium cations for silica fume (SF) hydration during the cement induction period. The advantages of calcium oxide is its high reactivity, no other ion contaminations and signs of exhibit the undesirable plastic behavior as it is in case of calcium hydroxide [9]. The utilization of soft-burnt (SBL) and dead-burnt lime (DBL) and curing conditions with respect to the reactivity and long term stability were studied. Srinivasan et al. studied the effect of lime water on OPC-SF binders [10].

2. Materials and methods

The white OPC CEM I 52.5 R (Aalborg Portland A/S, Denmark), white undensified silica fume RW-Füller Q1 (RW silicium GmbH, Germany), SBL CL 90-Q (Kotouč Štramberk, spol. s r. o., Czech Republic) sieved under the 63 μm, dead-burnt lime CL 90-Q (Vápenka Vitošov s.r.o., Czech Republic) sieved under the 63 μm, superplasticizer Master Glenium ACE 446 (BASF Stavební hmoty Česká republika s.r.o., Czech Republic), coarse aggregate (Filtrační písky, spol. s r.o., Czech Republic) and deionised water were used for UHPC preparation.

All the components were gradually mixed for 1 min, and then the ¾ of gauge water with the superplasticiser dosage was added and mixed for 5 min. The ¼ of the missing gauge water was added and mixed for other 5 min. The homogenized mixture was moulded to moulds with the dimensions 4 × 4 × 16 cm intended for mechanical testing. After 24 h the experimental specimens were demoulded and the dry or water curing conditions were applied under the controlled temperature of 25°C. Hydration process of binder was analysed by using of isothermal calorimeter TAMair (TA Instruments) under the 25°C. All of the used components were storage before using under the same conditions for 24 hours, i.e. controlled 25°C. Analysed samples were prepared by hand mixing of bath contents of 40 g of binder, deionized water and additives for 2 min, and then 5 g of this mixture was placed into lockable plastic ampules. The measurements were performed against to reference sample (i.e. water) with similar heating capacity. The studied hydration process was precisely stopped by acetone washing [11]. Formed hydration products (50 mg) were observed by TG-DTA analysis (Q600, TA Instruments) up to 1000°C with heating rate 10°C/min under the dried air atmosphere. Table 1 summarizes the composition and designation of studied samples.

Table 1. Composition (g) of measured samples.

Sample	OPC	SF	SBL	DBL	SP	Sand	H_2O
R	188	62.0	0	0	7.5	750	67
1.a	186	61.5	2.5	0	7.5	750	67
1.b	184	61.0	5	0	7.5	750	67
1.c	180	60.0	10	0	7.5	750	67
1.d	176	59.0	15	0	7.5	750	67
1.e	169	56.0	25	0	7.5	750	67
1.f	0	0	250	0	7.5	0	67
2.a	186	61.5	0	2.5	7.5	750	67
2.b	184	61.0	0	5	7.5	750	67
2.c	180	60.0	0	10	7.5	750	67
2.d	176	59.0	0	15	7.5	750	67
2.e	169	56.0	0	25	7.5	750	67
2.f	0	0	0	250	7.5	0	67

3. Results and discussion

3.1. Setting of UHPC

With the increasing content of the SBL (see figure 1.a), the heat flow in pre-induction period was gradually increased, the difference between R and 1.e sample was 12.2 mW/g. If the un-influenced hydration process of SBL in UHPC occurs the difference in heat values supposed to be 35.7 mW/g. In case of DBL addition (see figure 1.b), the increase of the heat flow between R and 2.e samples was determined to 2.1 mW/g. From the obtained results is obvious that in the UHPC system the retardation of the CaO hydration occurred which was more substantial when the DBL was applied.

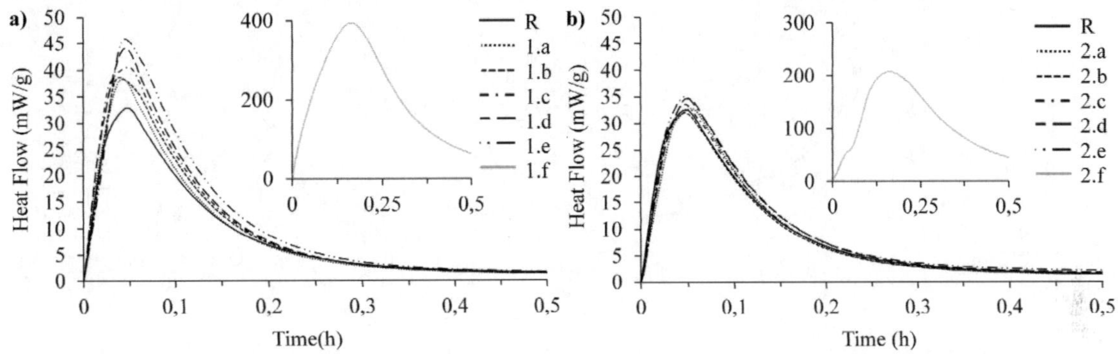

Figure 1. Heat flow in the pre-induction period a) SBL, b) DBL.

Figure 2 shows the retardation of main hydration peak when the SBL was applied. The retardation of the peak in the case of the sample 1.c is almost 1 hour against the R one. The hydration of samples with SBL addition was not significantly retarded too. Sample 1.e reached the maximum of main hydration peak in the same time as the reference R one. Without the dependence on the weight of the addition of SBL, the heat flow value of the main peak maximum was increased of 0.4 mW/g. The significant influence of soft-burned lime to hydration process was observed in part of induction period. The lowest heat flow 0.4 mW/g was determined in the case f R sample, with the increasing amount of lime addition the heat flow gradually increased to 2.3 mW/g in the case of 1.e sample. Samples 1.d and 1.e showed local minimum of heat flow after 1 hour of hydration. From this point the hydration rate was accelerated to the end of induction period. This phenomenon is clearly visible on the figure 2.b, where during first hour of hydration process the difference in total hydration heat, as well as the amount of formed hydrates, was between particular samples minimal. The total hydration heat increased proportionally with increasing lime content and proved the reaction of silica fume during the induction period and cement hydration process. Sample R reached for 60 mW/g heat more than the sample 1.e after 30 hours of hydration.

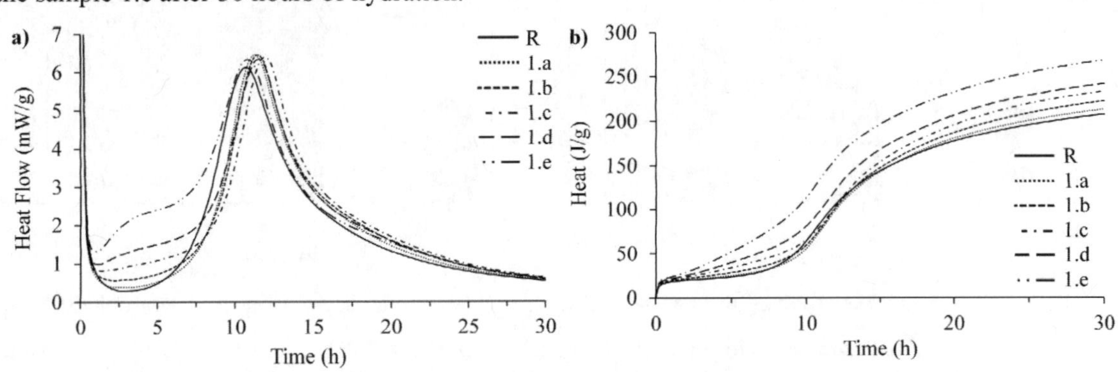

Figure 2. Calorimetric measurement of the UHPC with SBL a) heat flow, b) heat.

Samples with DBL addition showed similar hydration coarse as the one with SBL (see figure 3.a). However, during the first hour of the induction period the almost depletion of the previously observed local minimum occurred and the hydration rate after this minimum to the end of induction period was approximately constant. The retardation of the main hydration peak was similar in case of both types of the limes. Nevertheless, the maximum of the main hydration peak was with the DBL addition slightly decreased. Also the course of the main hydration heat was for the both types of the lime similar (see figure 3). The total hydration heat produced by sample R after 30 hours was or 57 mW/g less than the sample 2.e.

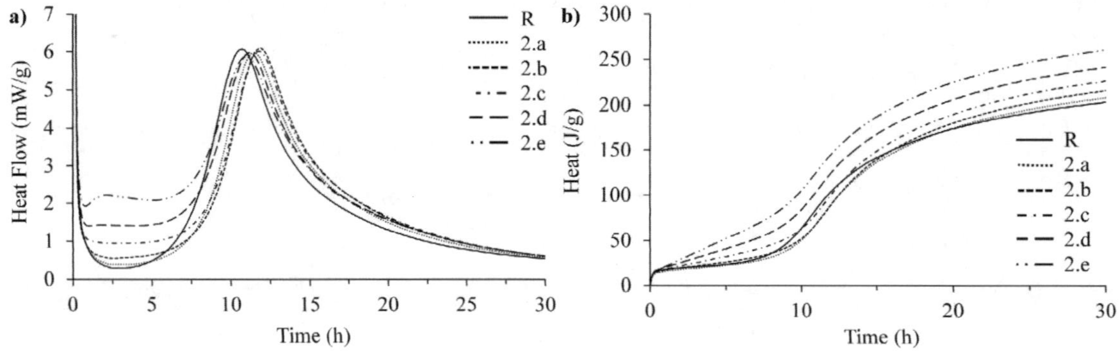

Figure 3. Calorimetric measurement of the UHPC with DBL a) heat flow, b) heat.

3.2. Testing of mechanical parameters

Tensile and compressive strengths for samples stored under the dry and water curing condition for hydration after 1 day were almost similar due to demoulding after 24 hours and subsequently set up of the different curing conditions for 24 hours. After 1 day of hydration (figure 1.a) the increase of tensile strengths values for 1 MPa was observed for the samples with low content of SBL. On the contrary the higher content of SBL caused the decrease of tensile strength values for 1 MPa against reference sample R. After 7 days of dry curing conditions the gradual increase of tensile strengths with increasing soft-lime content was observed from 9.5 MPa for sample R to 13.2. MPa for sample 1.d. Higher soft-lime content again caused the decrease of the tensile strength values. After 28 days of dry curing conditions the tendency in mechanical behaviour was identical as was observed after 7 days. The tensile strength value of sample R under the water curing conditions was 19 MPa, but the any addition of SBL caused the decrease in tensile strength values. Also the similar tendency after 28 days of water curing was observed, the value for reference sample R was even decreased to 17.5 MPa.

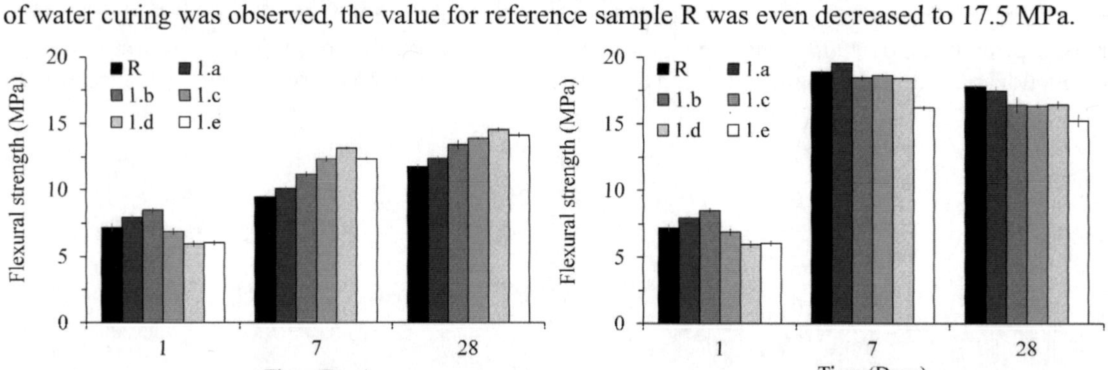

Figure 4. Flexural strength of the specimens with soft b. lime a) dry curing, b) water curing.

The compressive strength behaviour (figure 4.a) of the samples with the addition of SBL (R-1.d) and cured under the dry conditions was not significantly influenced and for 1.e the value even decreased. After 7 days of the water curing conditions (see figure 1.b) the compressive strength values decreased with the increasing lime content. The maximum was reached for sample 1.e. After 28 days of water curing, the values of compressive strength slightly increased for samples 1.a and 1.b with SBL addition. Sample 1.b reached the total maximum compressive strength's value of 131.3 MPa. On the contrary sample 1.e reached the maximum decrease against the reference R one.

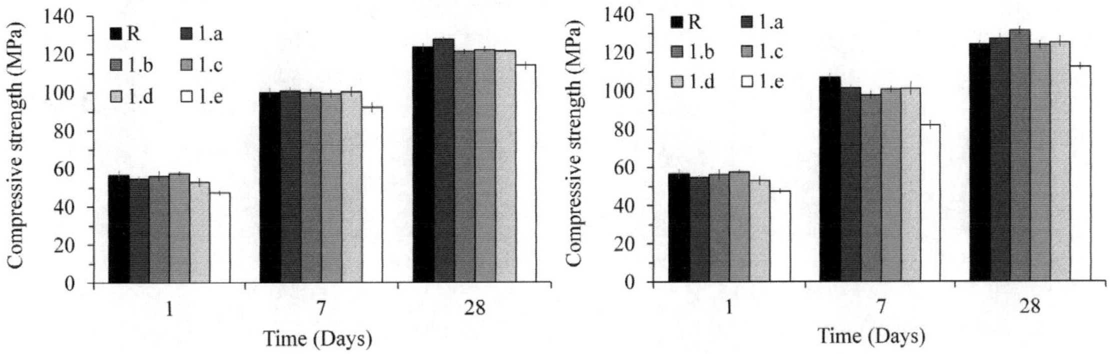

Figure 5. Compressive strength of the specimens with soft b. lime a) dry curing, b) water curing.

After 1 day of dry curing, the tensile strengths of the samples with dead-lime content reached the similar or higher values than the reference R one. In the case of the sample 2.a and 2.b (dead-burnt) the tensile strength increase was substantial then in the case of 1.a and 1.b (soft-burnt) after 1 day of the dry curing. The increasing tendency in tensile strength values with the increasing content of the lime was observed after 7 and 28 days of dry curing (see figure 6.a). The behaviour of the samples under the water curing conditions and dead-lime addition is given on the figure 6.b. The samples showed the minimal decreasing tendency after 7 and 28 days. Sample 2.d under the 28 day of the water curing conditions reached maximum value of tensile strength 19.3 MPa. The significant defects were observed for samples 2.e after the 7 and 28 day of water curing and the values were not able to determine. The compressive strength behaviour of the samples with DBL addition under dry and water curing conditions (figure 7) showed the identical tendency as the samples with soft-burnt one.

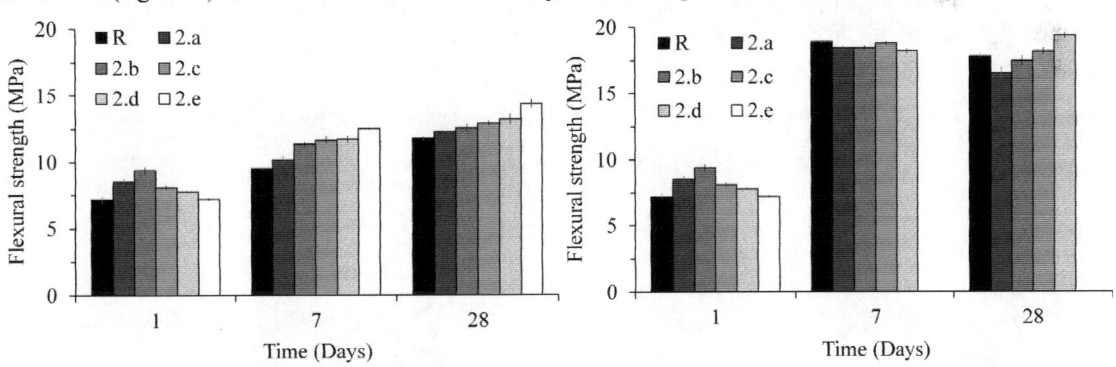

Figure 6. Flexural strength of the specimens with dead b. lime a) dry curing, b) water curing.

3.3. DTA analysis of hydration products in time

The thermal study of the hydration products formed after 1, 5 and 24 hours is given on figure 8. The mass loss visible on the TG curve in the region of 100°C can be assigned to free water content, decomposition of amorphous hydration products and ettringite. In the case of reference sample R the mass loss after 1 and 5 hours of hydration was minimal, however in the case of 1.d sample with the lime content was substantial. This finding proved the formation of hydration products in the presence of the lime even in induction period of cement hydration what was also proposed from the calorimetric data. Mass loss around 120°C can be assigned to gypsum decomposition, which was gradually depleted to the 5 hours of hydration and explain the absence of sulphate depletion peak in calorimetric curves with high intensity of heat flow within the main hydration peak [12]. Strong exothermic effect with low mass loss in the region from 250 to 300°C can be assigned to superplasticisers decomposition.

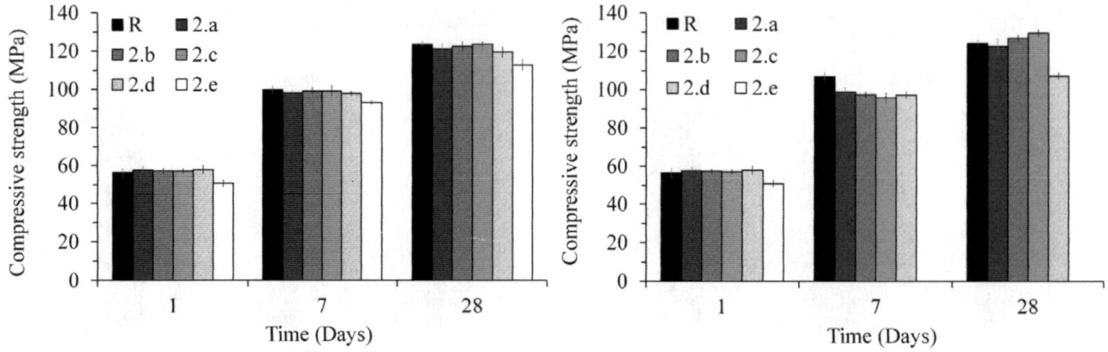

Figure 7. Compressive strength of the specimens with DBL a) dry curing, b) water curing.

This effect was not visible in the samples R and 1.d 0 h where the SP was not added. In the region around 450°C the decomposition of $Ca(OH)_2$ took place, which belonged to early stage hydration products.[13] In the case of reference sample R the portlandite peak was firstly depleted due to the portlandite consumption in reaction with silica fume species and afterwards formed due to the ongoing hydration process. In the case of sample 1.d the portlandite seemed to be slightly consumed within the first hour of hydration, further its content was increased due to the higher formation from the lime addition than the reaction rate with the silica fume. Decomposition of calcium carbonates took place above 650°C [14]. The addition of the lime to the system increased also the carbonate content even before the hydration process due to the not precise its firing and also carbonation process. During the hydration of the sample R the carbonate content was constant up to 5 hours of hydration, the significant increase was observed after the 24 hours of hydration. In the case of 1.d sample the content of carbonates during the whole hydration process was only slightly increased. Reaction of calcium ions with silica fume was observed by gradual depletion of portlandite in the end of cement induction period. In the case of the sample 1.d the significantly higher formation of hydration products during and also after induction period was observed.

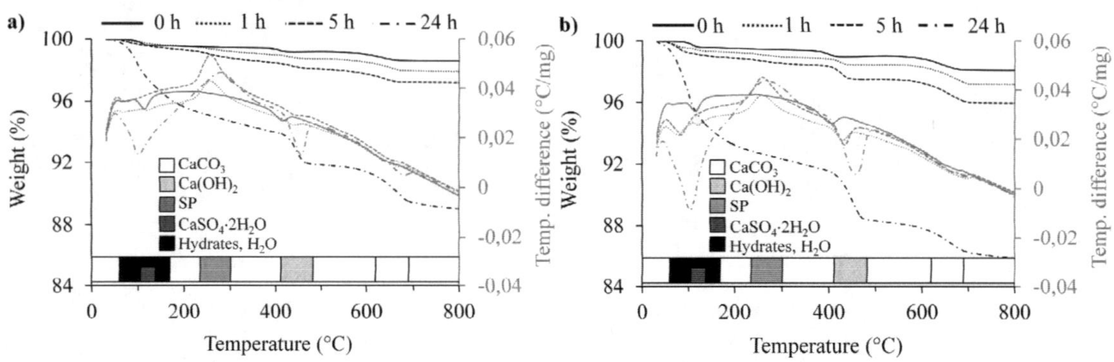

Figure 8. TG-DTA analysis of hydration products by means of time a) sample R, b) sample 1.d.

4. Conclusion

The presented data proved that the lime content can improve the silica fume hydration in binder systems even during the induction period of OPC hydration regardless the lime burning conditions. In the case of the content of dead-burnt lime followed by water curing conditions, mechanical strengths were not significantly affected, but there is high risk of product destruction with the high content of this lime.

The dry curing conditions applied to samples with both types of the lime remarkably enhanced the flexural strengths with minimal influence to compressive ones.

Acknowledgements
This outcome has been achieved with the financial support by the project: GA19-16646S "The elimination of the negative impact of zinc in Portland cement by accelerating concrete admixtures", with financial support from the Czech science foundation.

References

[1] Metha P K and Monteiro P J M 2014 *Concrete: Microstructure, Properties, and Materials* (New York: McGraw-Hill Education)
[2] Aǐtcin P C 2005 *High-Performance Concrete* (New York)
[3] Neville A M 2011 *Properties of Concrete* (New York: Pearson) chapter 6 pp 271–313
[4] Southararajan V M (2013) Micro filler effects of silica-fume on the setting and hardened properties of concrete *Research Journal of Applied Sciences, Engineering and Technology*
[5] Köksal F, Altun F, Yiğit İ and Şahin Y 2008 Combined effect of silica fume and steel fiber on the mechanical properties of high strength concretes *Con. Build. Mat.* **22** pp 1874-1880
[6] Kolani B, Buffo-lacarrière L, Sellier A, Scadeillas G, Boutillon L and Linger L 2012 Hydration of slag-blended cements *Cem. Concr. Com.* **34** pp 1009-1018
[7] Zhang Y, Kong X, Sellier A, Escadeillas G, Boutillon L and Linger L 2015 Correlations of the dispersing capability of NSF and PCE types of superplasticizer and their impacts on cement hydration with the adsorption in fresh cement pastes *Cem. Concr. Res.* **69** pp 1-9
[8] Wong H S and Razak H A 2005 Efficiency of calcined kaolin and silica fume as cement replacement material for strength performance. *Cem. Concr. Res.* **35** pp 696-702
[9] Ruiz-Agudo E, Rodriguez-Navarro C, Sellier A, Escadeillas G, Boutillon L and Linger L 2010 Microstructure and Rheology of Lime Putty. *Lang.* **26** pp 3868-3877
[10] Srinivasan M, Kumar V R, Saffiq R M, Gopinath S, Murthy A, Iyer N 2017 Effect of lime water on the properties of silica fume blended cementitious composite *I. J. E. Mat. Sci.* **24** pp 491-498
[11] Luz A P and Pandolfelli V C 2011 Halting the calcium aluminate cement hydration process. *Cer. Int.* **37** pp 3789-3793
[12] Novotný R, Bartoníčková E, Švec J and Mončeková M 2016 Influence of Active Alumina on the Hydration Process of Portland Cement *Pro. Eng.* **151** pp 80-86
[13] Esteves L P 2011 On the hydration of water-entrained cement–silica systems: Combined SEM, XRD and thermal analysis in cement pastes *Ter. Acta* **518** pp 27-35
[14] Boháč M, Palou M, Novotný R, Másilko J, Šoukal F and Opravil T 2017 Influence of temperature on early hydration of Portland cement–metakaolin–slag system. *J. Therm. Anal. Cal.* **127** pp 309-318

IOP Conf. Series: Materials Science and Engineering **583** (2019) 012005 doi:10.1088/1757-899X/583/1/012005

Hydrothermal reaction products of brick recyclates

J Palovcik[1], T Opravil[1], F Soukal[1] and J Masilko[1]

[1] Brno University of Technology, Faculty of Chemistry, Materials Research Centre, Purkyňova 464/118, Brno 612 00, Czech Republic

Email: xcpalovcik@fch.vut.cz

Abstract. This paper deals with possibilities of the brick-grind-dust application as a precursor for hydrothermal synthesis. Currently, this brick dust is partially reused in brick masonry or buried to the landfill site. The basic idea of this paper was to examine possibilities of the reaction of phases in brick dust with lime under hydrothermal conditions. The aim of the work was to find out an optimal ratio between lime and brick dust for maximal production of tobermorite. Physical and chemical properties of the brick-grind-dust were studied, and then a series of samples in autoclaves were prepared. The maximum amount of tobermorite after 24 hours of reaction was found in samples with addition of 25% lime. Phase composition of raw material and products of hydrothermal synthesis were analysed by powder X-ray diffraction analysis and differential thermal analysis. Structure of prepared samples was observed by scanning electron microscope.

1. Introduction

The tobermorite ($Ca_5Si_6O_{16}(OH)_2{\cdot}5H_2O$) is one of the basic mineralogical phase occurring in autoclaved aerated concrete and other naturally occurring minerals. Calcium silicate hydrates (CSH), or calcium silicate aluminate hydrates (CSAH) phases are quite explored area because their synthetical analogical structures are obtained as hydration products of cementitious building materials. The final structure and type of CSH or CSAH phases depend on the chemical and phase composition of the starting material and many other conditions such as amount of water, temperature, pressure, and reaction time [1].

The reaction conditions necessary for tobermorite formation are called hydrothermal conditions (environment of saturated steam) and the ideal temperatures range is between 173°C and 197°C roughly. This temperature range is given by the intersection of the temperature solubility dependence of CaO (decreasing with temperature) and the temperature solubility dependence of SiO_2 (increasing with temperature) [2]. The intersection of solubility CaO and SiO_2 is shown in the figure 1. Calcium to Silicone ratio (Ca/Si = 5/6) of forming tobermorite, comes from its chemical formula. The temperature range and Ca /Si ratio needed to form tobermorite and others CSH phases are shown in figure 2 [3].

Production of building materials requires their stability, which is closely linked with phase stability. In autoclaved aerated concrete tobermorite is the desired phase, but there is a possibility of conversion into the other CSH mineral, xonotlite ($Ca_6Si_6O_{17}(OH)_2$). The stability of tobermorite can be ensured by the presence of aluminum and iron cations, which affect the transfer of tobermorite to xonotlite [4, 5].

The formation of tobermorite is described in the literature by two possible mechanisms according to the presence of aluminum ions. In the first case, without aluminum ions, tobermorite crystalizes from CSH gel via reaction described in equation 1. In the presence of aluminum ions is tobermorite created by two paths that run simultaneously. One possibility is the reaction through the CSH gel as written above. The second possibility is that reaction proceeds through the mineral phase katoite (a type of hydrogarnet) and crystallization of tobermorite is solution controlled [4].

$$CaO + SiO_2 + H_2O \rightarrow CSH(II) \rightarrow CSH(I) \rightarrow C_5S_6H_5(tobermorite) \quad (1)$$

Traditionally used sources of SiO_2 in autoclaved aerated concrete described in literature are natural raw materials like a silica sand, kaolin, diatomaceous earth, crushed stone powder or same secondary raw materials represented by blast furnace slag, silica fume, fly ashes, ground glass waste, or porcelain [2, 6, 7, 8, 9]. Among the secondary raw materials brick grind dust can be included [10]. Burned brick can be used as a source of silicon oxide, aluminum oxide and iron oxide, in the form of mineral phases, if they are soluble. Soluble phases react under hydrothermal conditions to form new phases, potentially tobermorite, which they can stabilize to not to convert it into xonotlite [4, 5]. Brick-grind-dust is produced at the end of brick production when bricks are ground to achieve their high flatness. Brick dust is partially reused into the brick manufacturing process or buried to the land field site. Bricks are fired at a temperature of about 850-1000°C, so there is a possibility of a small amount of amorphous phase, depending on the feldspar content [11]. The amorphous content is potentially more reactive (more soluble under hydrothermal conditions) due to its disorder structure. A better solubility of the amorphous phase is shown in figure 1.

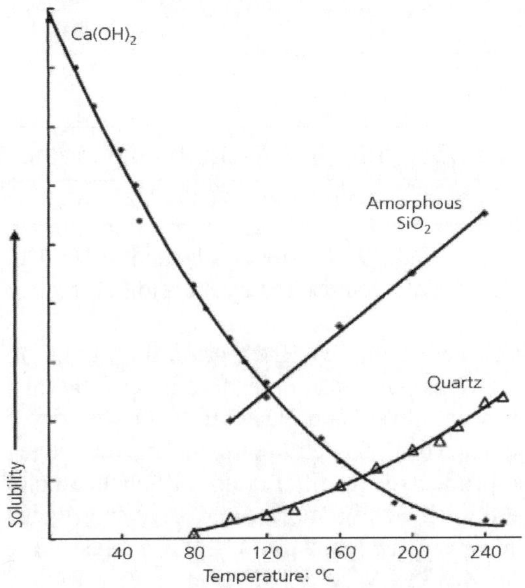

Figure 1. Temperature dependence of solubility of $Ca(OH)_2$ and SiO_2 [2].

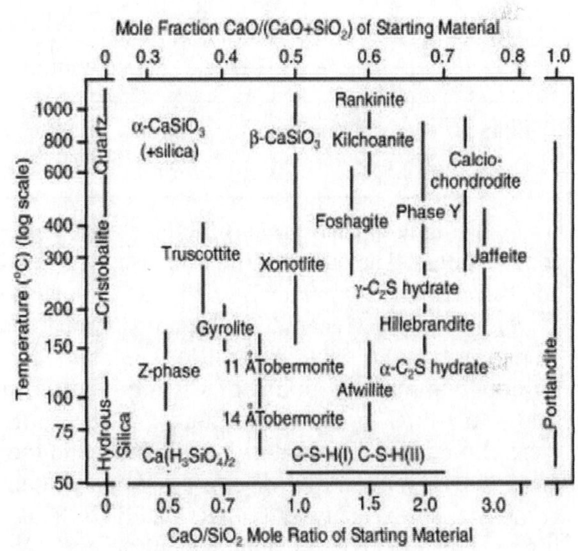

Figure 2. Phase changes in system $CaO/SiO_2/H_2O$ at elevated temperatures [3].

2. Materials and methods

Starting material was brick-grind-dust from brick factory Hevlín (company HELUZ). The specific surface of brick dust was 7300 $cm^2 \cdot g^{-1}$ (Blaine). Particle size composition of the fraction less than 0.63 μm (the content of this fraction was approximately 66%) was analysed by particle size laser diffraction analysis (HELOS Sympatec) (figure 3). D_{50} of this fraction was 4.19 μm. Chemical composition of brick-grind-dust was analysed by energy-disperse X-ray fluorescence analysis (by spectrometer VANTA VCR) and is given in table 1. Phase composition of brick-grind-dust was analysed by powder X-ray diffraction analysis (XRD) (Empyrean, Panalytical). For the determination of amorphous content the Rietveld analysis was conducted. Results of phase analysis of the brick-grind-dust are in table 2.

IOP Conf. Series: Materials Science and Engineering **583** (2019) 012005 doi:10.1088/1757-899X/583/1/012005

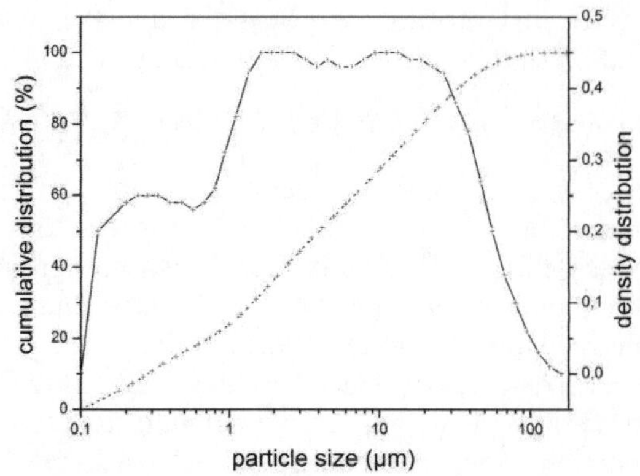

Figure 3. Particle size distribution of brick-grind-dust.

Table 1. Elemental composition of brick-grind-dust (LE = light elements).

Element	Si	Ca	Al	Fe	Mg	K	S	Ti	P	LE
(%)	22.82	8.21	8.19	4.4	2.84	2.38	0.66	0.51	0.06	49.93

Table 2. Phase composition of brick-grind-dust.

Phase	Amorphous	Quartz	Albite	Augite	Muscovite	Orthoclase	Akermanite	Hematite
(%)	28.5	30.4	17.1	5.6	11.2	2.0	4.3	0.9

As a source of calcium ions the limestone from Vitošov was used. Limestone was heated to 1000°C for two hours. The purity of the lime used was greater than 97% CaO determined by thermogravimetric and differential thermal analysis (TG-DTA analysis).

Samples were prepared in laboratory steel autoclaves with Teflon linings. Brick-grind-dust and lime were mixed in different Ca/Si ratios (see table 3). Amount of water was constant in all samples (100 ml). After two minutes of mixing of starting mixture, autoclaves were closed and placed to laboratory dryer set up to 180°C for a different time, 24, 48, 72 and 120 hours. After this heat treatment the autoclaves were cooled to the laboratory temperature and the reaction products were filtered on a Büchner funnel and dried at 60°C. After drying the prepared samples were crushed in a mortar and analysed. Phase composition was analysed by XRD analysis. Selected samples were analysed by TG-DTA analysis (TA Instruments Q600) and scanning electron microscope (SEM, ZEISS EVO LS 10).

Table 3. The starting composition of samples.

Sample	10_%_CaO	15_%_CaO	20_%_CaO	25_%_CaO	30_%_CaO	37,5_%_CaO	50_%_CaO
CaO (g)	2	3	4	5	6	7.5	10
Br.dust (g)	18	17	16	15	14	12.5	10
Ca/Si	0.71	0.91	1.14	1.40	1.70	2.24	3.49
H_2O (g)	100						

3. Results and discussion

According to chemical analysis the Ca/Si ratio in brick dust is 0.36. Stoichiometric ratio origine from the tobermorite formula is 0.83, but this value comes from a clear system which should contain only reactive CaO, SiO_2 and H_2O. Drochytka in his publication Pórobeton [12] describes a phase diagram

$CaO–SiO_2–H_2O$ by Bessey. There is a possibility to create tobermorite from ratio ranging from 0.8 to 1.0 in the system which contains only CaO, SiO_2 and H_2O. Brick dust is composed of at least eight mineral phases. Some minerals, containing calcium and silicon, do not react under hydrothermal conditions and could be found in the sample even after the reaction. To find the optimum lime addition to the system based on brick dust, an addition of 10 to 50% CaO has been proposed. Resulting Ca/Si ratios are given in table 3. The idea of the experiment was to obtain the highest amount of tobermorite as possible without the traces of unreacted portlandite.

The phase composition of prepared samples was analysed by XRD analysis. In figure 4, the diffractograms of samples hydrothermally cured for 24 hours are shown. There are same phases originated from the brick-grind-dust like muscovite, quartz, albite and akermanite. Additionally after hydrothermal synthesis the presence of tobermorite, katoite and hibschite were observed. There is an increase in the tobermorite peak intensity in samples prepared using up to 25% of CaO. With more added lime the intensity is decreasing. Calculated ratio for the sample with 25% addition of CaO comes from XRF spectra of brick and CaO addition results in 1.4. Along with the increasing amount of added lime, the intensity of portlandite peaks is increasing. For samples containing 15% and 20% lime, portlandite diffraction is almost insignificant. The intensity of calcite peak is also increasing with lime addition. With increasing lime content, a slight increase in hibschite peaks may be observed at the expense of katoite. This observation corresponds to the theory of the formation of tobermorite over the katoite which was mentioned by Matsui et. al. [4]. With the increase in lime content, the dissolution rate of the muscovite, quartz and feldspar increase. For the maximum possible yield of tobermorite after 24 hours, the most suitable addition of CaO is in the range 20–25% CaO.

A mixture containing 25% CaO was subjected to another experiments, which extended the reaction time to up to 48, 72 and 120 hours. The diffractograms are shown in figure 5. The diffraction spectrum shows that portlandite peaks are lost after 48 hours. The intensity of calcite diffraction is approximately the same after all reaction times. This amount of calcite is probably a product of portlandite carbonation because of CO_2 content in the air. The amount of tobermorite increases in time, while the amount of quartz decreases. The biggest change is apparent after 48 hours, and after prolongation of the reaction, the phase composition of the prepared samples is comparable. There is no conversion of tobermorite into xonotlite, even after the 120 hours of reaction. The influence of the aluminium and iron presence in the reaction system to prevent transformation of tobermorite to xonotlite is in good agreement with the results published earlier [4, 5].

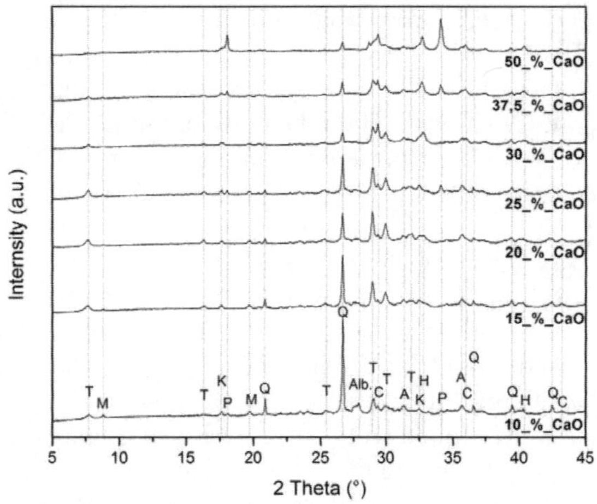

Figure 4. Comparison of diffractograms for different CaO concentrations, 24 hours of reaction.

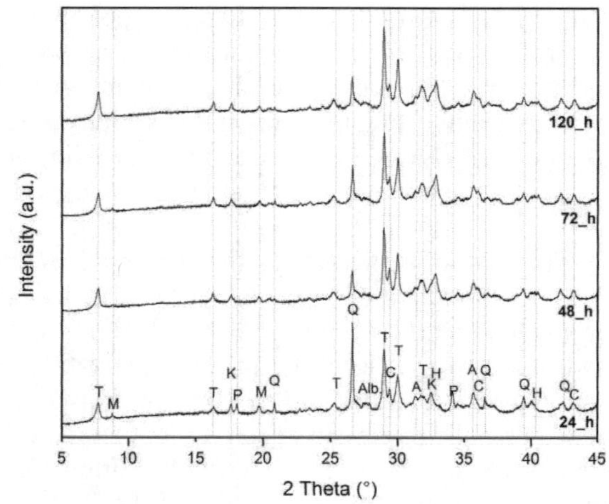

Figure 5. The diffractograms of samples prepared with 25 % of CaO for different time of reaction.

Markers: T – 11 Å tobermorite, M – muscovite, K – katoite, P –portlandite, Q – quartz, Alb. – albite, C – calcite, A – akermanite, H – hibschite.

The morphology of the brick-grind-dust can be described as angular fine particles with different shape and size (see figure 6). After 24 hours of hydrothermal conditions (figure 7), crystals on the surface of bricks-grind-dust particles are formed. Their morphology and shape of the crystals is difficult to describe even with a magnification of 20 000 ×. With increasing reaction time (figure 8), plates or flakes of tobermorite crystallized. This structure is very similar to the structure described in the article by Gutovic et al. [10], where brick recycles with Portland cement under hydrothermal conditions were studied.

Figure 6. Brick dust before reaction (mag. 15 000×).

Figure 7. Brick dust after 24 hours of the hydrothermal reaction (mag. 20 000×).

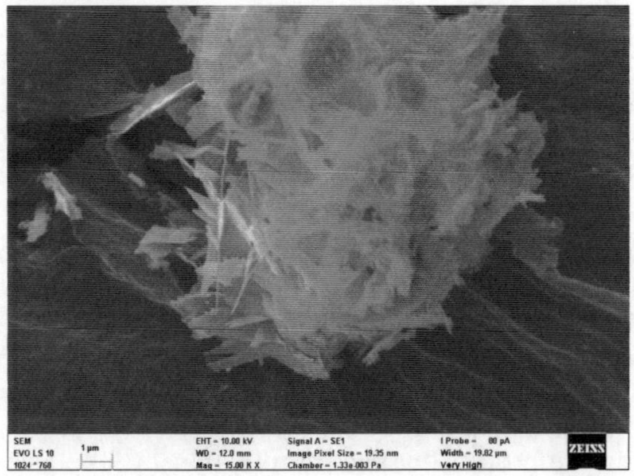

Figure 8. Brick dust after 120 hours of the hydrothermal reaction (mag. 15 000×).

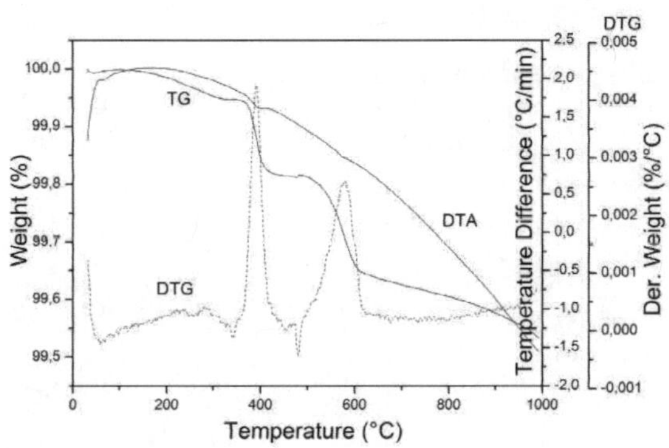

Figure 9. Thermogravimetric and differential thermal analysis of brick-grind-dust (exo up).

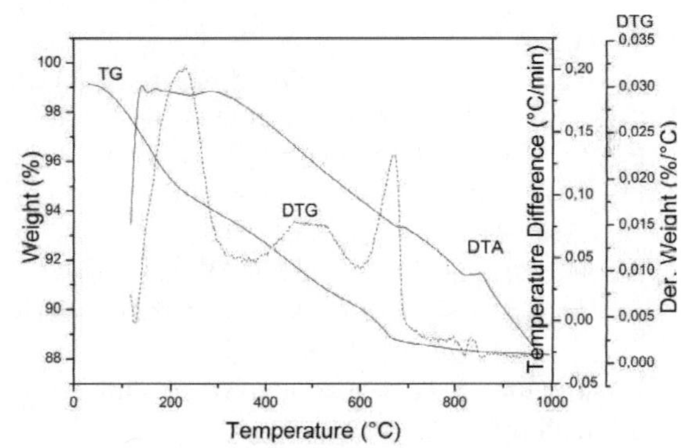

Figure 10. Differential thermal analysis after 120 hours of reaction (exo up).

Sample after 120 hours of reaction and unreacted brick dust was analyzed by TG-DTA analysis. From the figure 9 it is obvious, that brick dust contains little amount of water probably bonded in a C–(A)–S–H phases or in a structure of gypsum which is lost between 50°C and 250°C. With increasing temperature, two more peaks are observed on DTG curve. First, there is a peak of dehydroxylation of portlandite above 400°C, then above 600°C decomposition of carbonates occur. After 120 hours of reaction (figure 10) loss of water between 50°C and 250°C from C–(A)–S–H phases is observed because a part of the structurally bonded water is included in the tobermorite structure. Finally, tobermorite recrystallizes to wollastonite above 850°C, which is described by Huang et al. [13]. This is observable as the exothermic peak on DTA at this temperature. Shaw et al. [14] describe this conversion of tobermorite to wolastonite by the differential scanning calorimetry/thermogravimetric analysis. Shaw et al. describe this fact on the natural minerals and Huang et al. describe this effect on the system based on pure precursor. The amount of tobermorite which is prepared from brick-grin-dust is probably smaller, due to presence of non-reactive phases. Therefore, the phenomena on the DTA curve are not very pronounced and differential scanning calorimetry should be probably better method for tobermorite characterization.

4. Conclusion

The reaction of a fine fraction of brick recyclates with calcined limestone in hydrothermal conditions results in a production of new minerals: tobermorite, katoite and hibschite. The optimal amount of burned lime was found in range of 20–25% of CaO. When 25% of CaO or more was used, then some amount of portlandite was found, which is undesirable for potential application in autoclaved aerated concrete.

The presence of tobermorite was also observed by TG–DTA analysis, whereby at temperatures around 850°C the recrystallization of tobermorite into wollastonite occur. Reaction time plays the greatest role in terms of the morphology of prepared tobermorite. After 24 hours, the structure of the reaction products was not fully developed. After 120 hours of reaction, the platelet crystals of tobermorite were already well developed.

Subsequent aim for the research of this topic would be the reduction of amount of portlandite in final product because of stability. Next step is to prepare autoclaved aerated concrete with brick-grind-dust and test the mechanical properties of them.

Acknowledgment
This work was supported by the project of specific research: FCH-S-18-5194 "Inovativní pokročilé materiály – stadium vlastností a možnosti aplikace" with the financial support of Brno University of Technology.

References
[1] Taylor H F W 1993 *Adv. Cem. Based Mater.* (Cults, Aberdeen) **1** pp 38–46
[2] Arabi N, Chelghoun N, Jauberthie R and Molez R 2016 *Adv. Cem. Res.* vol 27(3) pp 153–62
[3] Nelson E B 1990 *Dowell Schlumberger Ed. Serv.* (Houston, USA)
[4] Matsui K, Kikua J, Tsunashima M, Ishikawa T, Matsuno S, Ogawa A and Sato M 2011 *Cem. Con. Res* vol 41 pp 510–19
[5] Haastrup S, Yu D and Yue Y 2018 *Mater. Chem. Phys.* **205** pp 180–5
[6] Melichar J, Černý V and Drochytka R 2018 *WSEAS Transactions on Environment and Development* **14** pp 408–17
[7] Jimenéz I, Pérez G, Guerrero A and Ruiz B 2017 *International Journalo of Mineral processing* **158** pp 8–12
[8] Song T H, Lee S H and Kim B 2014 *Constr Build Mater* **52** pp 105–15
[9] Li Z, Wang G, Yan Ch and Zhang A 2014 *Micro and nano Letters* **9** pp 536–8
[10] Gutovic M, Klimesh D S and Ray A 2004 *Constr Build Mater* **19** pp 353–8
[11] Pytlík P 1995 Cihlářství (Brno: CERM)
[12] Drochytka R 1992 Pórobeton (Brno: VUTIUM) vol 1
[13] Huang X, Jiang D and Tan S 2001 *Materiaals Research Bulletin* **37** pp 1885–92
[14] Shaw S, Henderson C M B and Komanschek B U 2000 *Chemical Geology* **167** pp 141–59

IOP Conf. Series: Materials Science and Engineering **583** (2019) 012006 doi:10.1088/1757-899X/583/1/012006

Stabilisation of belite prepared from modified limestone waste sludge

A Rybova[1], D Kubatova[1] and A Zezulova[1]

[1] Research Institute for Building Materials, Hněvkovského 65, 617 00 Brno, Czech Republic

E-mail: rybova@vustah.cz

Abstract. Presenting work deals with the use of waste materials - waste sludge from mining and washing of limestone and waste material from acetylene gas cylinders containing mainly tobermorite and xonotlite. Both materials were characterized by X-ray powder diffraction, laser granulometry and chemical analysis. After drying and milling of raw materials, they were mixed together and homogenised resulting in 3 sample series with different ratios of these materials. The mixtures were then burned at 1150°C and 1350°C and the main products formed were lime, dicalcium silicate, wollastonite, pseudowollastonite and gehlenite. To preserve hydraulically active belite, K_2O as stabilizing agent was added in amount of 1 and 3 wt. %. Burned products were studied by means of X-ray diffraction and optical microscopy. Also hydration was observed.

1. Introduction

Since cement manufacturing is responsible for 5% of global CO_2 emission [1], many researches have been done in order to find less energy consuming binders. The main sources of CO_2 during cement manufacturing are limestone which decomposes to CaO and CO_2 and also fuels used during burning. Very effective way to reduce CO_2 emission is to reduce clinker content in cements by using of supplementary cementitious materials (SCMs), to reduce energy demand by lowering the final temperature of burning or to prepare clinker with different ratio of clinker phases. Lower burning temperature can be achieved by preparing clinker, which main component is belite instead of alite. Many authors [2–5] prepared reactive belite clinkers with properties comparable to ordinary Portland clinker (OPC). Belite (β-C_2S, β-Ca_2SiO_4), that usually forms 15–30% of OPC, can be synthetized by burning at temperature around 1200°C while alite (C_3S, Ca_3SiO_4), that is the main component of OPC – content of 50–70%, requires high burning temperatures – around 1450°C [6, 7].

The main interest required on burning process is the stabilization of C_2S which has 5 polymorphic modifications – α, α'H, α'L, β and γ. All of them except γ are metastable at room temperature and could be preserved only by burning at high temperatures or by addition of stabilizing ions. Unfortunately, the thermodynamically stable modification – γ-C_2S usually formed from β-C_2S transformation at 630°C has not any hydraulic properties. This transformation is accompanied by volume increase which causes phenomenon called "dusting" [8, 9]. Polymorphic modifications α, α' and β are considered hydraulic, but reactivity of β-C_2S depends on many factors such as crystal imperfections, crystal size etc. [7]. To maintain hydraulic modification of belite, the burning process requires fast cooling and/or addition of dopants (e.g. Ba, Na, K, Fe, Mg, S in various compounds as $BaCl_2$, Na_2SO_4, NaF, K_2CO_3, Na_2CO_3, $FeSO_4$) [2, 10, 11]. However even hydraulically active belite modification has a disadvantage in very low early strength due to its high thermodynamic stability and dense structure that slows down reaction with water [12].

Apart from belite based clinkers, some alternative ideas have been published – cements based on completely different materials, secondary materials, binary or ternary binders made of SCMs and activators. Calcium sulphoaluminate binders or binders based on highly reactive C_2S are studied to greater extent nowadays [1]. These are made of conventional raw materials but could be based also on some industrial by-products, such as fly ash, blast furnace slag, metallurgical slags. Potential alternative raw materials for production of cementitious binders that have been studied are for example rice hush ash, sewage sludge, municipal solid waste incineration fly ash and metakaolin. [3]

The aim of the present work is to prepare belite-based binders with use of waste materials. In this paper, waste sludge from mining and washing of limestone was used for preparation of clinkers. Waste from acetylene storage cylinders was chosen as an additional raw material. Both these materials are stored in landfill and their amount is still increasing. Use of these products is possible, by burning of their mixtures to form potentially hydraulic binder, mainly based on belite. As a dopant for maintaining hydraulic modification K_2O in form of K_2CO_3 was used.

2. Experimental

2.1. Materials and methods

For the purpose of this project, two waste materials were chosen as they are stored uselessly in landfills and their composition could be suitable for preparing of hydraulic binders. The first material A is a waste sludge from mining and washing of limestone, the second one B is a waste from acetylene storage cylinders. Both materials were dried at 60°C, milled and characterized by XRD (X-ray diffraction) analysis (table 1), TG/DTA (thermogravimetry and differential thermal analysis) and also their chemical composition was determined (table 2). Specific surface before milling was evaluated according to British Standard B.S.12:1958 as follow – 262 m^2/kg for sample A and 1848 m^2/kg for sample B.

Table 1. Approximate phase composition of raw materials (wt. %).

	Calcite	Quartz	Albite	Illite/muscovite	Xonotlite	Tobermorite	Amorphous phase
A	81.0	14.0	2.0	3.0	-	-	-
B	18.0	26.0	-	-	17.5	19.5	19.0

Table 2. Chemical composition of raw materials (wt. %).

	SiO_2	Al_2O_3	CaO	Na_2O	K_2O	Fe_2O_3	SO_3	LOI
A	19.40	3.09	41.79	0.03	0.44	1.47	0.02	33.27
B	51.32	0.17	33.68	0.56	0.10	0.06	0.11	13.36

According to XRD, the major component of sample A is calcite ($CaCO_3$) following by quartz (SiO_2). Also illite/muscovite ($KAl_2(AlSi_3O_{10})(OH)_2$) and albite ($NaAlSi_3O_8$) are present in small amounts - under 5 wt.%. For sample B, all components are presented in comparable amounts. Sample contains quartz, calcite, tobermorite ($5CaO·6SiO_2·5H_2O$), xonotlite ($6CaO·6SiO_2·H_2O$) and amorphous phase.

TG/DTA results confirmed results of XRD – for sample A clay mineral dehydroxylation and calcite decarbonation (amount of calcite around 80%) are depicted. For sample B, insignificant dehydration can be seen. Calcite decarbonation showed amount of calcite around 20%.

The raw materials were mixed in weight proportions A:B – 75:25, 50:50 and 25:75 and resulting raw meals were denoted AB1, AB2, AB3. With and without addition of K_2CO_3 they were homogenized for

IOP Conf. Series: Materials Science and Engineering **583** (2019) 012006 doi:10.1088/1757-899X/583/1/012006

2 hours. K_2CO_3 addition was calculated so that resulting amount of K_2O in raw meals was 1% and 3 % respectively. These raw meals were then pressed into pellets 4 cm in diameter and burned in a super-kanthal furnace at 1150°C and 1350°C for 2 hours following by rapid cooling. Burned samples were milled in laboratory vibrating mill to fineness under 0.09 mm. All range of samples was analysed by XRD.

X-ray powder diffraction analysis was carried out on Bruker D8 Advance apparatus with Cu anode ($\lambda K\alpha = 1.54184$ Å) and variable divergence slits at Θ-Θ Bragg-Brentano reflective geometry. Quantitative phase analysis was done using Rietveld method. The amorphous phase content was calculated using internal standard method, as a standard 20 wt. % of fluorite (CaF_2) was used.

Monitoring of hydration of samples was done also by XRD analysis. Milled sample was mixed with water (water to binder ratio was 0.5), inserted into XRD sample carrier, covered with kapton foil and measured immediately. After the measurement that lasted 1 hour, sample carrier was uncovered and placed into high humidity environment and measured again after 1, 2, 3 and 7 days.

The structure of burned samples was studied by petrographic polarizing microscopy using Eclipse LV100ND by Nikon. Parts of burned pellets (before milling for XRD analysis) were crushed and mixed with epoxy resin. A polished section was prepared from samples with resin and was etched by vapour of acetic acid to get different interference colours of phases that are not resistant to the acid. For this test only samples AB1 and AB3 with 0% and 3% addition of K_2O were chosen.

Thermogravimetry and differential thermal analysis of hydrated samples was done on Perseus STA 449 (Netzsch) in dynamic atmosphere in temperature range 35–1150°C with temperature rise 10°C/min. Before testing, samples were milled under 0.09 mm and dried at 50°C for 24 hours.

3. Results and discussion

3.1. Burning of samples and phase composition

As the first stage, samples were burned with and without addition of K_2O at 1150°C and 1350°C. After milling, phase composition was evaluated by XRD.

Sample series AB1

In sample series AB1 dominating phase was β-belite (β-Ca_2SiO_4). The addition of 1% K_2O did not have significant effect on β-belite content. All samples in this series contained around 40% of β-belite, except AB1 3% K_2O 1350°C. This sample contained 97% of β-belite. Higher burning temperature and K_2O additions led to stabilization of β-C_2S, as proved in literature [2, 13, 14].

Samples burned at 1150°C contained also unreacted lime (CaO) formed from calcite ($CaCO_3$). Its amount decreased with increasing addition of K_2O. Samples burned at 1350°C contained significant amount of rankinite (Ca_2Si3O_8), which usually forms as accompanying phase of belite [8] and also γ modification of belite. γ-belite content also decreased with increasing K_2O content. Minor phases in samples were wollastonite (β-$CaSiO_3$), pseudowollastonite (α-$CaSiO_3$), quartz (SiO_2), gehlenite ($Ca_2Al_2SiO_7$), α′-belite (α′-Ca_2SiO_4) and cristobalite (SiO_2).

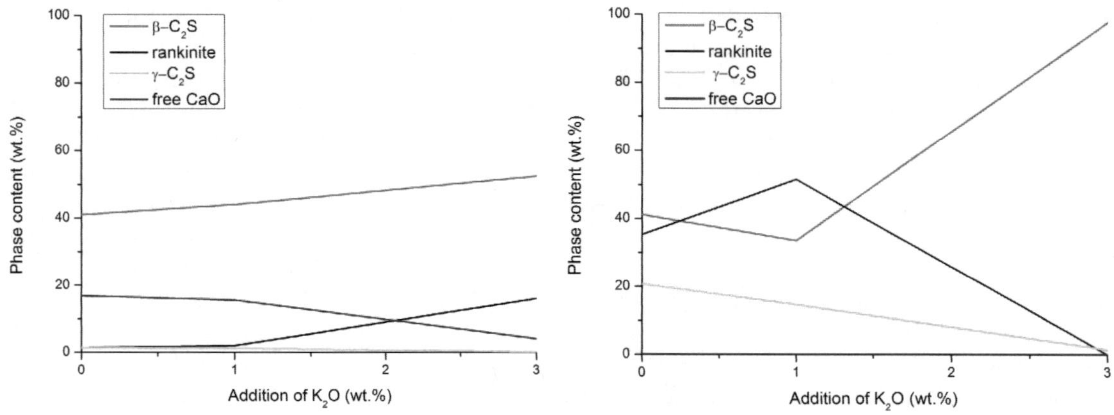

Figure 1. Content of β-belite, rankinite and γ-belite in samples AB1 with different ratio of K_2O burned at 1150°C (left) and 1350°C (right).

Sample series AB2

In these samples, dominating phase became pseudowollastonite. Its amount is increasing with higher temperature and also with higher addition of K_2O. Pseudowollastonite (α-$CaSiO_3$) is high temperature modification of wollastonite (β-$CaSiO_3$). Transformation temperature is 1125°C [8] but it is also affected by SiO_2 content. Higher SiO_2 content accelerates β to α transformation [15], which was proved also by our samples.

Samples burned at 1150°C contained also β-belite in amounts around 43%, for 3% K_2O it is 26% – it seems that K_2O supports SiO_2 reaction towards more pseudowollastonite formation. Accompanying phases in samples burned at 1150°C were wollastonite, rankinite, quartz, gehlenite, α´-belite and in small quantities also lime, cristobalite and γ-belite.

Burning at 1350°C led to the formation of around 70% of pseudowollastonite and 25% of rankinite. The rest was formed by α´ and γ modifications of belite.

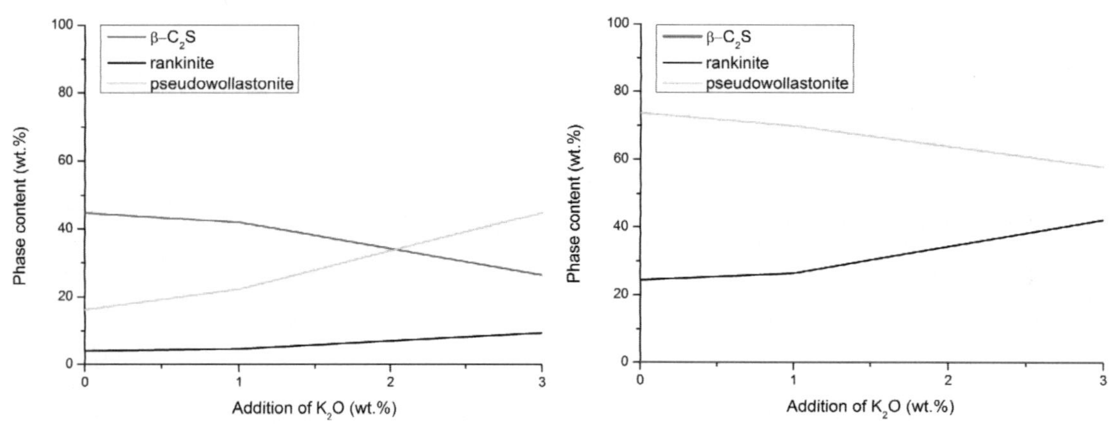

Figure 2. Content of β-belite, rankinite and pseudowollastonite in samples AB2 with different ratio of K_2O burned at 1150°C (left) and 1350°C (right).

Samples series AB3

Here, the major phase in all samples was pseudowollastonite. It formed more than 96% in samples burned at 1350°C. The rest was wollastonite. In samples burned at 1150°C pseudowollastonite also dominated, but its amount increased with amount of K_2O, as in previous sample series. High temperature and also high content of SiO_2 (for this series it was 43%, for series AB1 it was only 27%) stabilized pseudowollastonite. The same findings were reported by [8, 15]. However, pseudowollastonite is obviously stabilized also by K_2O addition, or at least transformation from wollastonite is faster if K_2O

is presented in system. Since no record about this phenomenon was found in literature, the process and its dynamics should be studied furthermore.

For samples burned at 1150°C similar dynamics were observed as in previous samples. β-belite and wollastonite content decreased with increasing K_2O. Additional phases were rankinite, quartz, gehlenite, cristobalite and γ-belite.

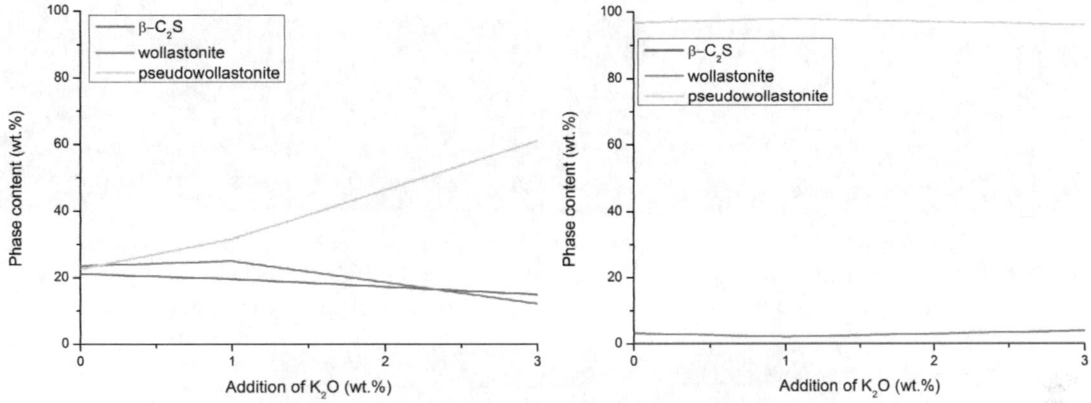

Figure 3. Content of β-belite, wollastonite and pseudowollastonite in samples AB3 with different ratio of K_2O burned at 1150°C (left) and 1350°C (right).

3.2. Microstructure

Microstructure of burned samples was studied by optical microscopy. Only samples AB1 and AB3 with 0% and 3% of K_2O were chosen, as other samples represent only the transition between these ones. For samples burned at 1150°C, details were not very well recognisable, as the structure was too fine for distinction, however, there were areas with distinguishable phases. In the sample AB1 0% K_2O 1150°C free lime and β-belite was observed that corresponds to XRD analysis results. In the sample AB3 0% K_2O 1150°C, areas of β-belite and probably wollastonite and pseudowollastonite were recognisable as well as a minor amount of free lime. Sample AB3 3% K_2O 1150°C contained areas of free lime and β-belite, although by XRD analysis free lime was not found. This is caused probably by very small overall amount of lime in the sample, that is under detection limit of XRD, especially when β-belite´s peaks are present (they overlap with lime´s peaks). The structure of AB1 3% K_2O 1150°C was not possible to distinguish.

The structure of samples burned at 1350°C is shown in figures 4–8. Grains of crystallised phases are visible. The areas between grains is solid amorphous melt formed during cooling. Solid melt with high SiO_2 content is amorphous probably due to inadequate chemical composition and cooling conditions for crystallization of other phases. Therefore, it is not possible to use XRD analysis for melt composition identification. During cooling, secondary overgrowth of lamellae occurs on the edge of the belite grain. In belite grains, cracks were formed that are probably connected with partial substitution of γ-C_2S modification (AB1 3% K_2O 1350°C) which occurs during cooling [16]. AB1 3% K_2O 1350°C contains only β-C_2S according to XRD. By optical microscope it can be seen that β-C_2S is bigger and more rounded than in the previous case. The melt is present as well. Microscope image of the sample AB3 0% K_2O 1350°C corresponds to XRD – grains of pseudowollastonite are present, the melt occurs in a smaller amount. On the section of the sample AB3 3% K_2O 1350°C, pseudowollastonite and a small amount of wollastonite are shown. Grains of wollastonite are needle-like, grains of pseudowollastonite are more angular. Both of them, wollastonite and pseudowollastonite, are resistant to vapour of acetic acid, they have no specific colour.

IOP Conf. Series: Materials Science and Engineering **583** (2019) 012006 doi:10.1088/1757-899X/583/1/012006

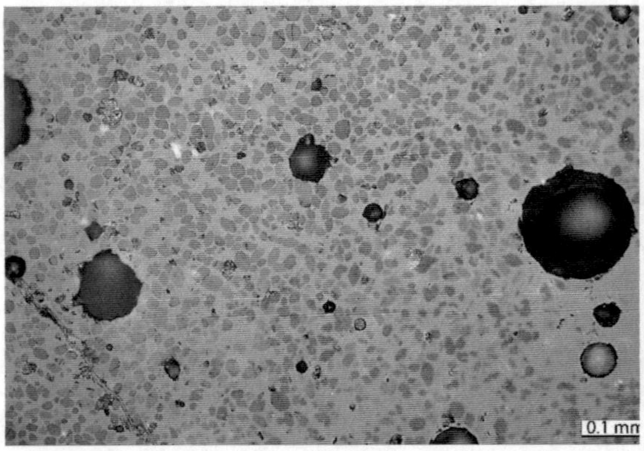

Figure 4. AB1 0% K_2O 1350°C, brown rounded grains - belite, light areas between grains - melt, dark diffused areas - open pores.

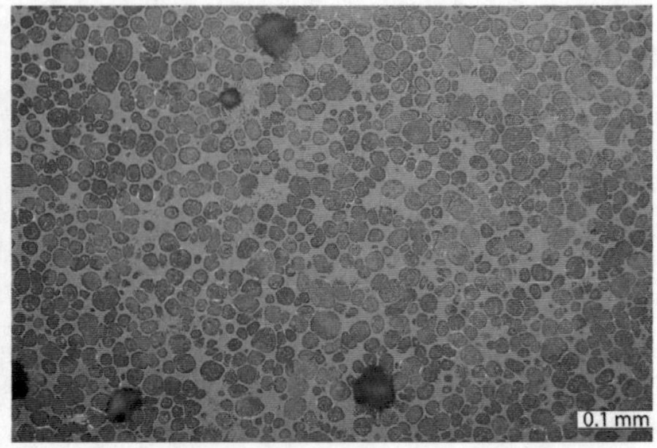

Figure 5. AB1 3% K_2O 1350°C, brown rounded grains - belite, light areas between grains - melt, dark diffused areas - open pores.

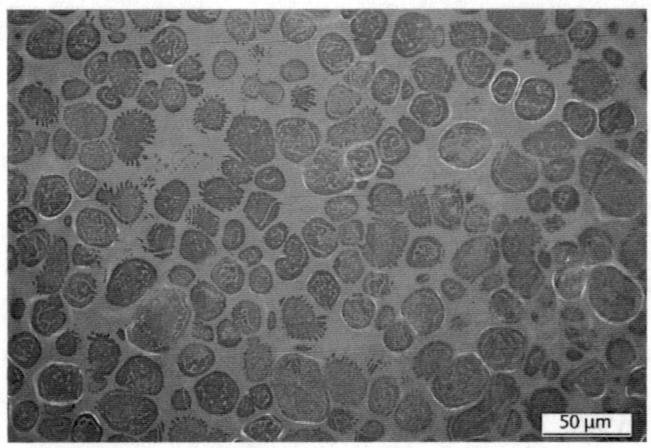

Figure 6. Detail of AB1 3% K_2O 1350°C, brown rounded grains - belite, light areas between grains - melt.

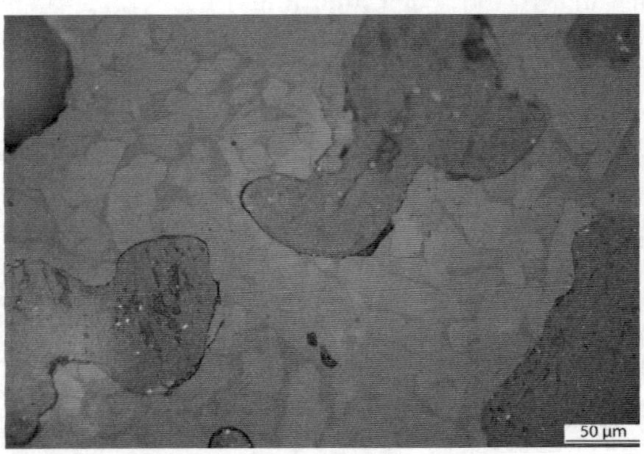

Figure 7. AB3 0% K_2O 1350°C, grey grains - pseudowollastonite, areas between grains - melt, dark diffused areas - open pores, large grey area -epoxide.

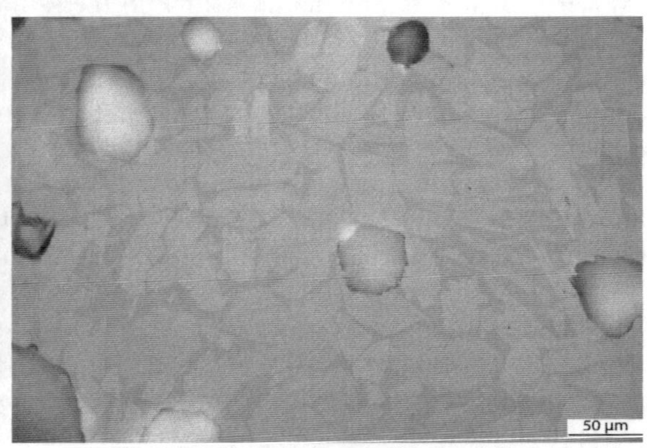

Figure 8. AB3 3% K_2O 1350°C, needle-like grains - wollastonite, other grains - pseudowollastonite, areas between grains - melt, dark diffused areas - open pores.

3.3. Hydration

Since results of XRD, confirmed by microscopy observations, showed that samples AB1 and AB3 with 3% K_2O burned to 1350°C contained around 97% of β-belite and pseudowollastonite respectively, the ability to hydrate was studied by means of XRD analysis.

In sample AB1 3% K_2O 1350°C, β-belite as a major phase did not hydrate, but carbonated. After 1 day of hydration significant amount of calcite occurred. Calcite peaks increase in time while β-belite peaks decrease (see figure 9). Belite is not probably hydraulically active, as stabilized β modification should be. The reason may be in its structure, as is reported in [7] or probably in its formation process, which needs further observations.

On the other hand, calcium silicates′ carbonation is very well known in literature. They react with CO_2 in aqueous conditions to generate calcium silicate hydrate (C-S-H) and calcite instead of portlandite ($Ca(OH)_2$). In adequate conditions, carbonation of alite (C_3S) and belite (C_2S) occurs immediately within a few minutes or hours [17, 18], as this was probably the case of our samples. Even hydraulically not active belite carbonates quite fast. During short-term carbonation all C_2S/C_3S grains do not fully react, therefore continued exposure to moist or humid environments post-carbonation allows for the subsequent hydration of residual unreacted portion [19]. For our samples, different environments of hydration will be studied. Also similar condition for long-term hydration/carbonation will be studied, since according to [20, 21] extensive carbonation leads to formation of silica gel and calcite.

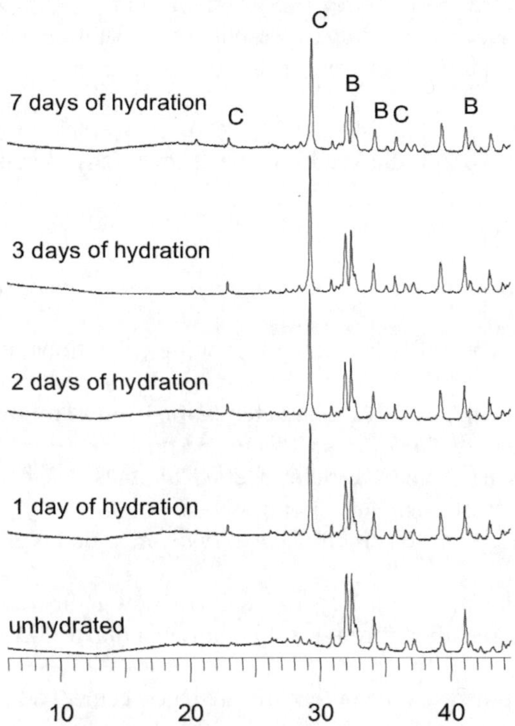

Figure 9. Hydration of sample AB1 3% K_2O 1350°C (C – calcite, B – β-belite).

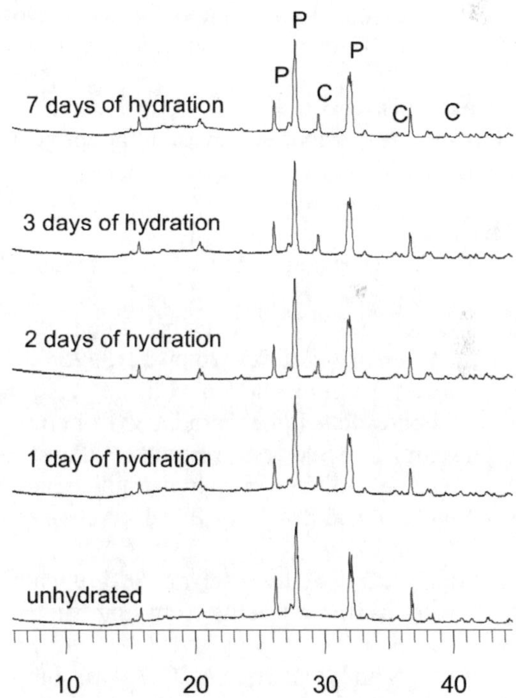

Figure 10. Hydration of sample AB3 3% K_2O 1350°C (C – calcite, P – pseudowollastonite).

The same experiment was done for sample AB3 3% K_2O 1350°C. It was confirmed that also pseudowollastonite carbonates, though not that intensively as β-belite. Calcite peak could be seen after 1 day of hydration and its intensity is slightly increasing (see figure 10).

Similar phenomenon is the base of non-hydraulic concrete system, calcium-silicate binder (SCS), known as Solidia Cement [22, 24], patented in 2015. Low-lime calcium silicate minerals such as wollastonite/pseudowollastonite and rankinite reacts with CO_2 to form a binder with low carbon footprint. SCS paste can store up to 17% of CO_2 by mass into a form of stable carbonates [23].

Considering lower production temperature and lower limestone use, the overall CO_2 footprint of carbonated concrete made of SCS has been reported [24] to be about 70% less than that of OPC concrete. Also, according to [22], concrete elements made of SCS have satisfactory freeze-thaw resistance and resistance to high temperature environment.

As we prepared pseudowollastonite from waste materials, the carbon footprint of potential product would be even smaller than that of the Solidia. The carbonation in humid environment will be studied furthermore, as well as mechanical and durability properties of these and previous samples.

4. Conclusions

Mixtures of two different waste materials (waste sludge from mining and washing of limestone and waste from acetylene storage cylinders) with and without addition of K_2O were burned at 1150°C and 1350°C. XRD analysis results showed that lower content of SiO_2 in mixture led to formation of β-C_2S while higher SiO_2 content supported the formation of wollastonite and pseudowollastonite. Higher temperature and higher K_2O addition stabilised β-belite in samples with higher CaO and lower SiO_2 content while for samples with lower CaO and higher SiO_2 content pseudowollastonite was stabilized.

Hydration of sample with β-C_2S showed no common hydration products. Only calcite peaks were detected after 1 day. Fast carbonation was observed at this sample. Similar trend could be seen also for sample containing pseudowollastonite. Calcite peaks were present after 1 day of hydration.

Both prepared materials require further research. However according to presented results tested waste materials could be conveniently used together for preparation of low-energy alternative binder.

Acknowledgements

This paper was elaborated within the project TJ01000158 financed by the Technology Agency of the Czech Republic.

References

[1] Ludwig H M and Zhang W 2015 Research review of cement clinker chemistry *Cem. Concr. Res.* **78** pp 24–37
[2] Chatterjee A K 1996 High belite cements — Present status and future technological options: Part I *Cem. Concr. Res.* **26** pp 1213–25
[3] Chen Y L, Lin Ch J, Ko M S, Lai Y Ch and Chang J E 2011 Characterization of mortars from belite-rich clinkers produced from inorganic waste *Cem. Concr. Comp.* **33** pp 261–66
[4] Kacimi L, Simon-Masseron A, Salem S, Ghomari A and Derriche Z 2009 Synthesis of belite cement clinker of high hydraulic reactivity *Cem. Concr. Res.* **39** pp 559–65
[5] Staněk T and Sulovský P 2015 Active low-energy belitė cement *Cem. Concr. Res.* **68** pp 203–210
[6] Maheswaran S, Kalaiselvam S, Arunbalaji S, Palani G S and Iyer N R 2015 Low-temperature preparation of belite from lime sludge and nanosilica through solid-state reaction *J. Therm. Anal. Calor.* **119** pp 1845–52
[7] Dutta N and Chatterjee A 2017 Hydrothermal synthesis of dicalcium silicate based cement *Mater. Sci. Eng.* **216**
[8] Hewlett P C 2004 *Lea's Chemistry of Cement and Concrete* (Oxford: Elsevier Science & Technology Books)
[9] Taylor H F W 1990 *Cement chemistry* (London: Academic Press)
[10] Ávalos-Rendón T L, Pastén Chelala E A, Mendoza Escobedo C J, Figueroa I A, Lara V H and Palacios-Romero L M 2018 Synthesis of belite cements at low temperature from silica fume and natural commercial zeolite *Mater. Sci. Eng.* **229** pp 79–85
[11] Koutník P 2017 Preparation of β-belite using liquid alkali silicates *Mater. Constr.* **67**
[12] Sinyoung S, Kunchariyakun K, Asavapisit S and MacKenzie K J D 2017 Synthesis of belite cement from nano-silica extracted from two rice husk ashes *J. Environ. Manage.* **190** pp 53–

60
[13] Kriskova L, Pontikes Y, Zhang F, Cizer Ö, Jones P T, Van Balen K and Blanpain B 2014 Influence of mechanical and chemical activation on the hydraulic properties of gamma dicalcium silicate *Cem. Concr. Res.* **55** pp 59–68
[14] Rodrigues F A 2003 Synthesis of chemically and structurally modified dicalcium silicate *Cem. Concr. Res.* **33** pp 823–27
[15] Rodriguez E T, Garbev K, Merz D, Black L and Richardson I G 2017 Thermal stability of C-S-H phases and applicability of Richardson and Groves' and Richardson C-(A)-S-H(I) models to synthetic C-S-H *Cem. Concr. Res.* **93** pp 45–56
[16] Petrog 2017 *Optical Microscopy of cement clinker* (*Online* http://www.petrog.com/ws/files /articles/ Microscopy_2017_Oct_Leica.pdf)
[17] Berger R, Young J and Leung K 1972 Acceleration of hydration of calcium silicates by carbon dioxide treatment *Nature* **240** pp 16–18
[18] Young J, Berger R and Breese J 1972 Accelerated curing of compacted calcium silicate mortars on exposure to CO_2 *J. Am. Ceram. Soc.* **57** pp 394–97
[19] Klemm W and Berger R 1972 Accelerated curing of cementitious systems by carbon dioxide: part I. *Cem. Concr. Res.* **2** pp 567–76
[20] Goodbrake C, Young J and Berger R 1979 Reaction of hydraulic calcium silicates with carbon dioxide and water *J. Am. Ceram. Soc.* **62** pp 488–91
[21] Šauman Z 1971 Carbonization of porous concrete and its main binding components *Cem. Concr. Res.* **1** pp 645–62
[22] Ashraf W, Olek J and Jain J 2017 Microscopic features of non-hydraulic calcium silicate cement paste and mortar *Cem. Concr. Res.* **100** pp 361–72
[23] Ashraf W, Olek J and Atakan V 2016 Carbonation reaction kinetics, CO_2 sequestration capacity, and microstructure of hydraulic and non-hydraulic cementitious binders *Sust. Constr. Mater. Technol.*
[24] Solidia Technologies 2018 Solidiatech (*Online* http://solidiatech.com)

IOP Conf. Series: Materials Science and Engineering **583** (2019) 012007 doi:10.1088/1757-899X/583/1/012007

Preparation and properties of Portland limestone cements

T Stanek[1], M Bohac[1], A Rybova[1] and A Zezulova[1]

[1] Research Institute for Building Materials, Hněvkovského 65, 617 00 Brno, Czech Republic

E-mail: stanek@vustah.cz

Abstract. Limestone is a suitable substitute for economically less favourable Portland clinker in Portland blended cements. Limestone in cement not only serves as a very fine filler, but can also act as a reaction component. In this work, the properties of laboratory Portland cements with limestone were studied. The cements were prepared from two limestone types and two clinker types with a graded limestone content of up to 30 wt. %. The limestones had a different $CaCO_3$ content. Clinkers were distinguished by a marked difference in C_3A content and hydraulic activity. The effect of limestones on the technological properties of cements, the hydration process and the impact of separate and together grinding of limestone and clinker were studied. In this case, the type of clinker had stronger impact than the type of limestone on the technological parameters. It was also confirmed that a low limestone content has a positive effect on cement hydration.

1. Introduction

Portland cement is the most important hydraulic binder and its worldwide consumption is constantly increasing. The production was around 4.1 billion tones in 2017 [1]. Cement production is characterized by high energy and material demands. When producing one ton of cement clinker, about 0.87 tons of CO_2 is released [2]. One way to reduce cement clinker consumption and CO_2 emissions is to replace it with suitable economical and environmentally-friendly materials, called supplementary cementitious materials (SCMs). One of these materials is limestone. According to the standard EN 197-1, cements with 6–35 wt. % of limestone are called Portland-limestone cement (PLC).

Some authors divided the action mechanism of limestone in cement-based materials into two broad mechanisms of filler effect (including filler effect, nucleation effect and dilution effect) and chemical effect, while other specified filler effect, nucleation effect, dilution effect and chemical effect as distinctive mechanisms. Limestone powder can positively or negatively affect the hydration and resulting parameters of the cement binder depend on its fineness and amount. [3]

Ground limestone works as a filler if its particles are smaller than cement particles. This improves the particle size distribution and the total packing density of the cement-based material [4]. As a result, water demand in concrete is reduced and its strength and durability may be improved [5].

Since Soroka and Setter [6] introduced the limestone nucleation effect, also other authors have confirmed that fine limestone particles provide spaces for the formation of seeds of hydration products [7]. This phenomenon speeds up the hydration [8] and improves the degree of cement hydration [5]. It has been reported that the increase in C-S-H precipitation on the surface of limestone particles is due to the similarity between the planar configuration of Ca and O atoms in calcite layers and in CaO layers in C-S-H [9].

The chemical effect of limestone is mainly related to its particles size and to the content of Al_2O_3 from C_3A and C_4AF in cement. Since the aluminate phase content of the clinker is limited, the chemical effect of limestone in the cement is small. The chemical effect can be increased by the addition of Al_2O_3 from other SCMs [10]. It is reported that limestone stabilizes ettringite and reacts with Al_2O_3 to calcium carboaluminates, thereby reducing the porosity of cement mass [11, 12].

The dilution effect of limestone is that it does not have pozzolanic properties and its excess does not participate in any reaction. Therefore, there is more available free water to react with cement particles. Consequently, the degree of cement hydration is increased, which has been confirmed in various studies [4, 5, 13].

In this work, the combined effect of two types of limestone and clinkers from two cement plants on the technological parameters of the PLCs was monitored. The impact of together and separate grinding was also verified.

2. Materials and methods

2.1. Grinding together

PLCs were prepared by combination of two industrial clinkers and two limestones. Limestones are from one quarry and both limestones, according to TOC content, meet the requirements for the LL category. The limestone labelled H has a high content of $CaCO_3$ (94.6 wt. %) and 2.1 wt. % SiO_2. The limestone labelled L is contaminated by chert and contain 79.0 wt. % $CaCO_3$ (with a high content of SiO_2 – 16.5 wt. % and also Al_2O_3 – 1.6 wt. %). The clinkers are from two cement plants, they have high content of alite (C_3S), however, the content of C_3A is very different (see table 1). The clinker C was burnt from high reactivity raw meal and has fine-grained structure. Raw meal for the production of clinker M has worse reactivity and the clinker is coarse-grained (see table 1). At first, two comparative reference cements were prepared only with the addition of gypsum as a set regulator. These cements were ground to a specific surface according to Blaine to 350 m^2/kg and the grinding time was measured. The addition of gypsum in case of clinker M was 5 wt. % and in case of clinker C 6 wt. %. Other cements with graduated amounts of limestone were ground for the same time as reference cements. Basic information about these cements are given in table 2 and 3.

Table 1. Quantitative phase composition of clinkers determined by optical microscopy in wt. %.

Clinker	C_3S	C_2S	C_3A	C_4AF	Cfree	ø size of C_3S
M	73.5	7.4	7.0	9.9	2.2	31 μm
C	70.2	7.6	18.5	3.5	0.2	20 μm

Table 2. Composition of PLCs with clinker M in wt. %, specific weight in kg/m^3 and specific surface in m^2/kg.

Cement	M	M-LL 10H	M-LL 20H	M-LL 30H	M-LL 10L	M-LL 20L	M-LL 30L
Clinker M	100	90	80	70	90	80	70
Limestone H	0	10	20	30	0	0	0
Limestone L	0	0	0	0	10	20	30
Specific weight	3087	3048	3009	2970	3052	3017	2981
Specific surface	351	382	410	451	397	439	466

Table 3. Composition of PLCs with clinker C in wt. %, specific weight in kg/m^3 and specific surface in m^2/kg.

Cement	C	C-LL 10H	C-LL 20H	C-LL 30H	C-LL 10L	C-LL 20L	C-LL 30L
Clinker C	100	90	80	70	90	80	70
Limestone H	0	10	20	30	0	0	0
Limestone L	0	0	0	0	10	20	30
Specific weight	3075	3037	2999	2961	3041	3006	2972
Specific surface	349	398	436	469	411	458	509

2.2. Grinding separate

In the second part of the experiment, PLCs were prepared by combination of two industrial clinkers used in the first part and industrially ground high-purity limestone of two different types of fineness (see table 4). The limestone type 7 had specific surface 372 m^2/kg and the limestone type 9 had 484 m^2/kg. At first, two comparative reference cements were prepared, only with the addition of set regulator. The cements were ground to 360 m^2/kg. The addition of gypsum was the same as in the first case. Other cements with graduated content of limestone were prepared by homogenization. Basic information about these cements are given in table 5 and 6.

Table 4. Granulometric characteristics of industrially produced finely ground limestones according to ČSN 72 1220 in wt. %.

Clinker	Max. residuum on sieve 0.5 mm	Max. residuum on sieve 0.2 mm	Max. residuum on sieve 0.09 mm
Type 7	10.0	-	40.0
Type 9	-	0.1	5.0

Table 5. Composition of PLCs with clinker M in wt. %, specific weight in kg/m^3 and specific surface in m^2/kg.

Cement	M II	M-LL 10-7	M-LL 20-7	M-LL 30-7	M-LL 10-9	M-LL 20-9	M-LL 30-9
Clinker M	100	90	80	70	90	80	70
Limestone type 7	0	10	20	30	0	0	0
Limestone type 9	0	0	0	0	10	20	30
Specific weight	3087	3048	3009	2970	3048	3009	2970
Specific surface	362	363	364	365	374	386	399

Table 6. Composition of PLCs with clinker C in wt. %, specific weight in kg/m^3 and specific surface in m^2/kg.

Cement	C II	C-LL 10-7	C-LL 20-7	C-LL 30-7	C-LL 10-9	C-LL 20-9	C-LL 30-9
Clinker C	100	90	80	70	90	80	70
Limestone type 7	0	10	20	30	0	0	0
Limestone type 9	0	0	0	0	10	20	30
Specific weight	3075	3037	2999	2961	3037	2999	2961
Specific surface	365	365	366	367	377	389	401

2.3. Testing methods

Compressive strengths were measured in mortars from PLCs prepared according to EN 196-1 standard, water-cured for 1, 2, 7, 28, 56 and 90 days. Mortars were prepared according to standard, i.e. with binder to sand ratio 1:3 and water to cement ratio w/c = 0.5.

The setting process of prepared PLCs was observed by a Tussenbrock method. Tussenbrock method is performed on a device with a needle similar to Vicats. Depth of submersion of six cylindrical probes of decreasing diameter in the cement paste is measured in time dependence. The depth of indentation of particular probe indicates a threshold of shearing stress (in kPa) which corresponds to consistency achieved at the moment of measurement. This method allows to capture several anomalies during the setting process such as false setting.

The influence of limestone on hydration process was observed by calorimetric measurement. Heat released during hydration process was measured within 24 hours. Observation of the course of the temperature during hydration process of prepared cement pastes was performed on semiadiabatic calorimeter [14]. From the calorimetric curves, the specific heat of cements after 24 hours hydration was calculated according to the following formula:

$$Q_{24} = \left(\frac{C_{sum}}{m}\right) \times (t - t_0) \quad (1)$$

Where: C_{sum} – total heat capacity of calorimeter in J.K-1 (cement, water, container);
m – weight of cement in g;
t, t_0 – final and initial temperature of the mixture.

The hydration heat was determined for selected samples also after 7 days of hydration using dissolution method according to EN 196-8.

3. Results and discussion

As you can see in table 2 and 3, in case of grinding together with clinker, limestone fineness significantly increases and therefor resulting cements fineness increases with increasing amount of limestone. From these values it can be also seen that limestone L has better grindability than limestone H and clinker M has better grindability than clinker C.

The most important property of the cements is their strength. The development of compressive strength of the prepared PLCs is shown in figures 1–5.

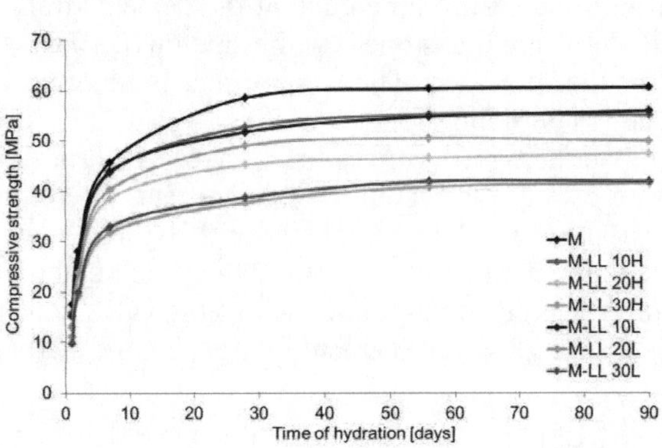

Figure 1. Development of compressive strengths for cements prepared from clinker M (grinding together).

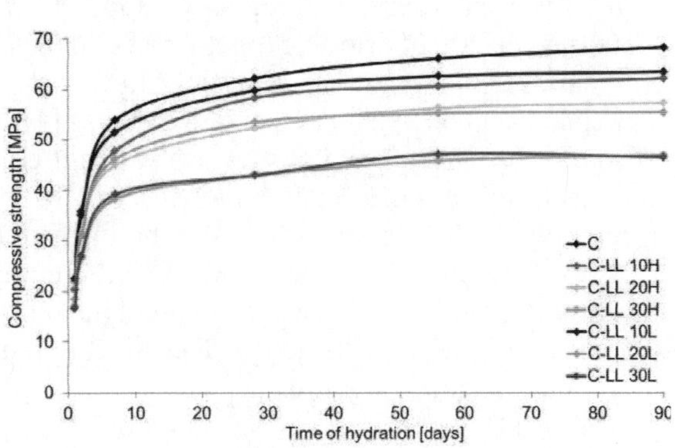

Figure 2. Development of compressive strengths for cements prepared from clinker C (grinding together).

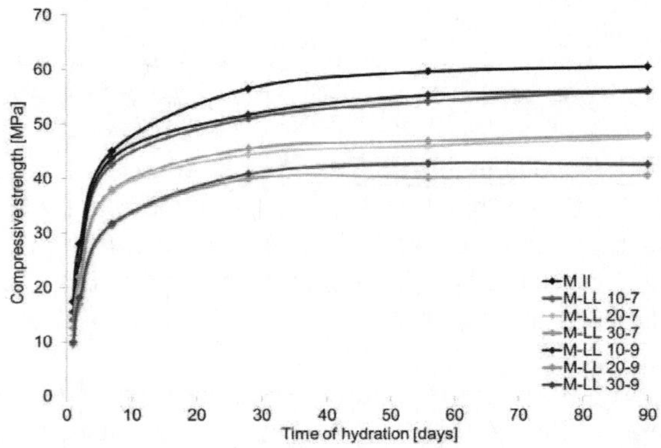

Figure 3. Development of compressive strengths for cements prepared from clinker M (grinding separate).

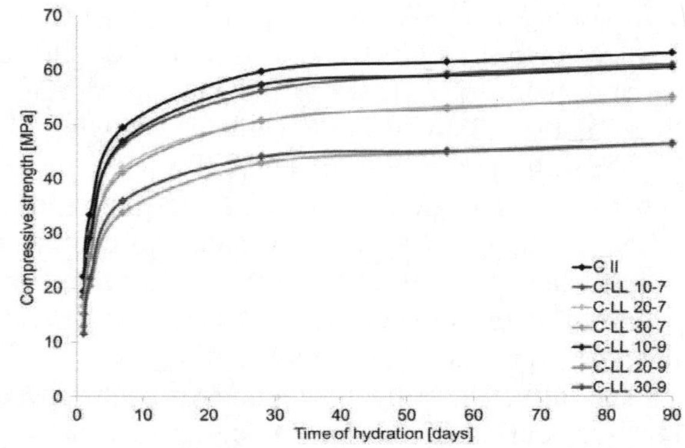

Figure 4. Development of compressive strengths for cements prepared from clinker C (grinding separate).

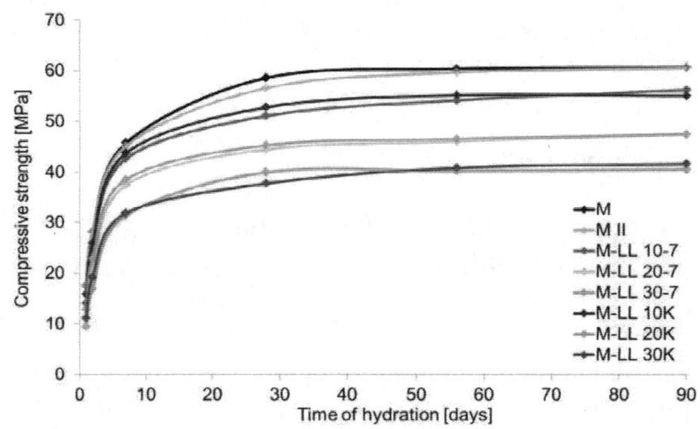

Figure 5. Comparison of compressive strengths for cements prepared from clinker M with different preparation.

It is evident from the strength development that with increasing lime content the cement strength decreases. In case of grinding together of clinker C and less pure limestone L in the amount of 10 wt. %, only short-term strengths are slightly higher than for the reference. The same effect is reported by Menéndez et al. [15] using 10 and 20 wt. % of similarly impure limestone.

In our case, the type of clinker has stronger effect than the type of limestone. Clinker M, burnt from low reactivity raw meal, has larger recrystallized alite crystals and medium C_3A content. Therefore, in terms of hydraulic activity, it can be characterized as slow. The strengths after 2 days of cement prepared from this clinker are lower and the increase in strength between 7 and 28 days is higher. On the contrary, clinker C is burnt from highly reactive raw meal, it is fine-grained with a high C_3A content and it is very hydraulically active. The increase in initial strengths of the cement is large, but the increase after 7 days is not as significant.

These differences are visible on calorimetric curves as well (see figure 6 and 7). Cements from clinker C are characterized by their narrower main hydration exotherms with steeper beginning. Compared to reference C, the addition of limestone and increasing of specific surface area by grinding at a constant time led to acceleration of hydration. This corresponds to work of other authors who reported the acceleration of cement hydration by addition of fine limestone powder due to the nucleation effect [6, 16, 17]. When using impure limestone L the hydration peak is higher compared to samples with limestone H, probably due to higher fineness and increased Al_2O_3 content. This supports the chemical effect of limestone on cement hydration and the hydration temperature may increase [3, 18].

On the contrary, cements from clinker M had delayed hydration, and sample with 30% replacement had very low hydration temperatures with a few hours delay of the main exotherm. However, in case of these cements, the positive effect of 10% limestone addition on hydration is demonstrated by higher total hydration heat after 24 hours than that of the M reference. The values were calculated from the calorimetric curves and are shown in table 7 and 8.

From literature it is evident that the highest positive effect on cement hydration has a fine limestone powder in the amount up to 10 wt. % [3, 16, 17, 19].

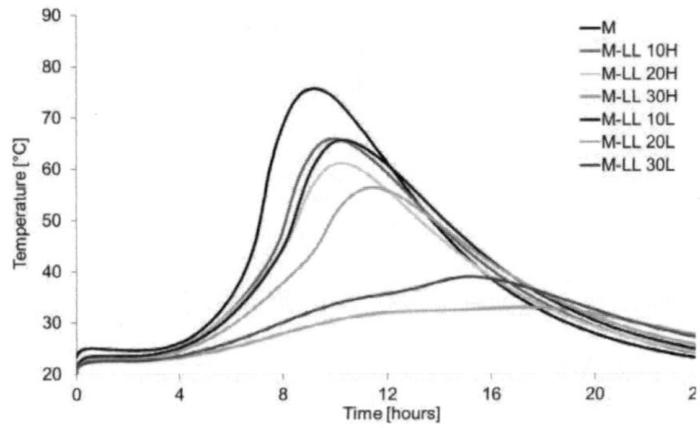

Figure 6. Hydration process of cement pastes M recorded by semiadiabatic calorimetry (grinding together).

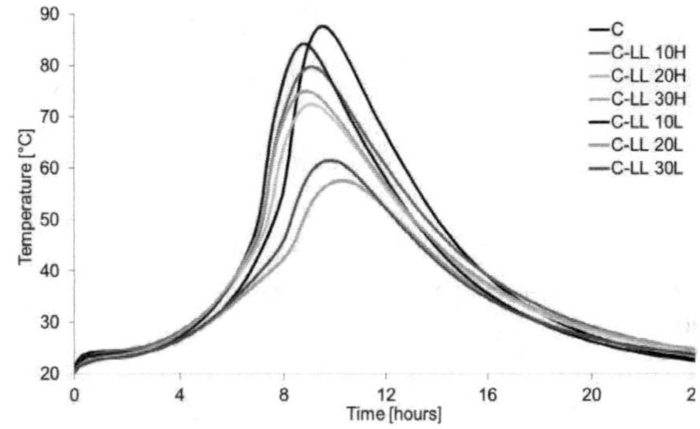

Figure 7. Hydration process of cement pastes C recorded by semiadiabatic calorimetry (grinding together).

Table 7. Heat of hydration of cements from clinker M released in 24 hours in the semi-adiabatic calorimeter in J/g (grinding together).

Cement	M	M-LL 10H	M-LL 20H	M-LL 30H	M-LL 10L	M-LL 20L	M-LL 30L
Q_{24}	367	371	350	193	398	334	231

Table 8. Heat of hydration of cements from clinker C released in 24 hours in the semi-adiabatic calorimeter in J/g (grinding together).

Cement	C	C-LL 10H	C-LL 20H	C-LL 30H	C-LL 10L	C-LL 20L	C-LL 30L
Q_{24}	450	456	409	329	442	431	341

Determination of the heat of hydration by the dissolution method (see table 9) shows that cements with 20% of limestone achieve a favorably low heat of hydration, and cement with 30% of limestone reaches values that classify them as cements with very low heat of hydration according to EN 14216.

Table 9. Heat of hydration of cements according to EN 196-8 with use of limestone H prepared by grinding together after 7 days of hydration in kJ/kg.

Cement	M	M-LL 10H	M-LL 20H	M-LL 30H	C	C-LL 10H	C-LL 20H	C-LL 30H
Hydration heat	343	295	246	196	345	280	233	186

The biggest difference in setting were found for PLCs prepared by grinding together. In a detailed observation of setting process of the cement pastes (see figures 8 and 9) it was found that with the increasing content of limestone, and hence also the increasing total surface area, the initial setting time is shorter. Cements prepared from more hydraulically-active clinker C have shorter initial setting time than cements from clinker M.

Voglis et al. [20] also stated shorter setting time while increasing the initial strength of cement with 15% addition of limestone compared to the reference OPC. The shortening of the cement setting time with increasing limestone content due to nucleation effect is also confirmed by other authors [21, 22, 23].

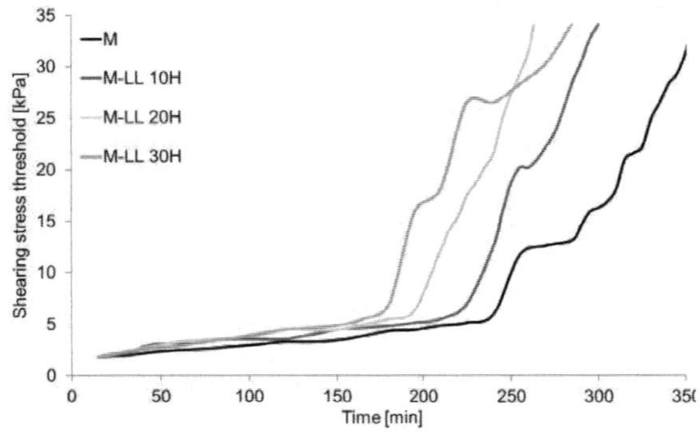

Figure 8. Setting process of cements from M clinker and limestone H (grinding together).

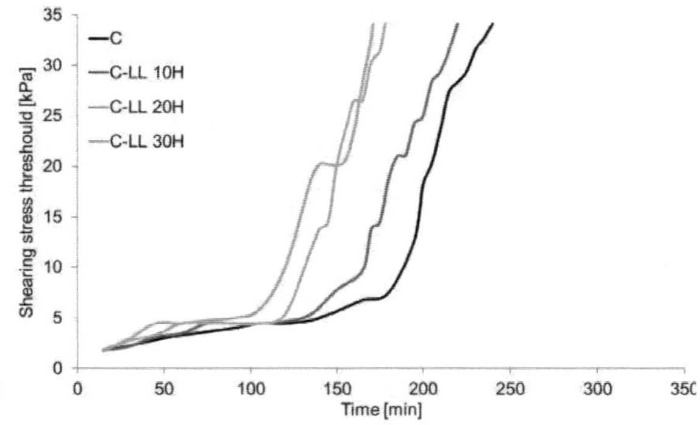

Figure 9. Setting process of cements from C clinker and limestone H (grinding together).

4. Conclusions

The effect of limestone content (up to 30 wt. %) on PLCs technology parameters was monitored. Two types of limestone and two types of clinker were used. Limestones varied in $CaCO_3$ content (94.6 and 79.0 wt. %), clinkers were from two different cement plants and differed in particular in C_3A content and in hydraulic activity.

It was found that the properties of the clinkers used had a much greater impact on the parameters of the cements than the quality of the limestones used.

In terms of the influence of two types of limestone, the addition of more contaminated limestone slightly increased the strength of the cements. This can be due to its higher fineness when using together grinding or the presence of impurities in the form of Al_2O_3 and amorphous SiO_2.

In the case of hydraulically more active clinker with a high C_3A content and a low limestone content, the positive effects on cement hydration in early stages are evident.

Different preparation of cements, i.e. the addition of separately ground limestone or the together grinding of limestone with clinker, did not show a significant impact on the strength, even though the total specific surface of the cements varied considerably. Significant differences were noted only in early stages of hydration.

Furthermore, it has been found that with the increasing content of limestone the initial setting time of cements is noticeably shorter.

Cements with a high limestone content significantly save less energetically and less environmentally suitable clinker and still achieve high-quality technological parameters. In addition, they exhibit very low hydration heat, which may be beneficial in some types of applications.

Acknowledgements

This paper was elaborated with the Institutional support for long-term development of research organizations by the Ministry of Industry and Trade of the Czech Republic.

References

[1] Zinke R K and Werkheiser W H 2018 *Mineral Commodity Summaries 2018* (U.S. Geological Survey, Reston, Virginia)

IOP Conf. Series: Materials Science and Engineering **583** (2019) 012007 doi:10.1088/1757-899X/583/1/012007

[2] WBCSD-CSI 2009 *Cement industry energy and CO_2 performance 'Getting the Numbers Right'* (Conches-Geneva, Switzerland)
[3] Wang D, Shi C, Farzadnia N, Shi Z, Jia H and Ou Z 2018 *Constr. Build. Mater.* **181** pp 659–672
[4] Schutter G D 2011 *Proc. 36th Conf. on our World in Concrete & Structures (Singapore)* (Singapore: CI-Premier Pte. Ltd.)
[5] Martin C, Phillipe L and Erick R 2006 *Cem. Concr. Res.* **36**
[6] Soroka I and Setter N 1976 *Cem. Concr. Res.* **6** pp 367–376
[7] Irassar E F 2009 *Cem. Concr. Res.* **39** pp 241–254
[8] Craeye B, Schutter G D, Desmet B, Vantomme J, Heirman G, Vandewalle L and Kadri E 2010 *Cem. Concr. Res.* **40** pp 908–913
[9] Rode S, Oyabu N, Kobayashi K, Yamada H and Kühnle A 2009 *Langmuir* **25** pp 2850–2853
[10] Taylor H F W 1997 *Cement Chemistry* (London: Thomas Telford Publishing)
[11] De W K, Ben H M, Le S G, Kjellsen K O, Justnes H and Lothenbach B 2011 *Cem. Concr. Res.* **40** pp 279–291
[12] Lothenbach B, Maschei T, Möschner G and Glasser F P 2008 *Cem. Concr. Res.* **38** pp 1–18
[13] Scrivener K L, Lothenbach B, Belie N D, Gruyaert E, Skibsted R, Snellings J and Vollpracht A 2015 *Cem. Concr. Res.* **48** pp 835–862
[14] Brandštetr J, Polcer J, Krátký J, Holešinský R and Havlica J 2001 *Cem. Concr. Res.* **31** pp 941–947
[15] Menéndez G, Bonavetti V and Irassar E F 2003 *Cem. Concr. Compos.* **25** pp 61–67
[16] Vance K, Aguayo M, Oey T, Sant G and Neithalath N 2013 *Cem. Concr. Compos.* **39** pp 93–103
[17] Thongsanitgarn P, Wongkeo W, Chaipanich A and Poon C S 2014 *Constr. Build. Mater.* **66** pp 410–417
[18] Aashay A, Gaurav S a Narayanan N 2016 *Constr. Build. Mater.* **102** pp 113–124
[19] Ramezanianpour A A, Ghiasvand E, Nickseresht I, Mahdikhani M and Moodi F 2009 *Cem. Concr. Compos.* **31** pp 715–720
[20] Voglis N, Kakali G, Chaniotakis E and Tsivilis S 2005 *Cem. Concr. Compos.* **27** pp 191–196
[21] Marzouki A, Lecomte A, Beddey A, Diliberto C and Ouezdou M B 2013 *Constr. Build. Mater.* **48** pp 1145–1155
[22] Valcuende M, Maco E, Parra C and Serna P 2012 *Cem. Concr. Res.* **42** pp 583–592
[23] Moon G D, Oh S, Jung S H and Choi Y C 2017 *Constr. Build. Mater.* **135** pp 129–136

IOP Conf. Series: Materials Science and Engineering **583** (2019) 012008 doi:10.1088/1757-899X/583/1/012008

The effect of zinc, water to binder ratio and silica fume on the hydration and mechanical properties of Portland cement mixtures

P Siler[1], I Kolarova[1], J Bednarek[1], M Janca[1], J Masilko[1], R Novotny[1] and T Opravil[1]

[1] Brno University of Technology, Faculty of Chemistry, Materials Research Centre, Purkyňova 118, CZ-61200 Brno, Czech Republic

Email: siler@fch.vut.cz

Abstract. The amount of zinc in the clinker or in the secondary raw materials has been increasing in recent years mainly due to the processing of waste materials containing zinc. The aim of this work is to show the effect of zinc on hydration of Portland cement and mechanical properties. These effects were studied by isoperibol calorimetry and mechanical properties testing. Zinc was added to the cement in the form poorly soluble compound ZnO. The concentration of zinc added was chosen as 1wt. percent. The newly formed compounds during hydration were identified by X-ray diffraction method (XRD). Another objective is to show the effect of silica fume (microsilica, SF) as the secondary raw material and water to binder ratio on hydration of cement in the presence of zinc.

1. Introduction

Concrete, usually Portland cement concrete is a composite material composed of fine and coarse aggregates bonded together with a Portland cement paste that hardens over time. The cement reacts chemically with the water and other ingredients to form a hard matrix that binds the materials together into a durable stone-like material that has many uses. Additives such as pozzolans are included in the mixture to improve the physical properties of the fresh mix or the finished material.

Portland cement is the most common type of cement in general use around the world as a basic ingredient of concrete, mortar, stucco, and non-specialty grout. It is a fine powder, produced by heating limestone and clay minerals in a kiln to form clinker, grinding the clinker, and adding small amount of supplementary ingredients.

The effect of zinc on the clinker properties was investigated in different works [1–6]. The presence of zinc affected hydration of C_3A and retards hydration of C_3S [2, 3]. $Zn_2(OH)_6{\cdot}2H_2O$ was described in literature [3] as product of hydration of Portland cement in the presence of zinc.

The zinc hydroxyl anions $Zn(OH)_3^-$ and $Zn(OH)_4^{2-}$ are found in the system between pH 12 and 13. Subsequently, they are transformed into insoluble $CaZn_2(OH)_6{\cdot}2H_2O$ before portlandite. Thus, hydration can be inhibited until these reactions are complete, since both OH^- and Ca^{2+} are present in low concentrations [4].

$$Zn^{2+} + 2OH^- \rightarrow Zn(OH)_2 \quad (1)$$

$$Zn(OH)_2 + 2OH^- \rightarrow 2H_2O + ZnO_2^{2-} \quad (2)$$

$$2ZnO_2^{2-} + C_3S/O - Ca^{2+} + 6H_2O \rightarrow C_3S/O - CaZn_2(OH)_6 \cdot 2H_2O + 2OH^- \quad (3)$$

Dashes in equation 3 indicate bonding of Ca through oxygen.

The impact of ZnO on the properties of Portland cement was published in [5–7]. Due to zinc retardation, mechanical strengths after 3 days were less than for samples without zinc. After one week, the strength was increasing and after 28 days it was higher than for the control sample. This effect was attributed to the action of ZnO nanoparticles as filler. Decreasing porosity with an increasing amount of ZnO was also found [6]. According to information published in [7], the mechanism by which ZnO retards hydration reaction could be nucleation and/or growth poisoning of C-S-H.

Silica fume (SF), also known as microsilica is an amorphous silicon dioxide. It is an ultrafine powder collected as a by-product of the silicon and ferrosilicon alloy production and consists of spherical particles with very small diameter. The main field of application is as a pozzolanic material for high performance concrete [8]. According to [7], for the effectiveness of secondary materials, their specific surface area is most important.

The aim of this work is to show the effect of zinc on hydration of Portland cement, mechanical properties, influence of water to binder ratio and influence of the presence of microsilica. These effects were studied by isoperibol calorimetry and mechanical properties testing. Zinc was added to the cement in the form of poorly soluble compound ZnO. The concentration of zinc was 1wt. percent. The amount of microsilica was 15 wt. percent. The newly formed compounds during hydration were identified by X-ray diffraction method (XRD).

2. Sample preparation

The mixtures were prepared from the cement CEM I 42.5 R (x_{10} = 0.58 μm, x_{50} = 11.11 μm, x_{90} = 40.51 μm) and CEM I 52.5 N (x_{10} = 0.45 μm, x_{50} = 7.93 μm, x_{90} = 28.88 μm) Heidelberg Cement Group Českomoravský cement, a.s. Mokrá, Czech Republic. Zinc was added in the form of insoluble ZnO in the amount of 1 wt. % of the cement substitution (the percentages of the replacement were always calculated on pure zinc in the binder). The pastes were mixed with distilled water. Used microsilica was Elkem Grade 920U (x_{10} = 0.52 μm, x_{50} = 1.98 μm, x_{90} = 8.33 μm) Elkem Silicon Materials, Norway.

For measurements on isoperibol calorimeter, 300 g of the mixture was weighed into a polystyrene crucible equipped by a thermo-insulating jacket and a thermocouple. Calorimetric measurements and calorimetric curves were obtained.

The pastes for compressive strength determination were embedded into 100 × 20 × 20 mm moulds and stored in wet conditions for sample analysis.

3. Description of the methods used

3.1. XRD (X-ray diffraction)

X-ray diffraction (XRD) is usually used to obtain information about the structure of crystalline materials. A primary use of the technique is the identification and characterization of compounds based on their diffraction pattern. X-ray diffraction analysis was performed with X-Ray diffractometer Empyrean (Panalytical) with Bragg–Brentano parafocusing geometry and using CuKa radiation. The measurements were done within the range from 5 to 120°2Θ with angular step 0.013°2Θ and 25 s duration using automatic divergence slits to maintain the constant irradiation of the sample area. The measurements were repeated four times and then summed.

3.2. Isoperibol calorimetry

The heat production is a result of the hardening processes in the cement paste. Very good representative values of the hydration process can be acquired using a multicell isoperibol calorimeter [9]. This simple method enables simultaneous monitoring of the progress of hydration of several samples and estimation of the beginning of the setting or hardening process. Unlike the isothermal calorimetry, the hydration curves show differences at constant ambient temperatures. By this way it is possible to simulate real

IOP Conf. Series: Materials Science and Engineering **583** (2019) 012008 doi:10.1088/1757-899X/583/1/012008

conditions of hydration. The data were collected using an isoperibol calorimeter constructed at the Faculty of Chemistry, Brno University of Technology (Brno, Czech Republic) and the resulting heat was obtained by numerical integration and multiplied by the integration constant of the calorimeter which was obtained from calibration. The calorimeter used was constructed based on the work published by Smrčková et al. [10]. This kind of calorimeter was primarily developed at the German Academy of Sciences (Halle, Germany). In this work, an upgraded version with 16 cells was used; its detailed description was reported by Brandštetr et al. [9].

4. Results and discussion

4.1. Influence of ZnO and microsilica

The effect of ZnO on the hydration of CEM I 42.5 R was investigated in literature [4]. There was found hydration retarding and maximum developed temperature decreasing thanks to a slower dissolution of ZnO. Further dissolution of ZnO leads to the incorporation of Zn^{2+} ions into the amorphous phase. A slight decrease of temperature and evolved heat can also be attributed to the formation of a crystalline compound of $Ca[Zn_2(OH)_6](H_2O)_2$ or to the binding of zinc into amorphous compounds [4, 6]. According to [8] the effect of ZnO is mainly in nucleation and/or growth poisoning of C-S-H. The effect of SF as a secondary material depends mainly on the size of their particles.

Figure 1 shows the effect of ZnO and microsilica on hydration of CEM I 52.5 N.

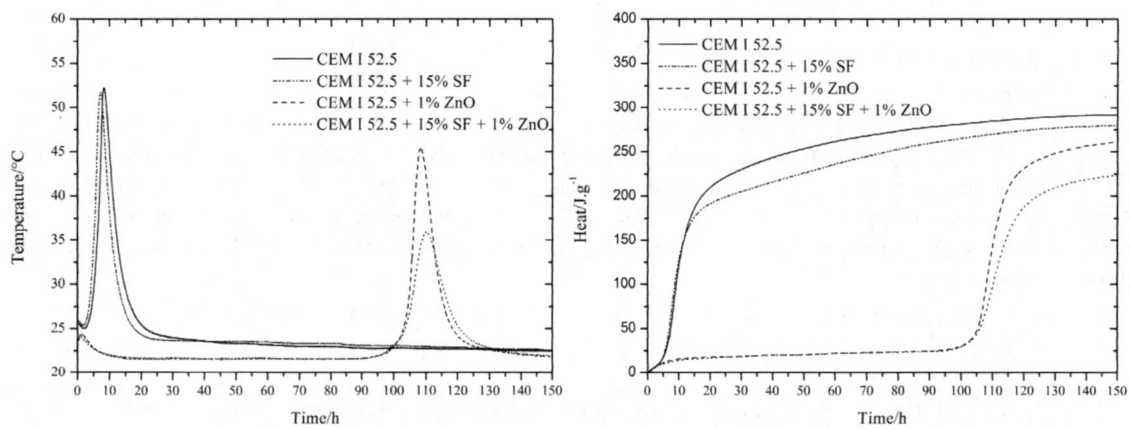

Figure 1. Isoperibolic calorimetry curves of cement CEM I 52.5 N.

The difference in maximum temperature achieved with cement without admixtures and with microsilica is very small (0.5°C). This difference is due to the fact that microsilica is not as reactive as cement during hydration and thus does not develop as high temperatures as for hydration of pure cement. The decrease of temperature in the presence of zinc was observed. With replacing cement with microsilica or zinc, less heat is released. This is probably due to the slow hydration of cement and the emergence of new zinc products. The pure cement curve CEM I 52.5 N is the same as for CEM I 42.5 R [4], which is mainly due to the same mineralogical composition of cement. Only cement particle size is different. The different particle size in this case does not affect the rate of dissolution of the cement and the rate of hydration reactions.

Only slight hydration acceleration when using microsilica and finer cement CEM I 52.5 N was observed. At a 1% zinc dose, no acceleration of hydration was observed but almost no delay of hydration was observed unlike CEM I 42.5 R [4]. Microsilica, thanks to its high softness, acts as a component allowing the precipitation of hydration products and thus acceleration of hydration. However, this acceleration is only small because particle diameters are not small enough (x_{50} = 1.98 μm).

A decrease in total released heat was also observed using CEM I 52.5 N. Due to the effect of zinc, another reduction was found. This reduction is more significant than for coarse cement. It can be assumed that the heat will increase during the next stages of hydration. This result shows that the fineness of the cement is important for stability of zinc hydroxyl membrane which is formed during hydration in the presence of zinc.

4.2. Induction period length

To illustrate the effect of zinc and microsilica, the lengths of the induction periods were determined and plotted in relation to the amount of zinc in the samples by the same way as in literature [4].

Table 1. Induction period lengths.

	Induction period lengths (hours)
CEM I 52.5 N	3.3
CEM I 52.5 N+SF	2.8
CEM I 52.5 N+Zn	97.4
CEM I 52.5 N+SF+Zn	97.4

Zinc and microsilica affect not only the maximum temperature but also the length of the induction period. Microsilica shortens the induction period (by 0.5 hours). It may be due to the fact that microsilica helps with nucleation due to its size (as shown in the literature [11]). Zinc, on the other hand, prolongs the induction period very significantly. As opposed to pure cement, the induction period was prolonged for zinc mixtures by 94 hours. This is probably due to gradual dissolution of zinc.

4.3. Influence of water to binder ratio

The influence of the water to binder ratio was tested on CEM I 42.5 R. Figures 2 and 3 shows this effect on the hydration of this cement. The addition of zinc was in the form of 1% ZnO.

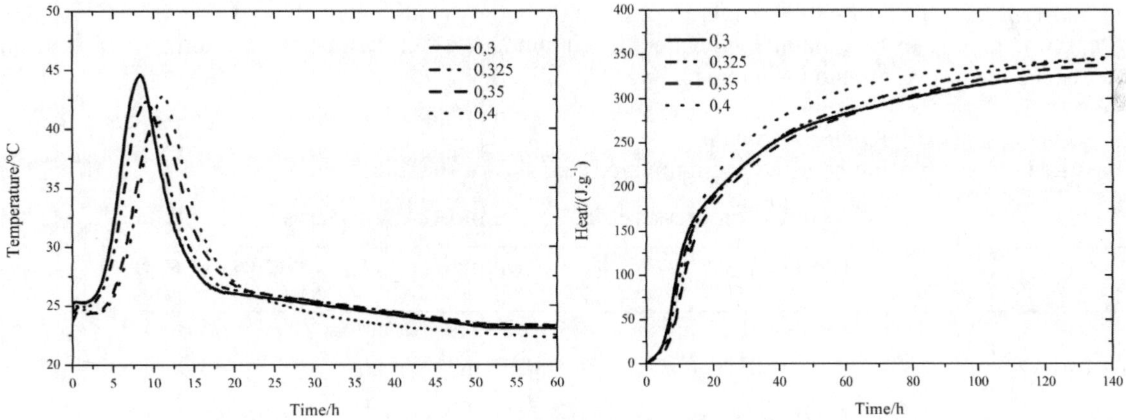

Figure 2. Influence of water to binder ratio on the pure CEM I 42.5 R.

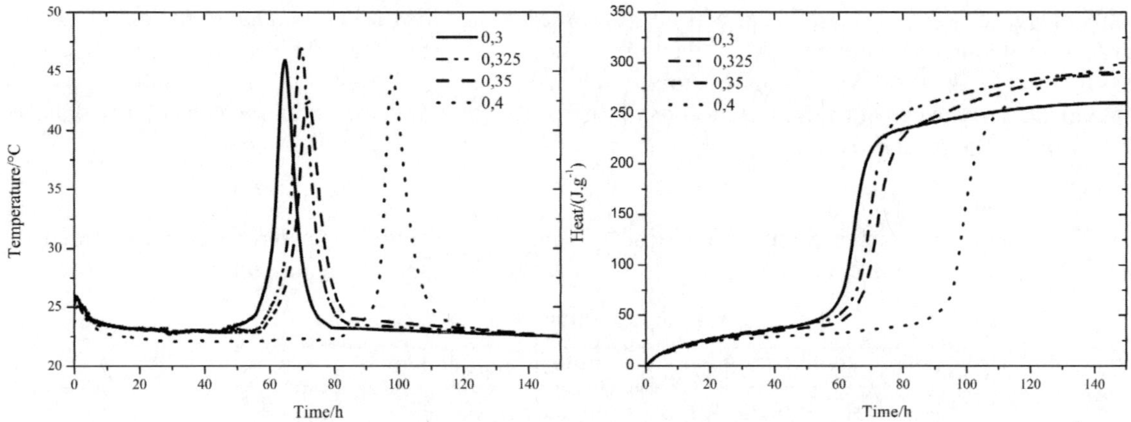

Figure 3. Influence of water to binder ratio on the CEM I 42.5 R doped with zinc.

When the water to binder ratio is decreasing, the length of the induction period is also decreasing. At low water content, water is saturated with Ca^{2+} ions faster and hydration will be accelerated. With lower amount of water, it is not possible to hydrate all phases, thus reducing of the developed heat occur. It can be assumed that storage of concrete will result in the continuation of hydration and release of additional heat. In case of higher amounts of water than is needed for hydration, unreacted water becomes as inert, which draws off heat. It has to be heated and reduces the maximum developed temperature. With higher amount of water, the phase's gets into better contact with water, the hydration is more complete and the developed heat increases slightly. Results also show retardation extending with increasing water content because of the greater distance of cement grains during hydration. These results are in line with the results published in the literature [12].

For zinc mixtures, retardation by zinc effect is indicated. With decreasing water to binder ratio, the rate of retardation is reduced due to the shortening of the induction period described for pure cement. Another reason is the less stable membrane from zinc hydroxyl anions of $Zn(OH)_3^-$ and $Zn(OH)_4^{2-}$due to less water. Due to less amount of water, lower amount of these anions will form. It also leads to a faster transformation into insoluble $CaZn_2(OH)_6 \cdot 2H_2O$.

4.4. Compressive strength measurement

The measured sample compressive strength values are shown in table 2.

Table 2. Compressive strength of prepared mixtures.

	Days of hydration	Flexural strength [MPa]	Compressive strength [MPa]
	1	6.7 ±1.2	36.6 ±4.0
CEM I 52.5 N	7	4.0 ±0.9	68.0 ±3.5
	28	4.7 ±0.8	71.3 ±2.8
	1	4.0 ±0.1	30.2 ±1.2
CEM I 52.5 N+SF	7	3.6 ±0.4	59.0 ±2.6
	28	6.3 ±0.6	67.7 ±6.8
	1	-	-
CEM I 52.5 N+Zn	7	2.8 ±0.5	87.5 ±4.3
	28	3.6 ±0.3	87.2 ±3.6
	1	-	-
CEM 52.5 I N+SF+Zn	7	3.8 ±0.8	58.0 ±4.1
	28	3.6 ±0.9	82.1 ±7.7

For cement with microsilica, both flexural and compressive strength are decreasing comparing with cement without admixture. This is probably due to the fact that the microsilica consumes a large amount of water due to its large surface, which is then missing when cement is hardened [13]. Another reason can be the loss of processability associated with the slowing of the hydration reactions. This effect would be avoided by using a superplasticizer. For samples with zinc, we can see a significant improvement in mechanical properties in later stage of hydration. This effect can be explained by the fact that zinc reduces the porosity of the structure and due to the filling effect of smaller nanoparticles of zinc oxide [6]. For samples with microsilica and zinc, the mechanical properties after 7 days are worse than that for cement without additives and almost comparable to sample only with microsilica, due to zinc retarding effects and water consumption by microsilica. The strengths after 28 days are again higher than those of zinc-free samples.

4.5. XRD results

Newly formed compounds during cement hydration in the presence of zinc were detected by X-ray diffraction analysis. The amount of these compounds is shown in table 3.

Table 3. Newly formed compounds detected by XRD.

	Days of hydration	$Ca(CO_3)$ Calcite	$C_6A\overline{S}_3H_{32}$ Ettringite	$Ca(OH)_2$ Portlandite	$Ca_2(Fe_2O_5)$ Brownmillerite	$Ca_2(SiO_4)$ Larnite (belite)	$Ca_3(SiO_5)$ Hatrurite (alite)	$Ca_4Al_2[CO_3][OH]_{12}[H_2O]_5$	$Ca_4Al_2O_6(CO_3)_{0,67}(SO_3)_{0,33}.11H_2O$	$Zn_4Al_2CO_3(OH)_{12}.3H_2O$ (Zaccagnaite)
CEM 52.5 N	1	5	4	23	7	10	50			
	7	4	2	37	7	10	39			
	28	12	3	32	4	14	31	3		
CEM 52.5 N + SF	1	6	4	28	6	11	44			
	7	11	5	33	6	13	30			
	28	8	5	32	6	12	36			
CEM 52.5 N + Zn	7	9	5	34	5	13	34			
	28	11	2	39	4	10	30	4		1
CEM 52.5 N + SF + Zn	7	8	0	24	6	18	43			
	28	7	8	24	4	11	41	2	1	

In addition to the compound $CaZn_2(OH)_6 \cdot 2H_2O$ which is described in literature [4] but was not found in our samples, other compounds have been discovered. The mechanism of their formation is not yet described in the literature. New compounds affect hydration. Further prolongation of hydration can be attributed to the formation of these new compounds.

5. Conclusion

The aim of this work was to show the influence of zinc on the hydration and mechanical properties of mixtures with Portland cement. Another objective was to show the effect of microsilica as a commonly used pozzolanic material and the effect of water amount as an important factor during hydration. By using of isoperibol calorimetry the effect of ZnO on the hydration of cement CEM I 52.5 N was investigated. The hydration retarding and maximum developed temperature decreasing thanks to a slower dissolution of ZnO was found. The different particle size of cement 42.5 and CEM I 52.5 N in this case does not affect the rate of dissolution of the cement and the rate of hydration reactions. Only slight hydration acceleration when using microsilica and finer cement CEM I 52.5 N was observed. The result shows that the fineness of the cement is important for stability of zinc hydroxyl membrane which is formed during hydration in the presence of zinc because there is more significant reduction of developed heat with the presence of zinc. When the water to binder ratio is decreasing, the length of the induction period is also decreasing thanks to faster saturation with Ca^{2+} ions. The same results were found for the presence of zinc. The reduction of hydration retardation in the presence of zinc was more significant thanks to lower stability of hydroxyl membrane with lower amount of water. Both compressive and flexural strength were decreasing especially during first 7 days of hydration thanks to presence of zinc but there were higher strengths then for zinc-free samples after 28 days due to filling effect of smaller nanoparticles of zinc oxide. The other reason can be thanks to formation of new surface for support of start of nucleation. Newly formed compounds during cement hydration in the presence of zinc were detected by X-ray diffraction analysis.

Acknowledgments
This outcome has been achieved with the financial support by the project: GA19-16646S "The elimination of the negative impact of zinc in Portland cement by accelerating concrete admixtures", with financial support from the Czech science foundation.

References

[1] Gineys N, Aouad G and Damidot D 2011 Managing trace elements in Portland cement - Part II: Comparison of two methods to incorporate Zn in cement *Cem. Concr. Compos.* **33** pp 629–36
[2] Olmo I F, Chacon E and Irabien A 2001 Influence of lead, zinc, iron (III) and chromium (III) oxides the setting time and strength development of Portland cement *Cem Concr Res.* **31** pp 1213–19
[3] Chen Q Y, Tyrer M, Hills C D, Yang X M and Carey P 2009 Immobilisation of heavy metal in cement-based solidification/stabilisation: A review *Waste Manage* **29**(1) pp 390–403
[4] Šiler P, Kolářová I, Novotný R, Másilko J, Pořízka J, Bednárek J et al. 2017 Application of isothermal and isoperibolic calorimetry to assess the effect of zinc on cement hydration *J. Therm. Anal. and Calorim.* **133** pp 27–40
[5] Weeks C, Hand R J and Sharp J H 2008 Retardation of cement hydration caused by heavy metals present in ISF slag used as aggregate *Cem. Concr. Compos.* **30** pp 970–78
[6] Nochaiya T, Sekine Y, Chooun S and Chaipanich A 2015 Microstructure, characterizations, functionality and compressive strength of cement-based materials using zinc oxide nanoparticles as an additive *J. Alloys Compd.* **630** pp 1–10
[7] Aitcin P-C 2005 Vysokohodnotný beton (High performance concrete) (Praha: ČKAIT)
[8] Ataie F F, Juenger M C G, Taylor-Lange S C and Riding K A 2015 Comparison of the retarding mechanisms of zinc oxide and sucrose on cement hydration and interactions with supplementary cementitious materials *Cem Concr Res.* **72** pp 128–136
[9] Brandštetr J, Polcer J, Krátký J, Holešinský R and Havlica J 2001 Possibilities of the use of isoperibolic calorimetry for assessing the hydration behaviour of cementitious systems *Cem. Concr. Res.* **31** pp 941–47
[10] Smrckova E, Palou M T and Tomkova V 1996 Application of conduction calorimeter for study of the reactivity of C_2S in the system $C_2S - C_4A_3$ *J. Therm. Anal.* **46** pp 597–605
[11] Siler P, Kratky J and De Belie N 2012 Isothermal and solution calorimetry to assess the effect of superplasticizers and mineral admixtures on cement hydration *J. Therm. Anal. Calorim.* **107**(1) pp 313–20
[12] Šiler P, Kolářová I, Krátký J, Havlica J and Brandštetr J 2014 Influence of superplasticizers

on the course of Portland cement hydration *Chem. Pap.* **68**(1) pp 90–7
[13] Šiler P, Krátký J, Kolářová I, Havlica J and Brandštetr J 2013 Calorimetric determination of the effect of additives on cement hydration process *Chem. Pap.* **67**(2) pp 213–20

IOP Conf. Series: Materials Science and Engineering **583** (2019) 012009 doi:10.1088/1757-899X/583/1/012009

Rheological study on influence of hydroxypropyl derivatives of guar gum, cellulose, and chitosan on the properties of natural hydraulic lime pastes

T Zizlavsky[1], **M Vysvaril**[1] **and P Rovnanikova**[1]

[1] Faculty of Civil Engineering, Brno University of Technology, Veveri 331/95, Brno 602 00, Czech Republic

Email: zizlavsky.t@fce.vutbr.cz

Abstract. Viscosity enhancing admixtures are widely used to improve properties of concrete and ready-mix mortars. Guar gum derivatives and chitosan ethers are reported to have water-retaining effect, which is beneficial for the adhesion of the mixture as well as for the hydratation of hydraulic particles in the binder phase of concrete or natural hydraulic lime (NHL) mortars. Influence of addition of admixtures in doses of 1, 5, and 10‰ of the binder weight has been studied on NHL based pastes with constant water:binder ratio. The measurements were carried out on the Discovery HR-1 rheometer with DIN concentric cylinders geometry. The flow properties and viscoelastic properties were studied on the specified mixtures. Pastes expressed shear-thinning behaviour with thixotropic character of flow curves. Admixture addition caused increase of yield stress, fluidity index and consistency coefficient varying on the amount and type of the admixture. Differences in the effectivity of admixtures were observed, the dosage-dependency of the admixtures varied.

1. Introduction

In the building industry the viscosity enhancing admixtures (VEAs), especially the cellulose-based ones are widely used to improve properties of ready-mix mortars and reduce the risks of bleeding and segregation in concrete [1]. Nowadays different biopolymeric admixtures such as guar gum derivatives, diutan gum, and xanthan gum have occurred [2–4]. Most of these admixtures are currently mainly used in the food industry, where they are used as stabilisers, thickeners and emulgators. The studies on their use, especially in concrete, are lately being published [1–6].

The main focus of the paper is on rheological properties of NHL-based pastes (grouts) modified by addition of hydroxypropyl- derivatives of three different biopolymers: cellulose, chitosan, and guar gum. Even though the usage of cellulose ethers is widely spread, the studies of effect of its addition on the properties of grout, or mortar based on non-cementitious binder are scarcely to be found. Water retaining admixtures are in general reported to have a thickening effect, thus increasing the yield stress of the mixture [7–16]. The exceptions of this statement occur in the literature, namely the hydroxypropylmethyl cellulose (HPMC) is reported to decrease yield stress with growing dosage after initial increase in low dose of admixture while used in cement based [12, 14, 15] as well as lime based [8, 17] mixtures. On the contrary, the hydroxypropyl guar (HPG) derivatives act as expected from the general conclusions [15, 16], increasing the yield stress with growing dosage of admixture in cement [11–13, 15] and also lime [10, 17] mortar. There are very few studies on chitosan derivatives found, it seems, that chitosan addition acts similarly to other VEAs in cementitious environment [18],

IOP Conf. Series: Materials Science and Engineering **583** (2019) 012009 doi:10.1088/1757-899X/583/1/012009

whereas in the lime mortar, hydroxypropyl chitosan (HPCH) decreases the yield stress in low doses followed by increase with growing amount of admixture from the certain dosage [9]. All of the admixtures showed thixotropic behaviour.

2. Materials and methods

2.1. Materials and mixing procedure

The dry mixture consisted of commercial natural hydraulic lime (Otterbein Calcidur) of the NHL 3.5 class according to EN 459-1, and the admixture of dosage 1‰, 5‰, and 10‰ of binder weight. The admixtures used and their specifications are presented in table 1. Dry mixture of lime and admixture was prepared, and then the specified amount of water (water/binder radio of 0.7:1) was added. All constituents were mixed together for one minute. Then the paste was introduced into the measuring cup.

Table 1. Properties of biopolymeric admixtures used.

Abbrev.	Chemical composition	Degree of substitution.	Viscosity (1% solution, 20°C) [mPa s]	Manufacturer
HPG H	hydrohypropyl guar	medium/high	1654	Lamberti s.p.a
HPG L	hydroxypropyl guar	low	3668	Lamberti s.p.a
HPCH	hydroxypropyl chitosan	high	22	Kraeber & Co GmbH
HPMC	hydroxypropylmethyl cellulose	low	3904	Lotte Fine Chemicals Co. Ltd

2.2. Methods of rheological measurements

Rheological measurements were executed on the hybrid rheometer Discovery HR-1 from TA instruments. The DIN concentric cylinders geometry has been chosen for all the measurements. The measuring apparatus with sample was tempered to 20°C. The results should not be reproduced quantitatively.

2.2.1. Flow properties. Five minutes after the start of mixing pre-shear was performed for 60 s at 100 s^{-1} to re-homogenize the sample. After pre-shear the sample rested for another 60 s, after that the measuring procedure was carried out. The detailed description of procedure has been published lately [19]. The results have been expressed as flow curves (shear rate vs. shear stress) and the Herschel-Bulkley model was applied to downward curves to describe rheological behaviour:

$$\tau = \tau_0 + k\dot{\gamma}^n \tag{1}$$

where τ - shear stress, $\dot{\gamma}$ - shear rate, τ_0 - yield stress, k consistency coefficient and n - fluidity index which characterizes shear-thinning ($n < 1$) or shear-thickening ($n > 1$) behaviour of a material.

2.2.2. Viscoelastic properties. To determine viscoelastic properties of lime pastes, the small amplitude (0.005%) oscillation tests at 1 Hz frequency was carried out. Using this method the critical strain γ_C which determines the end of the linear viscoelastic region (LVR), where the decrease in G' is observed, and the flow point γ_F, where $G'=G''$ (from this point onward the sample shows the character of a liquid without a consistent chemical or physical network-of-forces [20]) were measured. The frequency sweep tests in linear conditions (strain value was set to be lower than the critical strain of any of tested samples) were conducted. The material behaviour was observed by increasing the frequency from 0.1 Hz to 10 Hz. The results were expressed as complex modulus G^* and loss tangent tan(δ) defined by following equations:

$$G^* = \sqrt{(G')^2 + (G'')^2} \tag{2}$$

$$\tan(\delta) = G''/G' \tag{3}$$

3. Results and discussion

3.1. Flow properties

Downward flow curves of pastes with different amount of HPG H are shown in figure 1. The shape of expressed curves indicates pseudoplastic (shear-thinning, $n < 1$) behaviour of NHL-based pastes, which was also observed on the lime and cement-based mortars [10–13]. The presented curves were analysed using TRIOS software and Hershell-Bulkley (1) analysis to obtain values for yield stress (τ_0), consistency coefficient (k), and fluidity index (n). The data obtained are compared using bar charts in figure 2.

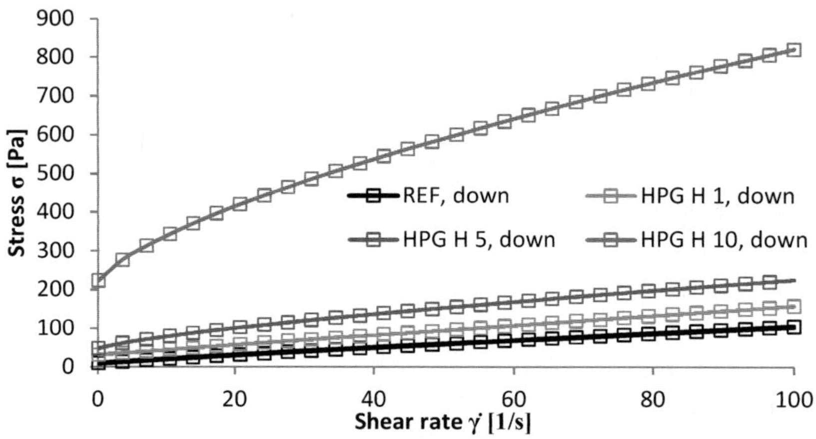

Figure 1. Downward flow curves of NHL pastes with different HPG H addition.

The admixture addition caused increase of yield stress followed by further growth of its value with increasing dosage. The non-monotonic behaviour of HPG L and HPMC was reported by Cappellari et al. [15] on HPG and methylhydroxyethyl cellulose (MHEC) and is supposed to be caused by the competition between two effects of biopolymers, the dispersing and lubricating effect, which becomes dominant within high dose of HPG L and HPMC, and the associative property which predominates in the other cases. The yield stress increase was reported for most of the applications of water-retaining agents in traditional building materials [8–19], the non-monotonic behaviour seems to be typical for most cellulose derivatives in all types of environments [8, 14, 15]. The increase of consistency coefficient (k) values shown in figure 2 is desirable, because it improves the adhesive and anti-sagging properties of the grout/mortar. Growth with higher doses in most cases with exception of HPCH, the especially significant increase in the case of HPMC was expected, for the main application of this product according to the manufacturer should be for tile adhesives. These results are in accordance with the ones obtained on cementitious materials [14, 15] and contrary to the ones observed on lime mortars [8]. This is due to a different concentration of $Ca(OH)_2$ in the binder solution, where cellulose ethers have a great ability to be adsorbed on $Ca(OH)_2$ crystals [17, 21, 22]. The fluidity index (n) describes shear-thinning ($n < 1$) or shear-thickening ($n > 1$) behaviour of the material. Most of the studied pastes showed shear-thinning characteristics, which is typical and desirable for building materials. The value for HPG H in lowest dosage is close to 1, which means, that it's almost Newtonian fluid. The lowering value of fluidity index with the growing dosage of admixture is common for most of the VEAs on cementitious materials [11, 13–15] while lime mortars exhibit increase of flow index with addition of VEA [8–10]. The notable increase its value within the lowest dosage was observed also by Cappellari et al. [15] on MHEC.

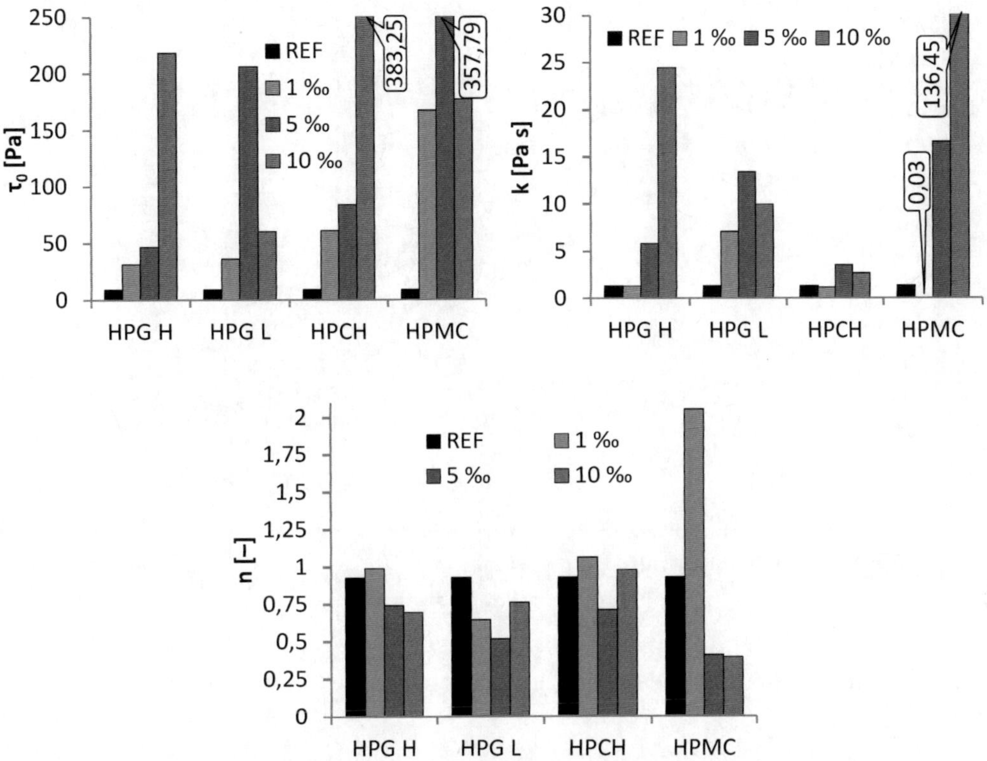

Figure 2. Comparison of yield stress (τ_0), consistency coefficient (k), and fluidity index (n) of NHL-based pastes with biopolymer admixtures.

3.2. Viscoelastic properties

3.2.1. Amplitude sweep test. Oscillation test was carried out to determine the end of the linear viscoelastic region (LVR), which is represented as critical strain (γ_c). The critical strain is represented by the end of horizontal part of the storage modulus (G') curve on figure 3.

The critical strain of the tested pastes overcome the value of reference paste in most cases, so the biopolymer addition increase the stability of the paste (the LVR is the same or longer as reference), this effect is dosage-dependent as can be seen either in figure 3 or 4, with increasing dosage of admixture grows the stability of the paste. The HPMC showed same non-monotonous behaviour as was observed within flow properties. All the pastes tested had the storage modulus values (G') larger than the loss modulus (G''), which means that the material consists of a strongly multi-connected network of elements; this is quite typical for lime or cement based pastes but also for aqueous solutions of some of the biopolymers [9, 10, 20, 23, 24]. The flow strain γ_F expresses how hard it is to make the sample become liquid. It is represented by the intersection of the moduli curves in figure 3, thus the point of equivalence between the moduli. In figure 4, there can be two basic dosage-dependent trends observed. HPG H and HPCH with growing dosage increase the flow strain, thus making it more difficult to make sample liquid, while HPG L and HPMC decrease the strain; taking into account the critical strain results, the addition of HPG H and HPCH widens the region in which the sample has gel-like behaviour, while HPMC and HPG L are making this region shorter. The flow strain results correlate with the degree of substitution of used admixtures (table 1).

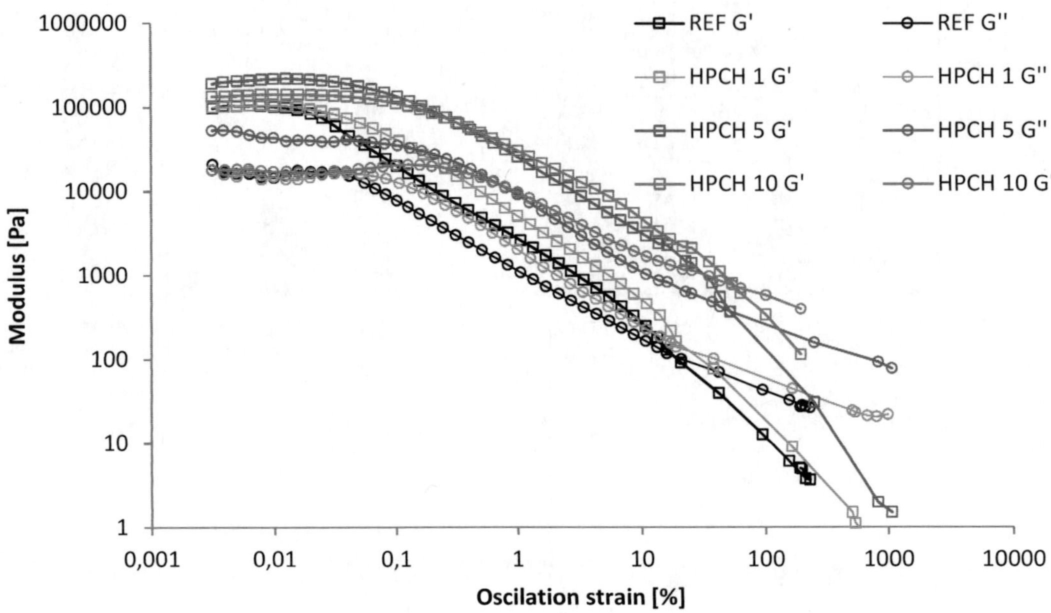

Figure 3. Storage modulus (G') and Loss modulus (G'') of HPCH pastes.

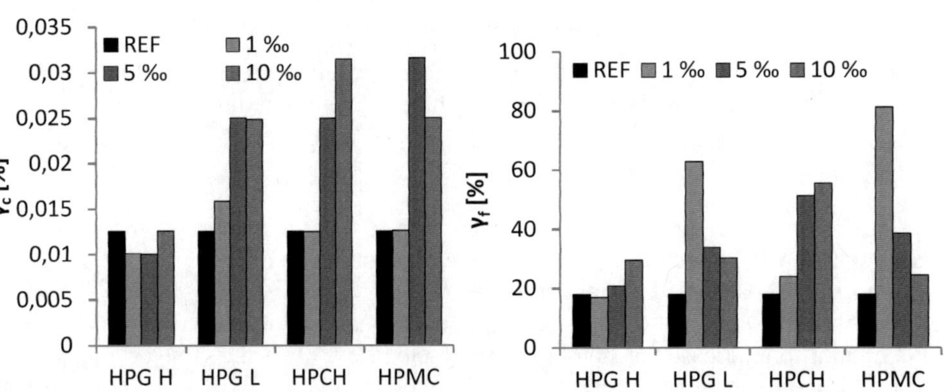

Figure 4. Comparison of critical strain (γ_C) and flow strain (γ_F) values of tested samples.

3.2.2. Frequency sweep test. Frequency sweep test at 0.005%. strain was performed to determine the viscoelastic properties of pastes. In figure 5 complex modulus, complex viscosity (η^*), and tan(δ) of HPMC are presented depending on frequency,while in figure 6, the charts show the values of complex modulus (G^*) and loss factor (tan(δ)) of the tested samples measured at 1 Hz frequency. The trends of G^* in the tested pastes are similar as for the flow strain, reporting the increasing resistance to deformation of the tested materials for HPG H and HPCH. The trend is opposite for HPG L and HPMC. The growing G^* correponds with the decreasing loss tangent, whereas the trends are not so clear as for the flow strain. The diminishing loss tangent reports more elastic behaviour of the material, thus greater resistance to external interference [15]. The results optained are comparable to the ones of lime mortars in the case of HPG, but contrary to the ones of HPCH [9, 10]. Generally, results and mainly the trends, obtained on NHL 3.5 pastes, correspond more likely with the results obtained on cementitious mixtures [12, 14, 15].

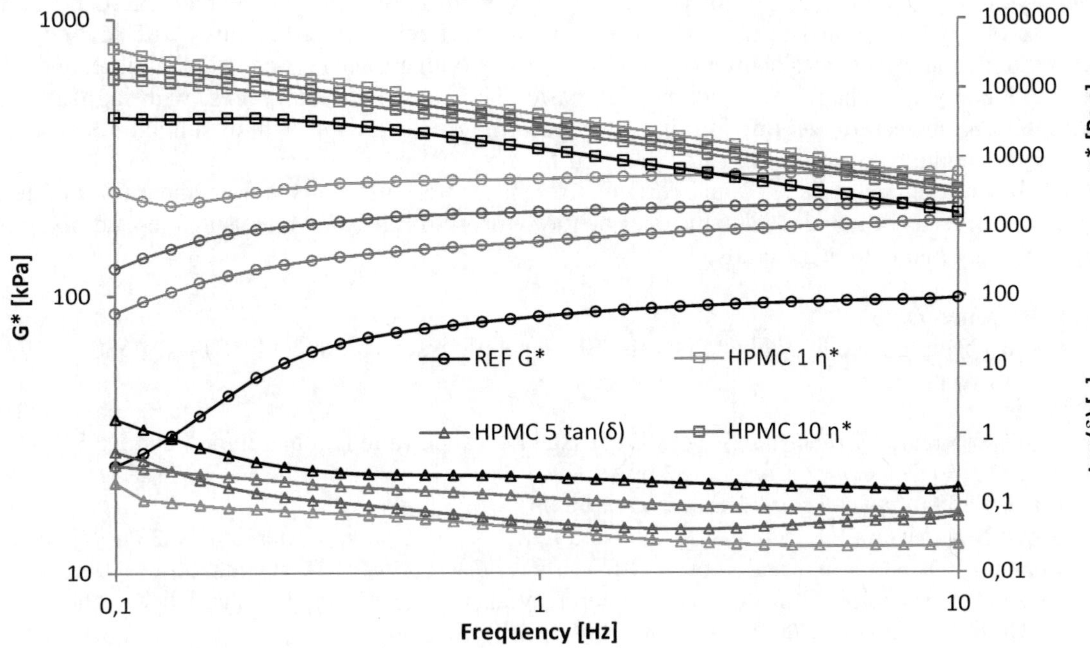

Figure 5 Results of frequency sweep test for NHL paste with HPMC.

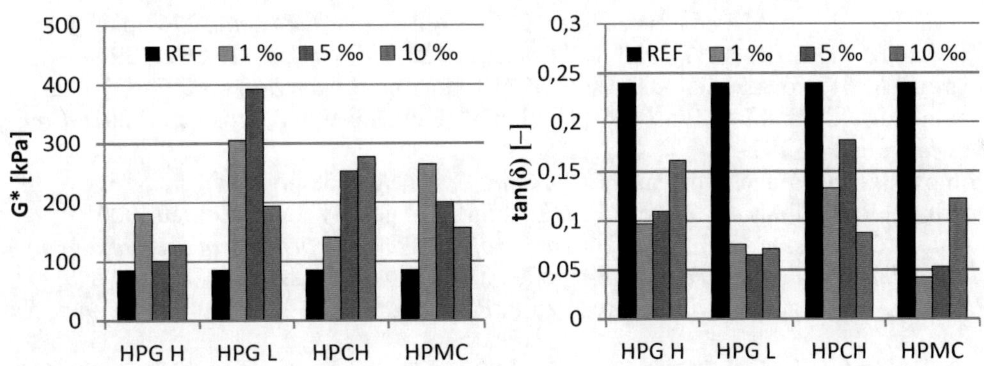

Figure 6. Complex modulus (G^*) and loss factor ($tan(\delta)$) comparison of tested pastes.

4. Conclusions

The study reports the influence of derivatives of chitosan, cellulose, and guar gum containing hydroxypropyl group on rheological properties of natural hydraulic lime based pastes.

The addition of hydroxypropyl derivatives of above mentioned biopolymers led to an increase of yield stress of tested paste, the hydroxypropyl guar and hydroxypropylmethyl cellulose, which both have low degree of substitution, expressed non-monotonic behaviour while yield stress of paste with highest dosage was inferior to the one of the paste with lower dose. The admixture addition led to increase of consistency coefficient, which is closely connected with the adhesive and anti-sagging properties of the mortars. All the pastes with exception of hydroxypropyl chitosan and hydroxypropylmethyl cellulose in the lowest dose reached fluidity index $n < 1$ which implies their shear-thinning behaviour.

The influence on viscoelastic properties mostly copied the flow properties. Addition of the admixtures with exception of hydroxypropyl guar with high degree of substitution led to an extension of linear viscoelastic region thus increased the stability of tested paste. Addition of biopolymeric admixture led to an increase of flow strain with two different dosage-dependent dependencies. Hydroxypropyl guar and hydroxypropylmethyl cellulose (admixtures with low degree of substitution) showed decreasing trend of flow strain value with growing dosage of admixture, but still reaching higher values than the nonadmixtured paste. The other two admixtures, with high degree of substitution, increased the flow strain value accordingly to the growth of admixture dosage. The complex modulus expressed the same trends as the flow strain.

The behaviour of the studied natural hydraulic lime pastes was analogous to the cementitious pastes, so similar effects can be anticipated to the ones on the cementitious mixes while using these admixtures in the natural hydraulic lime materials.

Acknowledgements
The work has been financially supported by the BUT specific research project FAST-J-19-5803.

References

[1] Van Der Vurst F, Grünewald S, Feys D, Lesage K, Vandewalle L, Vantomme J and De Schutter G 2017 *Cem. Concr. Compos.* **82** pp 190–201
[2] Plank J 2004 *Appl. Microbiol. Biotechnol.* **66** pp 1–9
[3] Singh N K, Mishra P C, Singh V K and Narang K K 2003 *Cement Concrete Res.* **33** pp 1319–29
[4] Zhang Y, Zhao Q, Liu C and Zhou M 2016 *Constr. Build. Mater.* **102** pp 648–53
[5] Mignon A, Snoeck D, D'Halluin K, Balcaen L, Vanhaecke F, Dubruel P, Van Vlierberghe S and De Belie N 2016 *Costr. Build. Mater* **110** pp 169–74
[6] Bezerra U T 2016 *Biopolymers and Biotech Admixtures for Eco-Efficient Construction Materials* (Duxford: Woodhead Publishing)
[7] Seabra M P, Labrincha J A and Ferreira V M 2007 *J. Eur. Ceram. Soc.* **27** pp 1735–41
[8] Vyšvařil M, Hegrová M and Žižlavský T 2018 *Solid State Phenomena.* **276** pp 69–74
[9] Vyšvařil M and Žižlavský T 2017 *IOP Conf. Ser.: Mater. Sci. Eng.* **251** 012039
[10] Vyšvařil M, Hegrová M and Žižlavský T 2018 *Key Eng. Mater.* **760** pp 257–65
[11] Govin A, Bartholin M-C, Biasotti B, Giudici M, Langella V and Grosseau P 2016 *Const. Build. Mater.* **122** pp 772–80
[12] Poinot T, Govin A and Grosseau P 2014 *Cem. Concr. Res.* **58** pp 161–8
[13] Govin A, Bartholin M-C, Biasotti B, Giudici M, Langella V and Grosseau P 2015 *Proceedings of the 27th Biennial National Conference of the Concrete Institute of Australia in conjunction with the 69th RILEM Week.* pp 845–57
[14] Patural L, Marchal P, Govin A, Grosseau P, Ruot B and Devès O 2010 *Cem. Concr. Res.* **41** pp 46–55
[15] Cappellari M, Daubresse A and Chaouche M 2013 *Constr. Build. Mater.* **38** pp 950–61
[16] Leemann A and Winnefeld F 2007 *Cem. Concr. Compos.* **29** pp 341–9
[17] Izaguirre A, Lanas J and Álvarez J I 2011 *Cem. Concr. Compos.* **33** pp 309–18
[18] Lasheras-Zubiate M, Navarro-Blasco I, Fernandéz J M and Álvarez J I 2011 *J. Appl. Polym. Sci.* **120** pp 242–52
[19] Žižlavský T, Vyšvařil M and Rovnaníková P 2018 *IOP Conf. Ser.: Mater. Sci. Eng.* **385** 012067
[20] Mezger T G 2014 *The Rheology Handbook* 4th ed. (Hanover: Vincentz Network)
[21] Knapen E and Van Gemert D 2009 *Cem. Concr. Res.* **39** pp 6–13
[22] Liu H, Zhao Y, Peng Ch, Song S and López-Valdivieso A 2016 *J. Mater. Sci.* **51** pp 9279–86
[23] Boháč M and Nečas R 2016 *Procedia Eng.* **151** pp 34–41
[24] Martín-Alfonso J E, Cuadri A A, Berta M and Stading M 2018 *Carbohydr. Polym.* **181** pp 63–70

IOP Conf. Series: Materials Science and Engineering **583** (2019) 012010 doi:10.1088/1757-899X/583/1/012010

Influence of soil-cement composition on its deformability

T Tracz[1], K Brasse[1], T Zdeb[1] and J Bielecka[1]

[1] Cracow University of Technology, Faculty of Civil Engineering, ul. Warszawska 24, 31-155 Kraków, Poland

Email: krystian.brasse@pk.edu.pl

Abstract. The paper discusses the results of deformability tests of soil-cement composites made with cohesive and non-cohesive soils. The compositions of the materials analysed were diversified in terms of their cement paste volume fractions and water-cement ratios. The aim of the research was to determine the secant modulus of elasticity in uniaxial compression of soil-cement composites in accordance with the EN 12390-13 standard dedicated to concrete testing. Due to the lower level of strength of the materials in question in comparison with ordinary concrete, nominal stress was lowered and the loading and unloading rates were modified. The number of loading cycles required for the determination of stabilised deformations was verified. The test results obtained and their analysis indicate the significant impact of the type of soil on the deformability of soil-cement composites. It was also found that the deformability of the materials tested is sensitive to changes in their compositions.

1. Introduction

In the practice of designing engineering structures, e.g. ones made of soil-cement, the key mechanical parameter is the building material's compressive strength. Owing to the ease of determining this feature, the high homogeneity and repeatability of the results obtained as well as the fact that it is possible to correlate these values with other characteristics, compressive strength is a basic and universal parameter of materials. According to the author of [1, 2], one of the fundamental parameters which can be correlated satisfactorily with the compressive strength of stabilised soil is its modulus of elasticity. However, the absence of an established methodology for testing the deformability of soil-cement composites and the strong dependence of this parameter on the type of reinforced soil translates into considerable variation with respect to the empirical relationships between the modulus of elasticity and compressive strength presented in literature. The insufficient homogeneity of the investigated soil-cements, which are produced e.g. using the DSM (Deep Soil Mixing) soil-cement pile construction technology, also has an equally fundamental effect on the discrepancies between the results obtained.

According to the author of [1, 3], when designing soil reinforcement based on the Deep Soil Mixing technology, the secant modulus of elasticity for soil-cement E_{50}, which is determined at 50% of the destructive stress value, is most often used. The expected value for soil-cement is the correlation coefficient between the secant modulus of elasticity and unconfined compressive strength E_{50}/UCS, which falls within different ranges depending on the UCS value. According to Topolnicki [1], for UCS < 2.0 MPa this range is from 50 to 300 UCS, and for UCS > 2.0 MPa it is from 300 to 1,000 UCS. A similar range for the E_{50}/UCS ratio, i.e. from 75 to 1,000, can be found in [4]. The authors claim that for design purposes, E_{50} values are most often assumed at 300 UCS. Other relationships can be found in [4], where for UCS < 1.0 MPa E_{50} = 120 UCS, and for UCS > 1.0 MPa E_{50} = 380 UCS. In [3], Topolnicki presents the results of tests of several dozen soil-cement specimens which were made using cohesive soil (loam) and differed in terms of their cement content. On the basis of those studies, it was found that the E_{50}/UCS correlation coefficient depends on the type of soil, cement content and the degree of homogeneity of soil-cement. However, the author of [1, 3] draws attention to the fact that during uniaxial compression in laboratory tests, soil-cement specimens are subjected to the different conditions than geotechnical structures. These tests do not take account for the presence of lateral stresses, which have a positive effect on the mechanical properties of soil-cement. In the presence of lateral stresses, soil-cement becomes more elastic-plastic and less brittle. In addition, under laboratory conditions a more favourable degree to which all soil-cement composite components are mixed is achieved. Therefore, secant moduli of elasticity of in-situ soil-cement are lower; according to [3], it can be assumed that in relation to the designed average compressive strength f_{cm} such moduli are in the following range: E_{50} = 300–500 f_{cm}. The same author narrows the range to E_{50} = 300–400 f_{cm} in [5].

A relationship concerning the correlation between the modulus of elasticity and UCS is also proposed in [6] on the basis of tests of moduli of elasticity of soil-cement specimens in the form of cores sampled from a structure. The stress levels applied ranged from 10% to 30% of the estimated UCS. The specimens tested differed in terms of the reinforced soil type (sand, clay and loam) as well as in terms of their curing time (30 to 200 days). As a result of those studies, empirical relationship in the form $E = 1482\ UCS^{0.8}$ (E and UCS in [MPa]) was determined, with the coefficient of determination close to 0.8. The range of correlation vs. UCS ranged from 1.5 to 35 MPa.

A broad review of literature concerning the search for correlation between the modulus of elasticity and compressive strength of soil-cements can be found in [7]. The authors state that the range between $E_{soil\text{-}cemen} = 12900\ UCS^{0.41}$ from [8] and $E_{soil\text{-}cement} = 30000\ UCS^{0.5}$ from [9] is appropriate for calculating modulus of elasticity values.

2. Study objective

This paper presents the results of studies aimed at determining the impact of the composition of soil-cement on its secant modulus of elasticity in compression. Another purpose of the research conducted was to verify whether the methodology contained in the EN 12390-13:2014 standard, which concerns tests of secant module of concrete, can be used in the testing of soil-cement composites after minor modifications. The results obtained are intended to be used in order to determine the relationship between the secant modulus of elasticity of soil-cement and its compressive strength. This relationship could be used in designing the structures produced using the Deep Soil Mixing technology.

3. Materials

The research programme involved producing a series of soil-cement composites whose compositions varied in terms of soil type, the amount of cement paste V_z used and water-cement (w/c) ratios. Three types of native soil, which are characterised in table 1, were selected for the production of the composites concerned. The soil-cements tested were produced by mixing those soils with water and with CEM II/B-S 32.5 R cement, which met the requirements of the EN 197-1:2012 standard.

Table 1. Properties of the soils selected.

Properties	**Medium sand**	**Clay**	**Loam**
Particle size distribution [% by mass]			
- gravel fractions 2/40 mm	5	-	-
- sand fractions 0.05/2 mm	95	40	30
- dust fractions 0.002/0.05 mm	-	42	42
- loam fractions <0.002 mm	-	18	28
Bulk density ρ [g/cm³]	1.60	2.10	2.08

For each soil, three different compositions of soil-cement mixtures were designed so that they exhibited adequate workability during mixing and consistency after mixing. Mixtures were required to be self-compacting with no segregation of ingredients. These requirements determined the amount of cement paste used and the adopted w/c ratios of the mixes produced. Fresh mix consistency tests were carried out in accordance with the standard [10]. Analysing the measurements conducted, it can be stated that in the case of the soils tested, mixtures with the appropriate consistency exhibited flow diameter values ranging from 150 to 270 mm. The liquidity of all mixtures guarantees the possibility of proper incorporation of these composites by the DSM method. All compositions are summarised in table 2.

Table 2. Compositions of soil-cement mixtures.

Soil	**Medium sand**			**Clay**			**Loam**		
Paste content V_z [dm^3/m^3]	400	450	450	600	600	650	700	750	800
w/c ratio	1.6	1.2	0.8	2.0	1.4	1.2	2.6	2.0	1.6
Cement content [kg/m^3]	208	296	401	258	348	427	240	323	416
Mixture designation	400/1.6	450/1.2	450/0.8	600/2.0	600/1.4	650/1.2	700/2.6	750/2.0	800/1.6

4. Methods

Mixing of composite components was carried out in a laboratory mixer with a capacity of 50 cubic decimeters. The mixer was equipped with two agitators working counter-rotating. The mixing time was determined by the type of soil and lasted until a homogenous mixture was obtained. The formed samples were stored for 28 days at + 20°C and 95% RH humidity until the mechanical properties were tested.

The modulus of elasticity was determined on the basis of the EN 12390-13 standard, which is dedicated to concrete testing. Owing to the lower compressive strength of the materials in question, test parameters were modified. The initial stress level was reduced by a factor of 10: from 0.5 MPa o 0.05 MPa. Depending on the type of soil used and the strength level achieved by the soil-cements produced from it, the rate at which the stress level changed was reduced from the standard value of 0.6 ±0.2 MPa/s to the values presented in table 3. The table also includes upper stress values σ_a which are equal to 1/3 of the compressive strength f_c of the soil-cements tested.

Table 3. Average compressive strength of soil-cements (f_c) and the technical parameters used in modulus of elasticity testing.

Soil	**Medium sand**			**Clay**			**Loam**		
Mixture designation	400/1.6	450/1.2	450/0.8	600/2.0	600/1.4	650/1.2	700/2.6	750/2.0	800/1.6
Compressive strength f_c after 28 days [MPa]	2.9	7.2	17.1	3.6	4.8	8.1	2.1	2.7	3.3
$\sigma_a = 1/3\ f_c$ [MPa]	1.0	2.4	5.7	1.2	1.6	2.7	0.7	0.9	1.1
Loading rate [MPa/s]	0.1	0.1	0.1	0.05	0.05	0.05	0.025	0.025	0.025

The EN 12390-13 standard recommends the use of cylindrical specimens with a diameter of 150 mm and a height of 300 mm. Since soil-cement composites do not contain large grain inclusions, the tests were conducted using representative cylindrical specimens with a diameter of 100 mm and a height of 200 mm whose opposite planes were ground before the test. Extensometers (figure 1) with a maximum measuring range of ±2.5 mm and an accuracy of 0.1 μm were used to measure longitudinal strain. During the test, those sensors were installed on the side surfaces of the cylindrical specimens in a suitable frame with a measurement base length of 100 mm.

Each test consisted of three specimen loading and unloading cycles. At the beginning of the test, the specimen is loaded to the stress value σ_p, for which the preload strain ε_p is measured along the compression axis. Subsequently, the load is increased to the stress level σ_a and the strain value ε_a is read.

Figure 1. Specimen during the test – measurement of strain (ε) using two extensometers.

The secant modulus of elasticity $E_{c,s}$ was calculated using formula (1), which uses mean strains at preload stress and at upper stress during the third cycle. Strain is measured each time after around 20 seconds of constant load value at each load level.

$$E_{c,s} = \frac{\sigma_{a,3} - \sigma_{p,3}}{\varepsilon_{a,3} - \varepsilon_{p,3}}, \tag{1}$$

where:

$E_{c,s}$ – secant modulus of elasticity in compression [MPa],
$\sigma_{a,3}$ – measured stress corresponding to nominal upper stress [MPa],
$\sigma_{p,3}$ – measured stress corresponding to nominal preload stress [MPa],
$\varepsilon_{a,3}$ – average strain at upper stress on loading cycle 3,
$\varepsilon_{p,3}$ – average strain at preload stress on loading cycle 3.

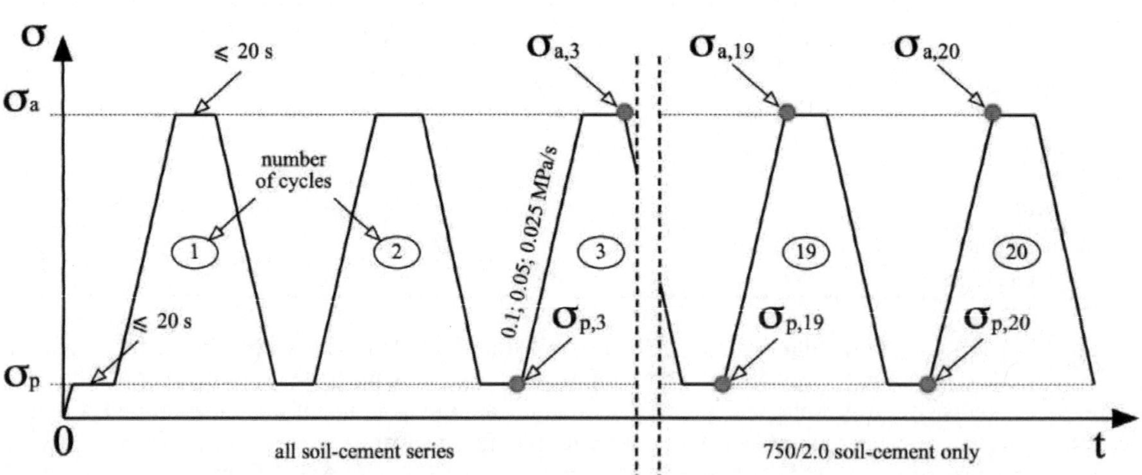

Figure 2. Diagram presenting stress level changes during the modulus of elasticity ($E_{c,s}$) test.

In the first place, before the basic deformability tests of all the specimens analysed were conducted, the number of cycles required for the assessment of the modulus of elasticity was verified. To this end, one series of specimens was selected, which was designated 750/2.0 (see table 2), and a total of 20 loading/unloading cycles were carried out, with strains at individual stress levels being measured at each time (see figure 2). Subsequently, the modulus of elasticity during individual cycles (from 3 to 20) was calculated; the results are presented in figure 3. Due to the instability of deformations in the first two cycles, they were omitted from figure 3 in accordance with the recommendations of the standard [11].

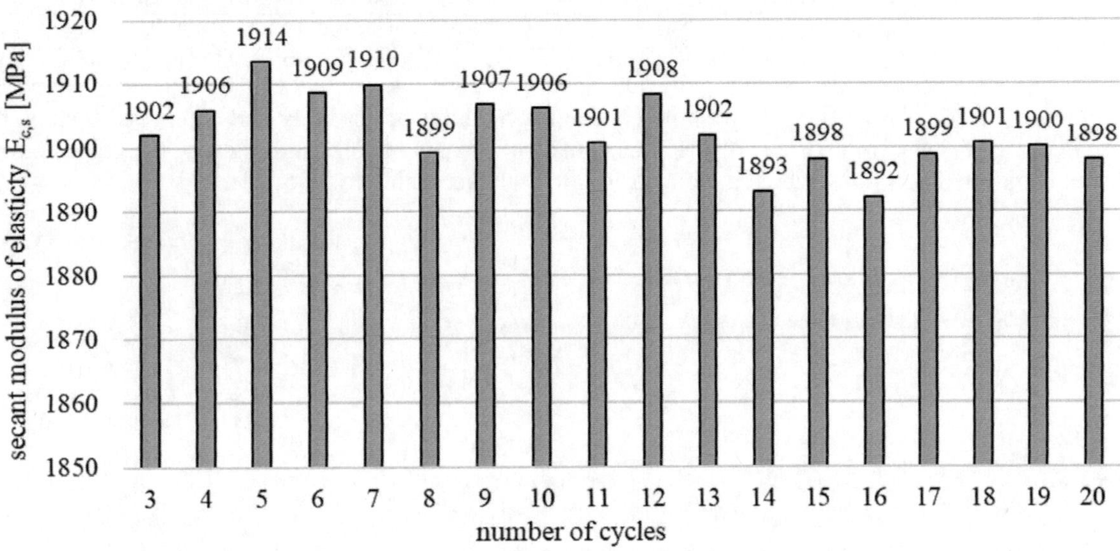

Figure 3. Variability of the modulus of elasticity ($E_{c,s}$) determined during 20 cycles for the soil-cement series designated 750/2.0.

The results presented indicate that the variation in the modulus of elasticity value in individual cycles from 3 to 20 is small: the results ranged from 1892 to 1914 MPa. Therefore, it can be assumed that this small variation results from natural measurement errors, in particular deformations and heterogeneity of the material tested. Very importantly, no upward or downward trends in the modulus of elasticity were observed as the number of cycles increased. Therefore, it was considered that the number of cycles (three cycles) adopted in the standard [11] is sufficient in order to reliably evaluate the modulus of elasticity in the soil-cements analyzed. Therefore, in the results presented below, modulus of elasticity values were calculated on the basis of strain and stress measurements during the third cycle (figure 2).

5. Test results

The test results obtained are presented in table 4. The values of secant modulus of elasticity presented in each case represent the average for 3 specimens. Additionally, table 4 shows the ratios of the modulus of elasticity $E_{c,s}$ to compressive strength f_c.

Table 4. Results of secant modulus of elasticity ($E_{c,s}$) tests after 28 days of curing [MPa].

Soil		Medium sand			Clay			Loam		
Mixture designation		400/1.6	450/1.2	450/0.8	600/2.0	600/1.4	650/1.2	700/2.6	750/2.0	800/1.6
Secant modulus of elasticity $E_{c,s}$ [MPa]	Average	10500	15200	21200	3000	4500	6900	1200	1900	3300
	Min	10120	14820	20760	2930	4360	6580	1170	1880	3150
	Max	10880	15730	21620	3070	4620	7100	1220	2080	3450
$E_{c,s}$ / f_c		3621	2111	1240	833	938	852	571	704	1000

Figure 4 below shows the relationship between the modulus of elasticity and compressive strength for all partial results from 9 series of specimens made of 3 types of soil. Additionally, the graph shows the range of the $E_{c,s}/f_c$ ratio, which is from ca. 100 to 1000 according to [3, 5, 12].

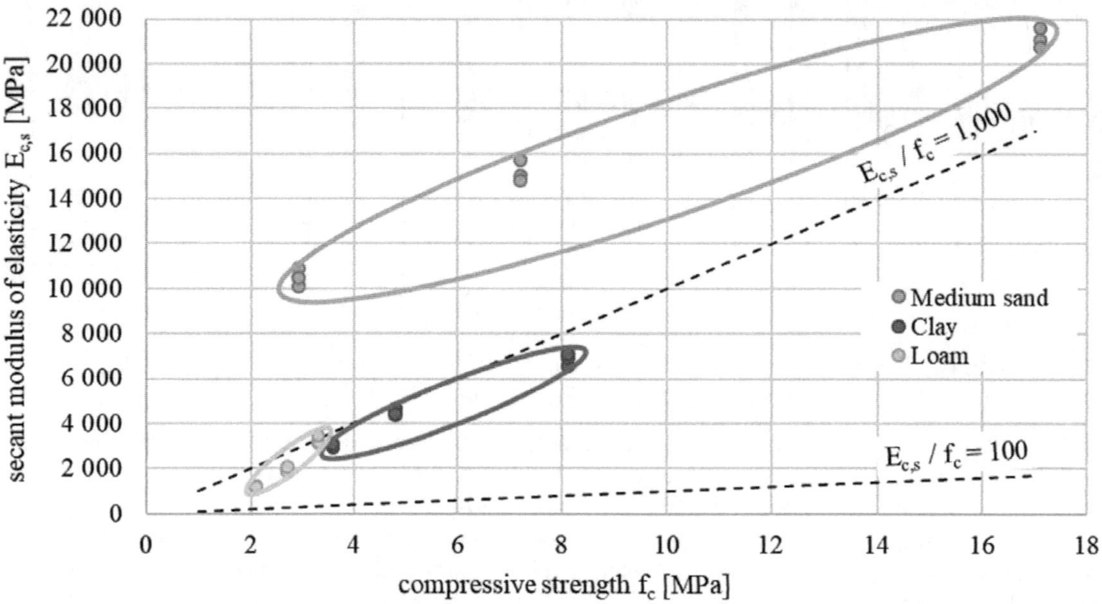

Figure 4. Relationships between compressive strength (f_c) and secant modulus of elasticity ($E_{c,s}$) after 28 days of curing.

The analysis of the above results demonstrates that the $E_{c,s}/f_c$ ratios determined on the basis of the results obtained differ significantly depending on the type of soil used. For soil-cements made using cohesive soils, the average $E_{c,s}/f_c$ values were as follows: 874 for clay soil and 758 for loam soil. It should be emphasised here that all partial values of these ratios are within the 100–1000 range which is mentioned in the literature [3].

However, for non-cohesive medium sand soil all $E_{c,s}/f_c$ ratio values significantly exceed the upper limit of the range stated. The average $E_{c,s}/f_c$ ratio for this soil type was 2324.

In the analysis of the impact of the composition of soil-cements on their deformability, the most notable phenomenon is the variability of the results obtained depending on the soil used. Soil-cements made from non-cohesive soil (medium sand) are characterised by a secant modulus of elasticity which ranges from 10500 to 21200 MPa. Much lower $E_{c,s}$ values were obtained for composites made from cohesive soils. For clay soil, the results range from 3000 to 6900 MPa, while in the case of loam soil this range is as low as 1200 to 3300 MPa.

In general, soil-cements include cement pastes with relatively high w/c ratios compared to the pastes present in ordinary concrete. This is related to the differences between aggregate grain size distribution in concrete and soil grain size distribution in soil-cement. Soil-cement mixes exhibit significantly greater demand for highly liquid cement paste. Analysing the secant modulus of elasticity results obtained for soil-cements with the same amount of paste within the same soil type, it can be concluded that in the case of medium sand soil and a V_z equal to 450 dm^3/m^3, an increase in the w/c ratio from 0.8 to 1.2 results in a decrease in the modulus of elasticity value by 28%, while for clay soil and a V_z equal to 600 dm^3/m^3, an increase in w/c from 1.4 to 2.0 results in a decrease in the $E_{c,s}$ value by 52%.

When analysing the impact of the composition of soil-cement composites on their deformability, tests concerning the effect of the amount of cement on the secant modulus of elasticity are of considerable value. The amount of cement results from the assumed w/c ratio and the amount of cement paste, thus it is a universal parameter of soil-cement composition [13]. For each soil type, an increase in the amount of cement increased the value of the secant modulus of elasticity. In the case of medium sand soil, an increase in cement content from 208 to 401 kg/m^3 resulted in an increase in the $E_{c,s}$ value by 102%. However, this impact is even more pronounced in the case of cohesive soils. For clay soil, an increase in cement content from 240 to 416 kg/m^3 resulted in an increase in the $E_{c,s}$ value by 130%, and for loam soil, an increase in cement content from 258 to 427 kg/m^3 made it possible to achieve an $E_{c,s}$ value higher by 175%. Obviously, this is due to the higher content of paste in soil-cement composition and, consequently, the greater impact of the amount of binder used on the properties of the entire composite.

6. Conclusions

The tests conducted have demonstrated that the application of the methodology contained in the EN 12390-13 standard for determining the secant modulus of elasticity of soil-cements is justified. It has been confirmed that during the third loading and unloading cycle, strains are already stabilised and no further cycles are required.

The fundamental factor in the composition of soil-cements which determines their deformability is the type of soil, which is the inclusion present in the composites. The use of loose (sand medium) soil in the composite, i.e. soil with coarser grains and lower demand for cement paste with the lowest possible w/c, enables a composite to be obtained with a relatively high compressive strength (f_c = 2.9–17.1 MPa) and a high modulus of elasticity ($E_{c,s}$ > 10000 MPa). The $E_{c,s}/f_c$ ratio determined for this group of soil-cements ranges from 1240 to 3621.

On the other hand, soil-cements made of cohesive (clay and loam) soils exhibit compressive strength values in the range from 2.1 to 8.1 MPa and moduli of elasticity in the range from 1200 to 6900 MPa. The $E_{c,s}/f_c$ ratio values calculated in this case range from 571 to 1000.

The correlations between soil-cement compressive strength and the modulus of elasticity proposed most frequently in the literature ($E_{c,s}/f_c$ = 100–1000) appear to hold for cohesive soils only. In the case of soil-cements made using loose soils, the $E_{c,s}/f_c$ ratio may even exceed 3000.

Summing up, estimates of modulus of elasticity values based solely on soil-cement compressive strength can be highly inaccurate. The $E_{c,s}/f_c$ ratio depends very much on the type of soil. Therefore, it is possible that soil-cements made from various soils may have significantly different moduli of elasticity despite exhibiting similar compressive strengths.

Acknowledgements

The authors also acknowledge the financial support of the Polish National Agency for Academic Exchange under the International Academic Partnership Programme: PPI/APM/2018/1/00027 *E-mobility and sustainable materials and technologies* EMMAT.

References

[1] Topolnicki M 2004 In situ Soil Mixing *Ground Improvement*, ed Moseley M and Kirsch K. (London and New York: Spon Press) pp 331–428

[2] Topolnicki M 2009 Design and Execution Practice of Wet Soil Mixing in Poland *International Symposium on Deep Mixing & Admixture Stabilisation* (Okinawa) pp 195–202

[3] Topolnicki M 2018 Zasady stosowania i projektowania wzmocnienia gruntu metodą wgłębnego mieszania na mokro (DSM) *Inżynieria i Budownictwo* **1** pp 3–16

[4] Bruce C, Collin J, Berg R, Filz G, Terashi M and Yang D 2013 Federal Highway Administration Design Manual: Deep Mixing for Embankment and Foundation Support p 244

[5] Topolnicki M 2017 Dobra praktyka stosowania i projektowania wgłębnego mieszania gruntu na mokro (DSM) *XXXII Ogólnopolskie Warsztaty Pracy Projektanta Konstrukcji* pp 1–29

[6] Tourrich J 2018 Handbook - Soil mix walls: Design and execution ed Denies N, Huybrechts N

[7] Fan J, Wang D and Qian D Soil-cement mixture properties and design considerations for reinforced excavation *Journal of Rock Mechanics and Geotechnical Engineering* vol 10 no 4 pp 791–797

[8] Briaud J-L, Nicholson P and Lee J 2000 Behavior of Full-Scale VERT Wall in Sand *Journal of Geotechnical and Geoenvironmental Engineering* vol 126 no 9 pp 808–818

[9] Wang D, Olowokere D and Zhang L 2014 Interpretation of Soil-Cement Properties and Application in Numerical Studies of Ground Settlement Due to Tunneling Under Existing Metro Line *Geotechnical and Geological Engineering* vol 32 no 5 pp 1275–1289

[10] PN-EN 1015-3: 2000 Methods of test for mortar for masonry - Part 3: Determination of consistence of fresh mortar (by flow table)

[11] PN-EN 12390-13:2014-02 Testing hardened concrete - Part 13: Determination of secant modulus of elasticity in compression

[12] Topolnicki M 2016 General overview and advances in Deep Soil Mixing XXIV Geotechnical Conference of Torino Design, Construction and Controls of Soil Improvement Systems pp 1–30

[13] Brasse K, Tracz T, Zdeb T and Rychlewski P 2018 Influence of Soil-Cement Composition on its Selected Properties *MATEC Web of Conferences* vol 163 pp 1–8

IOP Conf. Series: Materials Science and Engineering **583** (2019) 012011 doi:10.1088/1757-899X/583/1/012011

Influence of curing temperature on the mechanical properties of high-performance concrete

J Fladr[1] and I Broukalova[1]

[1] Department of Concrete and Masonry Structures, Faculty of Civil Engineering, Czech Technical University in Prague, Thákurova 7, 16629 Prague, Czech Republic

Email: josef.fladr@fsv.cvut.cz

Abstract. As part of the research focused on high-performance concrete (HPC) at our workplace, the impact of temperature curing on compression strength of HPC has been investigated. Four sets of samples (100 x 100 x 100 mm cubes, 30 samples in each set) were produced with the same mix design. Each set was cured with a different temperature: 20°C (reference samples), 50°C and 80°C (high and very high temperature curing) and 10°C (simulation of concreting in winter conditions). The curing was always carried out in a water bath with a given temperature during the whole time of maturing of the samples. Samples were tested in 1, 3, 7, 14 and 28 days of age. In each time step, three cubes were tested for compression strength and three cubes for tensile splitting strength. The results of the primary research are the time records of the increase of both strengths.

1. Introduction

Cement hydration is a chemical reaction in which heat is released. The elevated temperature influences the hydration rate, the development of the hydration heat, and the C-S-H gel structure [1]. The elevated temperature is exhibited mostly in the initial stages of hydration, especially during the first three days [2]. According to the standards, the curing temperature is 20°C; at the temperature of 30°C, growth rate of compressive strength is accelerated during the first few days. On the other hand, reducing the curing temperature to 10°C leads to a significant deceleration of hydration, as well as to a slow increase of compressive strength during the first days.

At elevated temperature a large amount of hydrates has not enough time for ideal growth, what may cause loss of final strength (at age of 28 days). This phenomenon is known as the "crossover effect" [2, 3, 4].

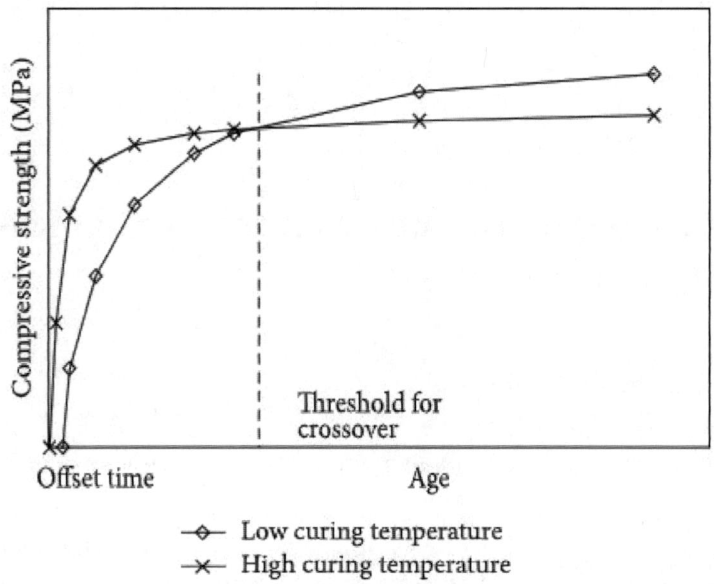

Figure 1. Crossover effect for low curing temperature and high curing temperature of concrete [2].

To avoid this effect, temperature of 70°C is considered safe maximum limit in wet environment, and 85°C in dry environment [5]. High curing temperature favourably affects behaviour of pozzolanic active additives as fly-ash and microsilica. During heating, their hydrating effect grows, what increases density of cement paste and consequently strength of concrete [6, 7].

In the treatment with high temperature a delayed ettringite formation may occur. In fact, it is an internal sulphate attack, in which monosulfate reacts with gypsum and the reaction product is ettringite. Ettringite is characterised by high hydration rate, at which increases its volume by up to 270% [8].

The ettringite crystal sizes are also temperature-dependent; the conventional crystals formed at a temperature of 25°C are smaller than crystals created at crystallisation around 60°C [6]. The recrystallization of ettringite takes place several days after the heat treatment, when the cement paste has already hardened, and recrystallization disturbs its structure. This is a fundamental difference against the ettringite crystals, which are created several hours after beginning of hydration. At this moment, the cement paste is still soft and therefore creating of ettringite crystals does not cause mechanical disturbance of the structure. At this stage, ettringite is in the form of a stable monosulfate and contributes to the initial increase in strength.

Delayed ettringite formation can be reduced by addition of fine pozzolanic admixtures, which, practically, are in all HPC recipes. If pozzolanic admixtures are used, the temperature treatment may reach 80°C without delayed formation of ettringite [9].

2. Experimental methods

The investigation of HPC behaviour at temperature treatment was performed on cubes of size 100 x 100 x 100 mm. The recipe shown in a table 1 was used during the whole experiment. Four sets of specimens were elaborated with the recipe, each set with 30 specimens. Half of specimens was used for compressive strength testing, the other half for tensile tests. Testing was performed at 1, 3, 7, 14 and 28 days.

Curing of specimens was always performed in water environment with temperature 10°C, 20°C, 50°C and 80°C with this procedure: after placing of fresh concrete in moulds the specimens rested 2 hours, and then they were immersed in water bath including the steel mould. The water had already been preheated to required temperature. The water level did not overtop the top edge of the mould.

IOP Conf. Series: Materials Science and Engineering **583** (2019) 012011 doi:10.1088/1757-899X/583/1/012011

The specimens were cured for 24 hours. After this time the specimens were demoulded, and first specimens were tested in compression and tensile. The rest od specimens were again immersed in water basin and the water level was raised to sink the whole specimens including the upper surface.

The temperature of water treatment was not changed during maturing of specimens. The authors are aware of the low efficiency of temperature treatment after 3 days of specimen age [10], however the presented results represent the pilot investigations, and for simplification the temperature remained constant during the whole experiment.

No enhancement of mechanical properties was expected for specimens treated in water with 10°C compared to reference specimens treated in 20°C. Curing with this temperature should only serve for better understanding relation of temperature and mechanical properties.

Table 1. Composition of the mixture.

Compound	**Specification**	**Dosage [kg/m^3]**
Cement	CEM I 42,5	650
Water		175
w/b		0.22
Aggregate 0/4	Basalt	780
Aggregate 4/8	Basalt	478
Aggregate 8/16	Basalt	360
Microsilica	Stachesil S	80
Plasticizer	Stachement	29

3. Results

3.1. Compressive strength

The main reference parameter was compressive strength, which was measured always on three cubes with dimensions 100 x 100 x 100 mm. Presented result is an average value from three measurements. The specimens were strained by constantly increasing load of rate 8 kN/s [11]. Results are depicted in figure 2.

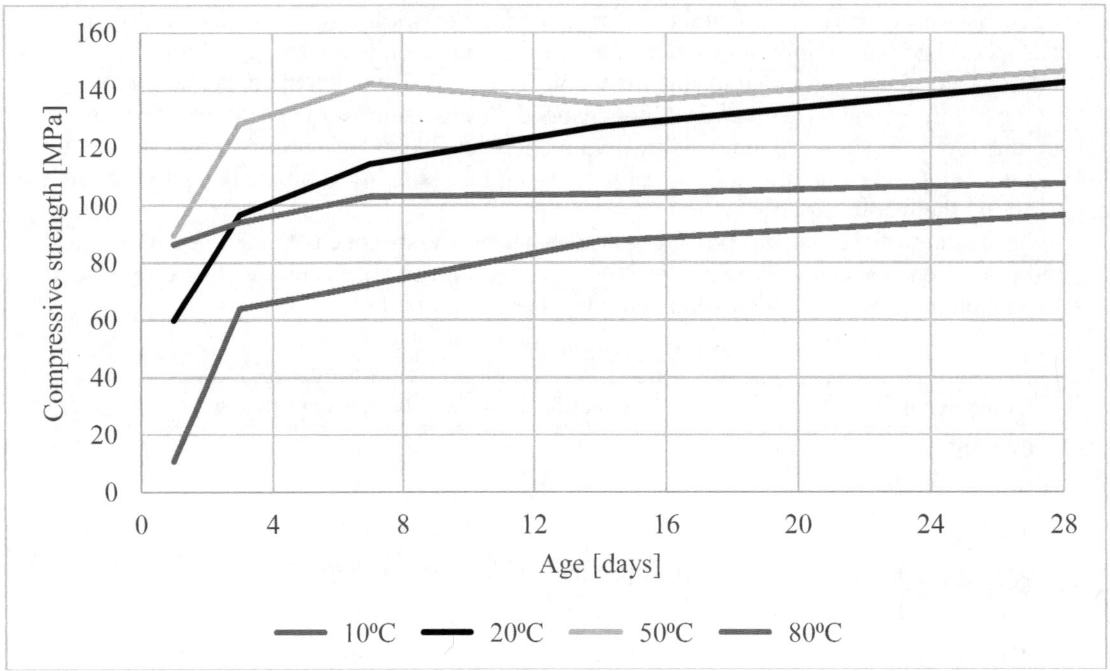

Figure 2. The development of compressive strength for all specimens.

Set of specimens treated at 80°C had the highest compressive strength (86.3 MPa) at age one day, but at age of three days the compressive strength was only 94.0 MPa, what is lower than compressive strength of reference specimens. The increase of compressive strength stopped at age 7 days for the specimens treated by temperature 80°C and the final compressive strength was by 32% lower than reference specimens′ strength. The authors are aware that 80°C is quite high temperature, but they based their investigations on precedent research [9], which showed that in case of application of latent hydraulic admixtures this temperature is acceptable. Presented investigations do not prove this theory.

3.2. Tensile splitting strength

The other investigated parameter was tensile splitting strength, which was also measured on cubes with dimensions 100 x 100 x 100 mm. Results depicted in figure 3 are the arithmetic average of three measurements. The specimens were loaded with constantly increasing load with 0,96 kN/s [12].

IOP Conf. Series: Materials Science and Engineering **583** (2019) 012011 doi:10.1088/1757-899X/583/1/012011

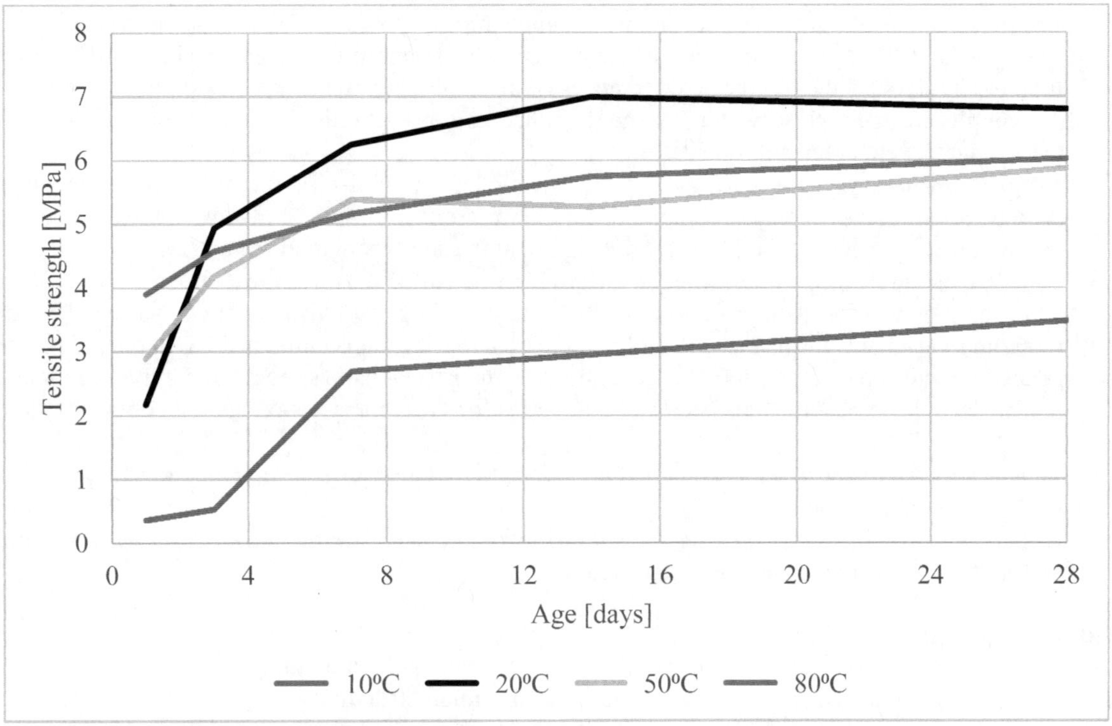

Figure 3. Distribution of tensile splitting strength for all kind specimens.

The samples that have been subjected to thermal curing reached lower splitting tensile strength than the samples that have been cured at standard room temperature (20°C). This is in accordance with another results from literature. It has been observed that the ratio between compressive and tensile strength can be changed from 1:10 up to 1:20 by thermal curing. The increase of the ratio depends mainly on thermal gradient [13].

Results of studies confirm our observations of effect of elevated temperature on compressive strength. The paper [14] came to similar conclusions as our team regarding measurement of compressive strength of specimens treated in different temperatures: "Comparison of the results for the same mixture at different curing conditions shows that, although higher temperatures improve the initial strength development, the value at a later stage, 14 or 28 days, seems to be penalized." Measuring of tensile strength of specimens cured at different temperatures is not in the focus of interest in the studies.

Several studies focused in investigation of autogenous shrinkage. The papers state that curing temperature affects autogenous shrinkage. Paper [15] concludes that the autogenous shrinkage of HPC increased with the increase of curing temperature. Since the decrease of tensile strength is directly linked to higher shrinkage, this may considered as a confirmation of our conclusions.

On the other hand, various other studies [16, 17, 18] show that undoubted conclusions on effect of temperature curing on autogenous shrinkage and consequently on tensile strength cannot be drawn.

4. Conclusions

The following conclusions can be drawn based on the obtained results:

- The positive effect of temperature treatment on the rate of compressive strength increase in the initial phase of concrete maturing has been demonstrated.
- Therewithal, an assumption was confirmed that temperature treatment may accelerate increase of compressive strength, not the final value of compressive strength. Compressive strength of reference specimens at age 28 days was 142.6 MPa and specimens cured in water of temperature 50°C had 28 days compressive strength 146.5 MPa.

- The presented results show that ideal curing temperature to achieve high short-term compressive strengths is 50°C. At this temperature, the compressive strength increased by about 30% within first 7 days, then the differences in the compressive strength between the samples treated in 50°C water and the reference samples decreased, and at the age of 28 days, the strengths were practically identical.
- Temperature treatment in 10°C was only a complement of the research. A slower increase of the compressive strength compared to the reference samples was expected. Likewise, the reference samples reached higher values of compressive strength at the age 28 days.
- Tensile strengths testing of investigated type of high-performance concrete confirmed assumption that heat treatment results in decrease of tensile strength. Highest tensile strength was reached for reference specimens. It was by 15% higher compared to specimens cured in 50°C and 80°C. Clear general conclusion on influence of temperature treatment on tensile strength apparently cannot be made. This phenomenon is unsystematic and varies for different types of cement composites and their composition.
- Tensile strength of specimens treated in 10°C reached half the value of reference specimens, what is also unpredicted result.
- Based on the results of the research, the authors recommend to use thermal curing with maximum temperature of 70°C for the given mixture.

Acknowledgments
This paper was prepared thanks to the support of the Science Foundation of the Czech Republic (GAČR), project "Analysis of the relations between the microstructure and macroscopic properties of ultra-high performance concretes" (No. 17-19463S).

References

[1] Šuman Z 1993 *Maltoviny I* (Brno: VUT) p 198
[2] Keun-Hyeok Y, Jae-Sung M and Myung-Sug C 2015 Effect of Curing Temperature Histories on the Compressive Strength Development of High-Strergth Concrete *Advance in Materials Science and Engineering* vol 2015
[3] Razak H A and Sajedi F 2011 The effect of a heat treatment on the compressive strength of cement-slag mortars *Materials and Design* vol 32 pp 4618–4628
[4] Collepardi M 2006 *Moderní beton* (Praha: ČKAIT)
[5] EN 13670: 2011 *Execution of concrete structures*
[6] Riad D and Benmalek M L 2014 Acceleration of the Hardening of Concrete Made with Mineral Admixtures by Using a Heat Treatment Process *Advances in Materials Science and Engineering* vol 4 pp 164–171
[7] Cwirzen A 2007 The effect of the heat-treatment regime on the properties of reactive powder concrete *Advances in Cemen Research* vol 19 pp 25–33
[8] Alexander M, Bentur A and Mindess S 2017 *Durability of Concrete* (USA: CRC Press) p 323
[9] Yasser A, Nordine L and Ouali A 2017 Effect of supplementary cementitious materials (scm) on delayed ettringite formation in heat-cured concretes *Enegy Procedia* vol 139 pp 565–570 (International Conference On Materials And Energy)
[10] Steven H, Kerkhoff B and Panarese W 2002 Curing Concrete *Design and Control of Concrete Mixtures* chapter 12 (*Online* http://www.ce.memphis.edu/1101/notes/concrete/PCA manual/Chap12.pdf)
[11] EN 12390-3:2009 *Testing hardened concrete – Part 3: Compressive strength of test specimens*
[12] EN 12390-6:2010 *Testing hardened concrete – Part 6: Tensile splitting strength of test specimens*
[13] Hela R 2005 *Technologie betonu* (Brno: VÚT Brno)
[14] Lura P, Breugel K and Maruyama I 2001 Effect of curing temperature and type of cement on early-age shrinkage of high-performance concrete *Cement and Concrete Research* vol 31 issue 12 pp 1867–1872

[15] Shen D, Jiang J, Shen J, Yao P and Jiang G 2016 Influence of curing temperature on autogenous shrinkage and cracking resistance of high-performance concrete at an early age *Construction and Building Materials* vol 103 pp 67–76
[16] Radocea A 1998 Autogenous volume change of concrete at very early age *Magazine of Concrete Research* vol 50 issue 2 pp 107–113
[17] Tazawa E, Matsuoka Y, Miyazawa S and Okamoto S 1995 Effect of Autogenous Shrinkage on Self Stress in Hardening Concrete *The international RILEM symposium: Thermal Cracking in Concrete at Early Ages E&FN Spon* (London) pp 221–228
[18] Jensen O and Hansen P 1999 Influence of temperature on autogenous deformation and relative humidity change in hardening cement paste *Cement and Concrete Research* vol 29 issue 4 pp 567–575

Utilization of basalt fabrics as reinforcement for alkali-activated blast furnace slag systems

P Hruby[1], L Kalina[1], J Masilko[1], M Kimm[2] and T Gries[2]

[1] Brno University of Technology, Faculty of Chemistry, Purkyňova 118, 612 00 Brno, Czech Republic
[2] Institut für Textiltechnik (ITA) of RWTH Aachen University, Otto-Blumenthal-Straße 1, 52074 Aachen, Germany

Email: xchrubyp@fch.vut.cz

Abstract. Alkali-activated materials based on the alkali-activated blast furnace slag (AABFS) seem to be a suitable alternative construction material due to the environmental and economic aspects. More widespread application potential could be supported by the utilization of fabric reinforcement into the matrix to form AABFS composite with improved mechanical properties. The basalt fibres or fabrics (BFs) represent a reinforcing material suitable for the AABFS due to its favourable price/properties ratio. However, the question of fibres stability under the alkaline conditions should be taken into account. The effect of BFs implementation in one or more layers into various types of matrices to the mechanical properties was studied with the regard to the compressive and flexural strengths. The used matrices were based on the alkali-activation of the BFS with various activators. The fabric/matrix adhesion same as the transition zone properties were studied using the SEM analysis and pull-out tests. Indication of appreciable improvement of mechanical properties utilizing one layer of BFs for some of the mixtures activated using sodium water glass and sodium carbonate was found out.

Keywords Alkali-activated materials, blast furnace slag, basalt fabric, pull-out test

1. Introduction

Alkali-activation of secondary raw materials like blast furnace slag (BFS) using the convenient alkaline activator allows to produce alternative construction materials e. g. AABFS with many favourable properties in contrary to the traditional ordinary Portland cement (OPC). [1, 2] Starting with the much-discussed environmental benefits of carbon dioxide emissions lowering along with the ability to achieve lower cost when the composition of the alkali-activated materials is optimised [3] and continued by many favourable properties like lower hydration heat in contrary to the OPC [4], high resistance o the aggressive environments [5] or good high temperature performances [6]. Still, there are some disadvantages like high shrinkage accompanied with formation of micro and macro cracks or formation of efflorescences as described in publication [7].

AABFSs represents suitable matrix for basalt fabric implementation to form inorganic composite with improved mechanical and thermal properties as well as durability. The main purposes of fibre/fabric reinforcement of the matrix are to provide a control of the cracking and to increase the fracture toughness of the brittle matrix through the stress transfer during both micro and macro-cracking. The ability of the stress transfer depends on the volumetric fibre fraction, fibres orientation and distribution. However, in order to achieve the desired properties, the existence of quality adhesion between the fibres and the matrix (F/M) same as formation of transition zone at their interface is necessary as discussed in studies [8–10]. Since the AABFS matrix forms highly alkaline environment for the reinforcing material and due to the price/parameter ratio basalt fibres have been chosen instead of the alkali resistant glass fibres in this paper.

There were no publications where the same type of matrix and basalt fabric were used, otherwise that is the reason why this paper has originated. Many authors used chopped basalt fibres instead of the fabric, hence possibilities of utilization of basalt fibres as a reinforcing material were tested. Timukal et al. [11] reported that reinforcement of Na_2SiO_3 and NaOH activated fly ash matrix utilizing 10 wt. % of basalt fibres led to the increase of compressive strength about 37% after 28 days but utilization of the dosage higher than 15 wt. % resulted in decrease of compressive strength in contrary to the reference because of lower precursor dissolution leading to the less binder content. Similar results were presented in Rill et al.'s paper [12] where the addition of 10 wt. % of the BFs to the matrix based on the metakaolin activated using the KOH resulted in the increase of flexural strength from 1.9 to 19.5 MPa. Masi et al. [13] noted that just a minimal chemical interaction between the AAM (based on fly ash activated with sodium aluminate solution) and the BFs had been found. They also reported improvement of the flexural strength between 600 and 1000°C due to the better F/M interaction after the sintering. These materials have the potential for applications at elevated temperatures up to about 800°C because they represent replacement of ceramic based composites for the midrange temperature applications. This issue was discussed more in study of Welter et al. [14].

This paper focuses on the preliminary studies with a goal to determine the influence of the basalt fabric utilization to the AABFS matrix on the mechanical properties and possibilities of characterization of the F/M interaction.

2. Experimental procedure

2.1. *Raw materials*

2.1.1. Blast furnace slag (BFS). Blast furnace slag was obtained from Kotouč Štramberk, Ltd. having a Blaine specific surface area of 400 $m^2 \cdot kg^{-1}$. Its chemical composition is given in Table 1. The BFS was 84% amorphous, while the rest was crystalline (akermanite – 9.5%, calcite – 3.7%, merwinite – 2.3% and quartz – 0.5%) according to the X-ray diffraction (XRD) using Rietveld method.

Table 1. Chemical composition of the BFS determined by X-ray fluorescence (XRF) analysis.

Chemical composition in wt. %									
CaO	SiO_2	MgO	Al_2O_3	SO_3	TiO_2	K_2O	MnO	Na_2O	Fe_2O_3
41.1	34.7	10.5	9.1	1.4	1.0	0.9	0.6	0.4	0.3

2.1.2. Alkaline activators. Liquid sodium water glass (WG) obtained from Vodní sklo, Inc. – Silicate modulus (M_s) = 2.24; Sodium hydroxide (SH) – PENTA Ltd. (pure); Sodium carbonate (SC) – INCHEMA Ltd. (pure).

2.1.3. Basalt fabric (BF). Basalt fabric (BF) in form of filament yarns connected into the square net was provided by the Institute for textile technique at RWTH Aachen University. BFs have been used in form of biaxial warp knitted textile but also as unique filaments yarns in same cases (pull-out test, SEM). Filament yarns have different tex in longitudinal (2.6) and transverse directions (1.3). Table 2 gives the information about the chemical composition of basalt fibres from Kammeney Vek determined by X-ray fluorescence (XRF). Amorphous structure of BFs was confirmed using the XRD.

Table 2. Chemical composition of the BFs from Kammeney Vek determined by XRF analysis.

Chemical composition in wt. %								
SiO_2	Al_2O_3	Fe_2O_3	CaO	MgO	Na_2O	K_2O	TiO_2	Other
55.7	15.4	10.8	7.4	4.1	2.4	1.5	1.2	1.4

2.1.4. Other materials. Standard siliceous sand according to ČSN EN 196-1 was used for the mortar preparation. STACHEMENT AC 600 produced by Stachema Kolín Ltd. was used as a shrinkage reducing admixture (SRA).

2.2. Tests and methods

2.2.1. Mechanical properties testing. Testing samples were prepared according to ČSN EN 196-1. Specimens with dimension 40 × 40 × 160 mm were used for compressive and flexural strengths (three-point bend arrangement) determination using the Desttest 4310 Compact A (Beton System, Ltd.). Water to binder (w/b) ratio (0.49) was kept the same along with the SRA dosage (0.75 wt. % by the BFS weight) and sand to binder ratio (3:1) for all mortars. Various alkaline activator mixtures were used for the alkali-activation of the BFS meanwhile. Starting with the WG/SC and WG/SH mixtures with silicate modulus (M_s) equal to 0.5 using activator dosage ranging from 6 to 12 wt. % by the BFS weight. Demoulded specimens were kept in wet chamber until further testing was done. These tests were performed on mortars (reference) and mortars reinforced with the basalt fabric(s). One layer of basalt fabric was added to the middle of the specimen commonly, if there were more layers the layers were distributed equally, so the same amount of mortar was between the layers always.

2.2.2. Single fibre yarn pull-out test. This method was used for a direct estimation of fibre/matrix (F/M) interfacial bond properties based on pulling out of a single fibre yarn of its surrounding matrix (mortars). Testing samples had shape of cubes (40 × 40 × 40 mm) where the fibre yarn was placed into the middle with embedment length (EL) of 5 mm. Pull-out tests were performed for the same EL after 72 hours of curing at wet conditions based on previous research. Fifteen samples were tested and evaluated for each matrix type. The measurement was done using the Universal mechanical testing machine Instron 5985 (testing speed 2 mm/min; distance between jaws 100 mm) with a maximal force (F_m) before the fibre broke down (fibre breakage about F_m = 333 N) or was pull out determination. Interfacial shear bond strength was then calculated based on the Choi´s study [15].

2.2.3. SEM analysis. The Zeiss Evo LS 10 microscope with EDX analysis was used for microstructure of interfacial zone properties study. Accelerating voltage was 10 kV. Samples were AABFS pastes (w/b = 0.36) reinforced with one yarn of basalt fibres. They were broken down after defined period (7/28 days of hydration) and excessive fibres were removed. These fragments were placed on the carbon tape, sputter coated with gold and underwent the SEM analysis.

3. Results and discussion

3.1. WG/SC and WG/SH activated BFS matrices

Figure 1 and Figure 2 show results of flexural and compressive strengths development as a function of alkaline activator dosage and BF reinforcement.

The highest compressive strength (83.8 ±4.2) MPa was obtained for 10% Na_2O WG/SC activated matrix without utilization of BFs reinforcement after 112 days. It should be noticed that increase of compressive strength for this mixture have continued up to 112 days while the other activator dosages did not show further gain of compressive strength after 56 days of curing. The highest flexural strength was found out for 8% Na_2O dosage. Differences in flexural strengths for 8 and 10% Na_2O dosages are less crucial than the ones for the compressive strengths, thus the 10% Na_2O seems to be preferable.

WG/SC activated systems showed lower initial mechanical strength when compared to the WG/SH activated ones due to the lower pH of initial solution. The most suitable activator dosages for WG/SH mixtures were 6% Na_2O (56.3 ±4.3 MPa) or 8% Na_2O (55 ±5.4 MPa) based on the results obtained after 112 days. Further increase of activator dosage up to 10 or 12 % led to the compressive and flexural strengths decrease. Comparison of the best mixtures of each activator type should be done as well. The difference in compressive strengths obtained after 112 days was 36.8% in favour of the WG/SC.

Addition of one layer of basalt fabric to the 10% Na_2O WG/SC led to the compressive strength improvement in most cases when compared to the reference. More preciously, it resulted in the compressive strengths increase about 24.7; 0.5; 17.5 and 14.5% after 7; 28; 56 and 84 days of curing. No beneficial effect of BFs addition in one layer on the mechanical properties was found out for other tested WG/SC and WG/SH mixtures in general observation of the entire interval.

Quality F/M adhesion same as the formation of transition zone is necessary for the effective stress transfer realisation, hence for the composite to be effective. No signs of quality F/M adhesion was found out for WG/SC. There were some indications of improvement in case of WG/SH after 7 days of hydration but no longer as will be described later. The deterioration of compressive and flexural strengths can be related to many factors beside the poor F/M adhesion. Starting with the presence of local fabric disorientations (leading to ineffective stress transfer, moreover local stress concentration and crack propagation) as described in Arnon´s publication [6]. The influence of poor constitution of the BFs used same as deterioration of fibres mechanical properties along with the decrease of their stress transfer capability due to the degradation under the highly alkaline environments should be considered as well.

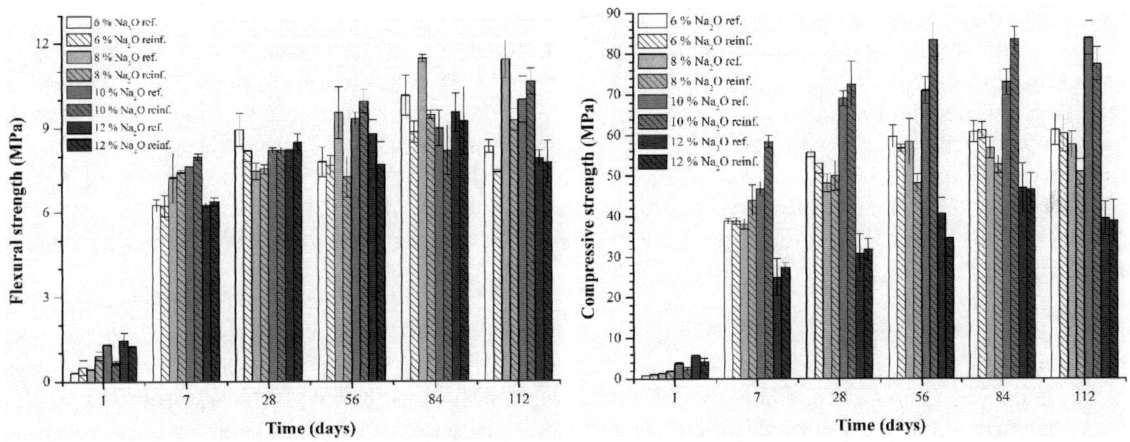

Figure 1. Flexural and compressive strengths development as a function of alkaline activator dosage (6–12% Na_2O) and basalt fabric reinforcement for WG/SC activated BFS systems.

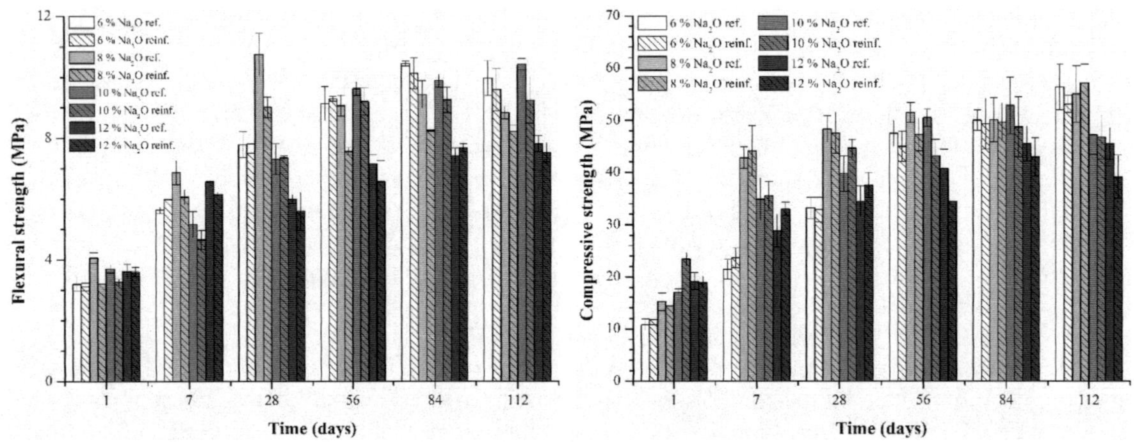

Figure 2. Flexural and compressive strengths development as a function of alkaline activator dosage (6–12% Na_2O) and basalt fabric reinforcement for WG/SH activated BFS systems.

3.2. Influence of number of fabric layers on the mechanical properties

Implementation of more fabric layers should be associated with improvement of mechanical properties as also mentioned in study of Du et al. [7]. Hence, the most suitable matrix (BFS activated using the 10% Na_2O WG/SC activator mixture) underwent further testing with increased number of basalt fabrics to two and three. The flexural and compressive strengths development is showed in Figure 3.

The results for samples reinforced with zero or one layer of BFs usually overlap each other considering the standard deviations. Still, one layer of the basalt fabric seems to be the most applicable when the reinforcement is required. Utilization of two or three layers caused lowering of compressive strength during the entire testing period. There was no significant difference between the results for reinforcement with two or three layers. Addition of basalt fabric to the matrix can be accompanied with the formation of lower amount of hydration products. It could explain the compressive strength drop against the non-reinforced reference sample same as the more significant drop with increasing number of fabric layers.

These results partially correlate with the ones presented in Du et al.´s paper [7]. Authors stated that in case of cementitious matrix reinforced with basalt fabric either one or two fabric layers had not exhibit significant reinforcement efficiency. Which is in a good correlation with the results obtained in this paper. On the other hand, Du et al. reported the strain-hardening behaviour when the cementitious matrix was reinforced with three or more layers of basalt fabrics. These results do not match the ones presented here. This difference should not be caused just by other matrix type. It is related to the deterioration of the fibres stress transfer capability due to their degradation. Existence of local stress concentrations caused by local bending of the fabric cannot neither be excluded.

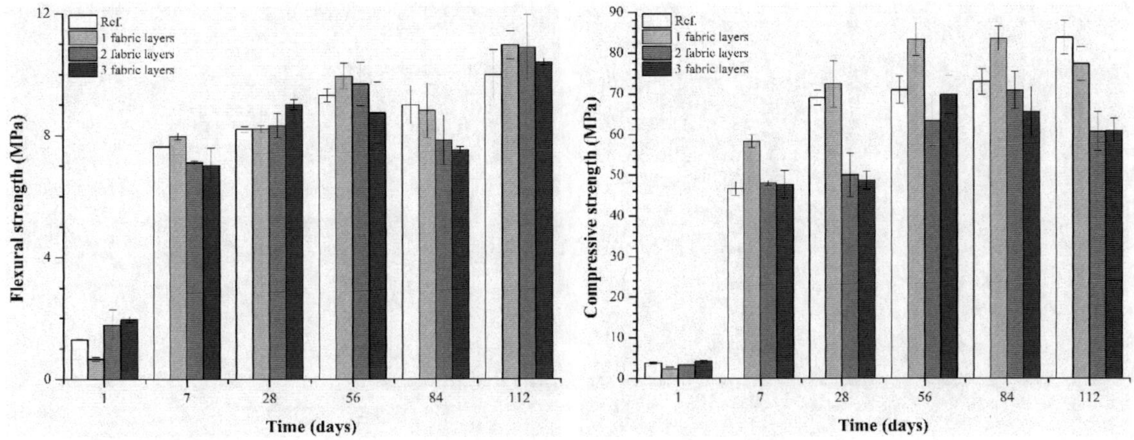

Figure 3. Flexural and compressive strengths development as function of fabric layers number for $M_s = 0.5$; WG/SC; 10% Na_2O based AABFS.

3.3. Fibre/matrix transition zone characterization

Nominal values of the interfacial shear bond strength (F_i) should be taken with reserve due to the limited instrumental equipment. Still, it brings interesting option for comparison of different matrices when various samples were tested under the same conditions. Embedment length about 5 mm and curing time of three days were selected as the most suitable to secure the fibre pulling out instead of its breakage while maintaining at least some time for the fibre/matrix interface formation.

The F_i results for various matrices (6–10% Na_2O WG/SC; WG/SH AABFS) reinforced with a single yarn of basalt fibres obtained using the pull-out tests are given in Figure 5. F_i increases up to 10% Na_2O, but beyond this dosage significant drop was noticed for WG/SC activator mixture, which is in a good correlation with the compressive strengths results. This was expected, because increase of the F_i is closely related to the increase of the compressive strength, respectively mechanical properties development at all. It indicates that fibre resistance against the pull-out is realized mainly due to the frictional bonds whose strength is related to the strength of the matrix itself and do not indicate such a significant contribution of chemical bonds formed between the F/M. The F_i increases with increasing activator dosage from 6 to 12% Na_2O for WG/SH activator mixture. It seems that chemical interactions beside the frictional forces are contributing to the total F_i in case of WG/SH AABFS, because the F_i should be lower for the 12% Na_2O dosage based on the compressive strength results.

The SEM images (Figure 4) showed that some sort of chemical F/M interaction could be realised for the 10% Na_2O WG/SH after 7 days of hydration since the etched surface of the fibres was visible along with the fact that matrix fragments extracted from surrounding matrix were adhered to the fibres surface. However, this was no longer visible after 28 days of curing. So, it cannot be excluded that the presence of protrusions on the fibres could be caused just by lower degree of hydration around the fibres. Still, the 10% Na_2O WG/SC AABFS did not show any signs of quality adhesion or F/M interaction at all after 7 or 28 days of curing (there was not observed presence of adhered paste on the surface of the fibres). The surface of the fibres sustained smooth without signs of their rapid degradation.

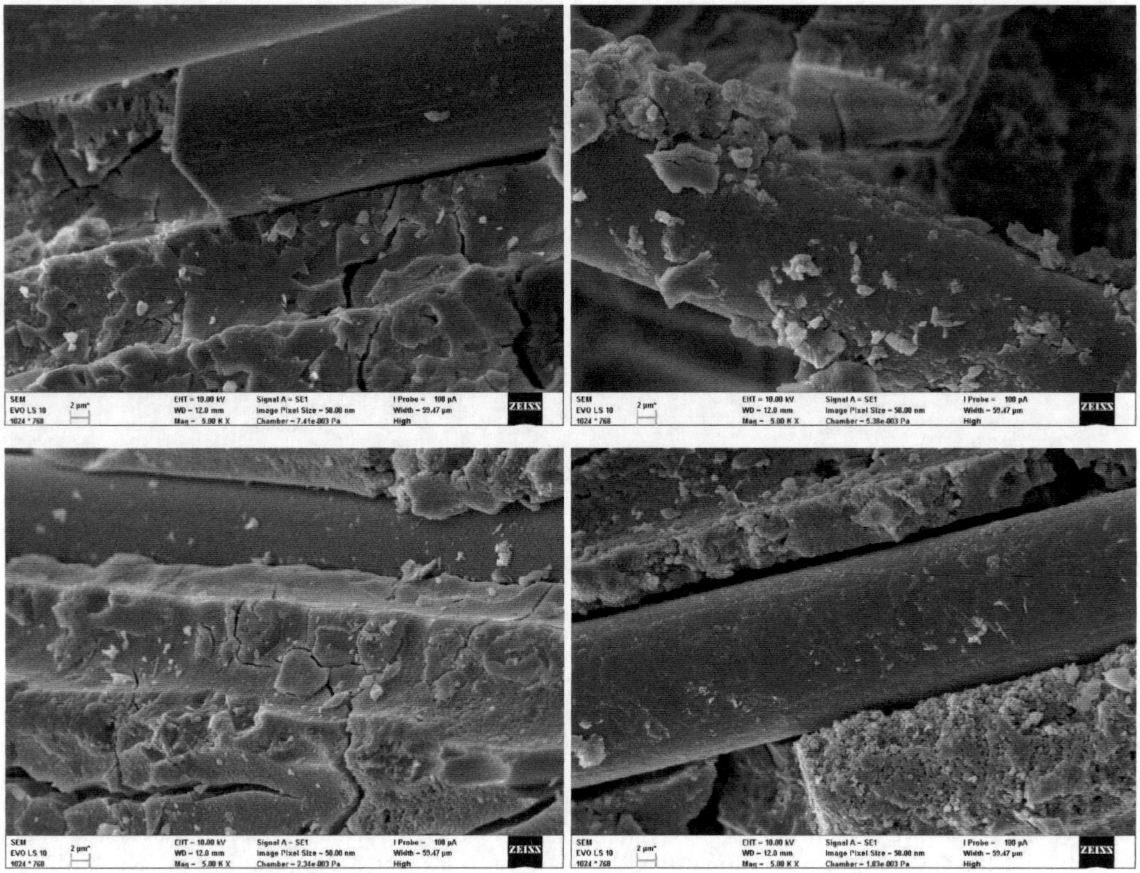

Figure 4. SEM images showing the detail of F/M interaction for 10% Na_2O WG/SC (left) and WG/SH (right) activated BFS pastes reinforced with the BFs after 7 days (top line) and 28 days (bottom line) of curing.

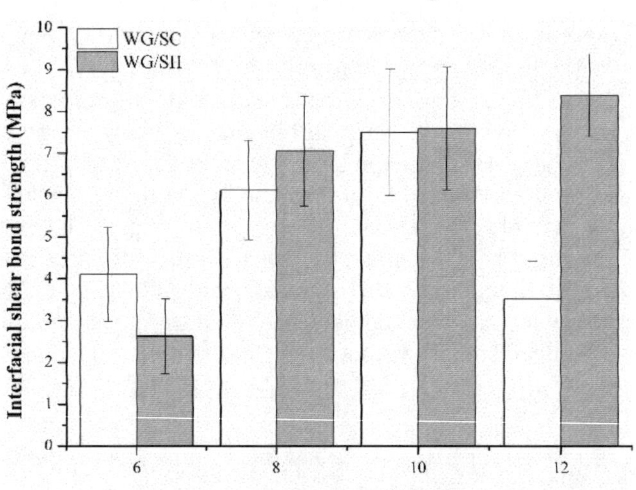

Figure 5. Interfacial shear bond strength development as a function of the activator dosage for WG/SC and WG/SH AABFS systems with fibre EL 5 mm after 72 hours of hydration.

4. Conclusion

There was not observed any significant improvement of tested mechanical properties when the various matrices were reinforced with one or more layers of basalt fabric. This is related to low F/M interaction associated with poor adhesion and ineffective stress transfer realisation or deterioration of fibres stress transfer capability due to their degradation. The presence of local fabrics disorientations cannot be excluded either. Some sort of noticeable improvement was noticed in case of $M_s = 0.5$; 10% Na_2O WG/SC AABFS matrix only. Despite these results utilization of basalt reinforcement in suitable form in the AABFS represents a promising possibility to increase the application potential of this material type however further and thorough research must be done to accomplish this goal.

Acknowledgments

This outcome has been achieved with the financial support by the project of specific research FCH-S-18-5194.

References

[1] Provis J L and Van Deventer J S J 2014 Alkali Activated Materials: State-of-the-Art Report RILEM TC 224-AAM

[2] Shi C, Krivenko P V and Roy D 2006 *Alkali-Activated Cements and Concretes* (New York: Taylor & Francis)

[3] Duxson P, Fernández-Jimenéz A, Provis J L, Lukey G C, Palomo A and Van Deventer J S J 2007 Geopolymer technology: the current state of the art *Journal of Materials Science* **42**(9) pp 2917–2933

[4] Bernal S A, Mejía de Gutiérrez R, Pedraza A L, Provis J L, Rodriguez E D and Delvasto S 2011 Effect of binder content on the performance of alkali-activated slag concretes *Cement and Concrete Research* **41**(1) pp 1–8

[5] Zhang J, et al. 2017 Durability of alkali-activated materials in aggressive environments: A review on recent studies *Construction and Building Materials* **152** pp 598–613

[6] Zhao R and Sanjayan J G 2011 Geopolymer and Portland cement concretes in simulated fire *Magazine of Concrete research* **63** pp 163–173

[7] Luukkonen T, Abdollahnejad Z, Yliniemi J, Kinnunen P and Illikainen M 2018 One-part alkali-activated materials: A review *Cement and Conrete Research* **103** pp 21–34

[8] Arnon B and Sidney M 1990 Fibre reinforced cementitious composites (Essex, England: Elsevier science publishers)

[9] Du Y, Zhang M, Zhou F and Zhu D 2017 Experimental study on basalt textile reinforced concrete under uniaxial tensile loading *Construction and Building Materials* 2017 pp 88–100

[10] Dias D P and Thaumaturgo C 2005 Fracture toughness of geopolymeric concretes reinforced with basalt fibers *Cement and Concrete Composites* **27**(1) pp 49–54

[11] Timakul P, Rattanaprasit W and Aungkavattana P 2016 Improving compressive strength of fly ash-based geopolymer composites by basalt fibers addition *Ceramics International* **42**(5) pp 6288–6295

[12] Rill E, Lowry D R and Kriven W M 2010 Properties of basalt fiber reinforced geopolymer composites *Proc. Int. Conf.: 34th International Conference & Exposition on Advanced Ceramics & Composites*

[13] Masi G, Rickard W D A, Bignozzi M Ch and Van Riessen A 2015 The effect of organic and inorganic fibres on the mechanical and thermal properties of aluminate activated geopolymers *Composites Part B: Engineering* (76) pp 218–228

[14] Welter M, Schmücker M and Mackenzie K J D 2015 Evolution of the Fibre-Matrix Interactions in Basalt-Fibre-Reinforced Geopolymer-Matrix Composites after Heating *Journal of Ceramic Science and Technology* **6**(1) pp 17–24

[15] Choi W, Jang S and Yun H 2015 Interface Bond Characterization between Fiber and Cementitious Matrix *International Journal of Polymer Science* 2015

IOP Conf. Series: Materials Science and Engineering **583** (2019) 012013 doi:10.1088/1757-899X/583/1/012013

Packing density modelling of non-spherical aggreg ates for particle composite design

O Koutny[1], J Kratochvil[1], M Drdlova[1] and E Bystrianska[1]

[1] Materials Research Centre, Brno University of Technology, Purkyňova 118, 612 00 Brno, Czech Republic

Email: xckoutnyo@fch.vut.cz

Abstract. The presented study follows the author's results of research where 3-parameter packing model for packing density determination was derived and calibrated for multimodal particular system. In this study previously derived model is extended by fourth parameter related to particle shape and its effect on bulk packing density. The model is primary derived for high performance cementitious composite design where compact structure and low binder content is preferred to maximise mechanical properties and durability. The effect of aggregate packing density on mechanical properties of ultra-high performance cementitious composites is investigated, using derived packing model

1. Introduction

Particle packing plays a key role in many fields of material engineering such as powder metallurgy, ceramics or polymer composites. Beside all of these, the cementitious composite industry utilizes the particle packing in the most common way into the production of the most world-spread construction material – concrete. The precise mixing of raw particular materials enables to achieve the structure with optimal or maximal packing density of particles. Such a dense structure improves many material properties such an increase of mechanical strength and durability, reduction of volume changes and overheating during the setting because of lower content of binder. The dense packing is the main requirement for concrete production in modern way into the form of material which is called high performance and ultra-high performance concrete (HPC and UHPC). In spite of large experiences of particle packing in the industrial fields mentioned above it is still relatively difficult to predict and effectively influence the particle packing density in modern concrete technology.

Packing density of particular system can be analytically determined via several existing models based on discrete packing theory. In this theory the particular system is divided to the discrete particle size classes where at least one class is dominant and it is packed preferably. The others particle size classes are then packed to this skeleton. First discrete packing model was introduced by Furnas and Westman on binary systems where particle size ratio was near to zero [1]. In this model there were defined two effects which increase packing density; namely filling effect of fine particles in gaps between coarse particle class and occupying effect of coarse particles which replace the porous space of fine particle size. When the particle size ratio is not in limit case (0 or 1), irregularities which decrease packing density take place. These irregularities, also called packing restrictions were firstly introduced by Stovall in the form of two structural effects which disrupt the regular packing and therefore decrease the packing density [2]. These effects are namely loosening effect of the fine particle class which disrupts the packing of dominant coarse particle class by squeezing themselves between them; and wall effect of the coarse particle class by disrupting of dominant fine particle class packing by forming wall-like structures on boundaries between particle classes. Participation rate of each effect depends on diameter ratio of interacted particle class and it is presented as interaction function of diameter ratio of two particle classes where one is dominant class. Several authors (de Larrard, Yu) derived their own interaction function and incorporated them to the form of so called Linear Packing Density Model [3, 4].

Dependence of packing density on volume percentage of fine particle class is actually not linear, as many authors suggest (de Larrard, Kwan) [5, 6]. Especially in the optimum composition point where linear packing theory overestimate the packing density i. e. there should be another effect which decrease the packing density especially around the optimum. This effect was called wedging effect and it was firstly introduced by Kwan [7].

In the original study of authors the Kwan extended 3-parameter packing model was calibrated to predict packing density of multimodal particle systems consisting of spherical particles. Calculated results were compared to experimentally determined packing densities for binary mixtures of multimodal fraction of spherical particles. Results from original formulation of 3-parameter packing model by Kwan show substantial deviation from experimental results. The reason is in formulation of interaction function for wedging effect which was not originally developed for multimodal particle systems and its contribution and the value of relevant interaction function have to be somehow related to the shape of granulometric curve. Therefore, interaction function for this effect has been adjusted [8].

Unfortunately, such a systems are not in coherence with real requirements of cementitious composites industry where the raw materials consist of particles with non-spherical shape. Kwan uses his original 3-parameter packing model for packing density determination of angular rock aggregate particles with correction of interaction functions for loosening effect and wall effect [9]. But there is still no direct connection between defined particle shape and change of packing. Therefore it is necessary to find out the parameter responsible for packing density change based on particle shape.

2. Experimental program

Presented study follows the results of [8], where the original Kwan extended 3-parameter packing model was calibrated to predict packing density of multimodal particle systems consisting of spherical particles of n particle size classes. The model can be described in its generalized and calibrated form by equation 1:

$$\frac{1}{\phi_i^*} = \sum_{i=1}^{n} \frac{y_i}{\phi_i} - \sum_{j=1}^{i-1} (1-a)\frac{y_j}{\phi_j}\left[1 - c\left(3.8^{y_j} - 1\right)\right] - \\ - \sum_{j=i+1}^{n} (1-b)(1-\phi_j)\frac{y_j}{\phi_j}\left[1 - c\left(2.6^{\sum_{j=i+1}^{n} y_j} - 1\right)\right] \tag{1}$$

where ϕ_C represents packing density of individual particle size class, y_i and y_i represents volume fraction of particle class size i and j respectively and parameters a, b and c represent interaction function for loosening effect, wall effect wedging effect according equation 2–5. The interaction function for wedging effect is solved separately for situation when $d_i < d_j$ using equation 4 and for situation when $d_i > d_j$ using equation 5.

$$a = 1 - \left(1 - \frac{d_i}{d_j}\right)^{3.3} - 2.6\frac{d_i}{d_j}\left(1 - \frac{d_i}{d_j}\right)^{3.6} \tag{2}$$

$$b = 1 - \left(1 - \frac{d_j}{d_i}\right)^{1.9} - 2\frac{d_j}{d_i}\left(1 - \frac{d_j}{d_i}\right)^{6} \tag{3}$$

$$c_1 = 0.322 \tanh\left(11.9\frac{d_i}{d_j}\right) k_1 \tag{4}$$

$$c_2 = 0.322 \tanh\left(11.9\frac{d_j}{d_i}\right) k_2 \tag{5}$$

where d_i is particle diameter of size class i, d_j is particle diameter of size class j (dominant class) and k_1 and k_2 are magnification factors for wedging effect interaction function. The real packing density is then defined as $\phi = \min\{\phi_1^*, \phi_2^*, ...\}$.

The raw materials for cementitious composites technology do not consist of spherical particles. Therefore the deviation can be expected in prediction of packing density of real particular systems using above derived model. This deviation has to be reflected and compensated by introduction of fourth parameter, as a function of packing density of individual size class ϕ_i instead of ϕ_C on parameter which characterises shape of particle in size class i.

The practical use of presented packing model for non-spherical particles is demonstrated on design of ultra-high performance cementitious composites. The effect of aggregate packing density on final mechanical properties is monitored on designed materials.

3. Materials and Methods

In this study there were used eight fractions of siliceaous aggregates as a non-spherical particle multimodal granulometric system, signed as S1, S2, S3, S4, S5, S6, S7 and S8 according to increasing median of particle size. Granulometric curves of each standard mix are shown in figure 1.

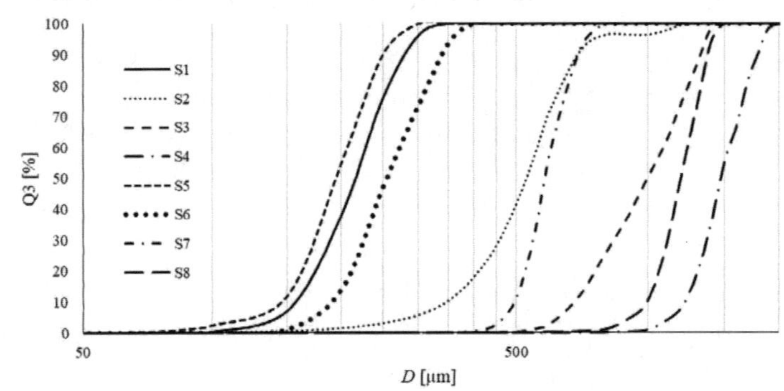

Figure 1. Granulometry of used materials.

The granulometric curves were determined using image analysis from optical microscopy record. For each curve there were analysed at least 1000 particles. The particle shape is described by circularity C of particle projection in optical microscopy record. The standardised circularity C^* is expressed as the median of C for relevant aggregate mix. From granulometric curves there were also determined the values of D_{min} and D_{max} as 1% quantile and 99% quantile. For each material mix there was also determined particle volume weight ρ. Material characteristics for each particle mix are summarized in table 1.

Table 1. Material parameters.

	S1	S2	S3	S4	S5	S6	S7	S8
D_1 [μm]	133	190	548	1034	89	147	414	962
D_{50} [μm]	201	528	997	1495	198	257	579	1215
D_{99} [μm]	309	1175	1408	1944	299	413	754	1449
ρ [g/cm^3]	2.67	2.67	2.65	2.64	2.66	2.64	2.63	2.64
C^*	0.834	0.876	0.884	0.883	0.816	0.842	0.846	0.855

From described materials there were prepared combined mixes of S1 and S4 (S1-S4), S1 and S3 (S1-S3), S2 and S4 (S2-S4), S2 and S3 (S2-S3), S5 and S7 (S5-S7) a S6 and S8 (S6-S8). These mixes were used for analysis and following correction of packing density. For each mix the packing density was experimentally determined in dependence on increasing volume ratio of finer standard fraction.

Experimental determination of packing density was provided according to EN 1097-3:1998 with additional compaction by standard vibration table. From measuring of compacted powder body weight *M*, the packing density ϕ_{exp} was determined according equation 6:

$$\phi_{exp} = \frac{1}{V} M \sum_{i=1}^{n} \rho_i y_i \tag{6}$$

where V is volume of testing container, ρ_i is the volume weight of constituent i and y_i is the volumetric ration of constituent i.

Mechanical properties, the compressive strength and the flexural strength, are monitored for samples of ultra-high performance mortars which were designed with different composition of aggregate skeleton and therefore with different packing density of aggregates. Two fractions of corundum aggregates were mixed in proper volume ratios (y K1 and y K2) to reach defined packing density ϕ_{agg} according the table 2.

Table 2. Composition of control mixes.

ϕ_{agg}	0.56	0.58	0.60	0.62	0.64	0.66	0.67
y(K1)	0.021	0.073	0.124	0.179	0.233	0.297	0.369
y_{agg}	0.536	0.557	0.577	0.597	0.618	0.638	0.648
ϕ_{agg}	0.66	0.64	0.62	0.60	0.58	0.56	
y(K1)	0.426	0.528	0.619	0.720	0.822	0.923	
y_{agg}	0.638	0.618	0.597	0.577	0.557	0.536	

The granulometry of used corundum fractions is presented in figure 2.

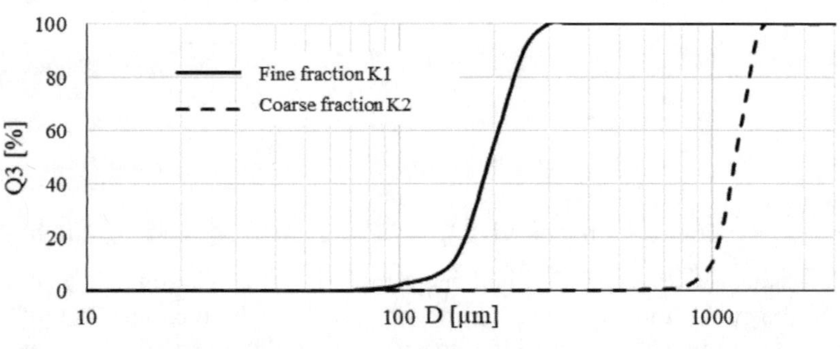

Figure 2. Granulometry of used corundum aggregates.

The aggregate volume fraction y_{agg} was determined as 1–0.05(1- ϕ_{agg}) where the expression 0.05(1- ϕ_{agg}) is equal to volume fraction of binder y_b. A binder based on Portland cement CEM I 52,5 R, undensified silica fume and ground limestone in mass portion of 7:2:1 was used for all mixes. The high-range water reducing agent based on polycarboxylate-ether copolymer was used for all mixes in portion 2.5% to binder. The water-to-binder ratio was set on 0.22. The test specimens were prepared in dimensions 40 x 40 x 160 mm for mechanical properties determination. Samples were compacted and cast using vibration table. The compressive strength was determined as a mean of six values and the flexural strength was determined as a mean of three values. After the mixing the workability was controlled using mini-spread flow test.

4. Results and discussion

Experimentally determined values of packing density for defined mixes were compared with values provided by calibrated modified 3-parameters model derived in previous study [8]. Two main criteria were monitored; the maximal value of packing density and corresponding composition of the mix. The results are shown in figure 3.

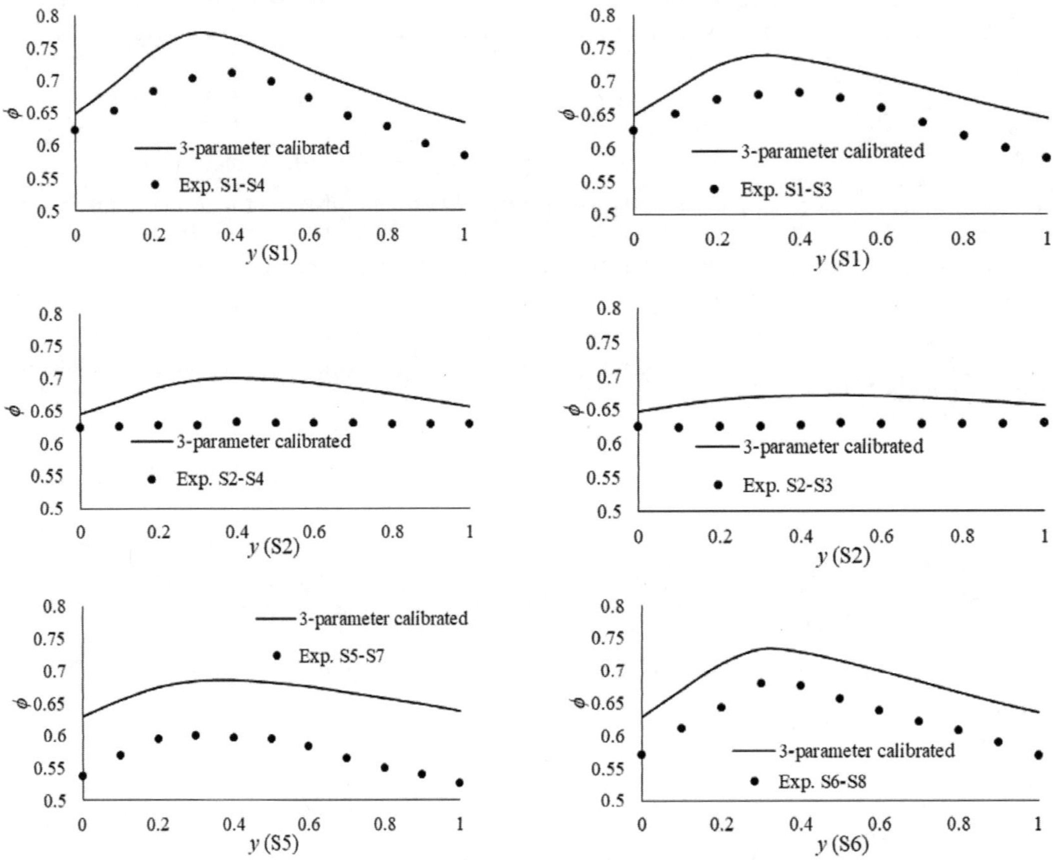

Figure 3. Packing density of mixes S1-S4, S1-S3, S2-S4, S2-S3, S5-S7 and S6-S8.

A common characteristic is visible on individual diagrams in figure 3. The significant overestimation of packing density predicted by model is visible in contrast to experimentally determined values. The effect of non-sphericity of particles on packing density was predicted and demonstrated by several authors [9, 10, 11]. Due to the fact that the similar materials are used according to granulometry, as in the previous study, it can be concluded that the deviation is caused by different packing density of individual particle size classes ϕ_i. The value of ϕ_i is independent on particle size and for systems of spherical particles has a constant value of ϕ_C. It seems to be that the deviation from spherical shape of particle causes the decrease of individual particle size class packing density ϕ_i and therefore of whole particular system. Providing an iteration of packing density using calibrated modified 3-parameter model where the value of ϕ_i is replaced by variable results in minimal deviation of modelled and experimentally determined packing density. When the newly determined value of ϕ_i is placed against parameter characterises particle shape, the circularity C^*, following dependence can be obtained (figure 4).

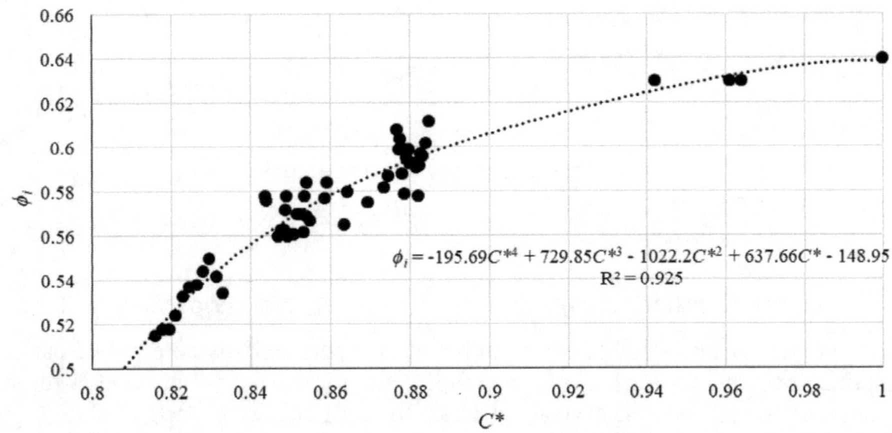

Figure 4. Influence of particle shape on packing density of individual particle size class ϕ_i.

The limit value 0.64 for spherical particle ($C = 1$) is set according [12]. In addition, the samples with C^* in the range of 0.94 to 0.96 are included to the context with determined results of packing density from previous study [8]. It is evident that the packing density of individual particle size class ϕi can be described as a function of particle shape, in this case as standardized circularity C^* and can be fitted by fourth order polynomial equation. Based on the regression curve the value of ϕ_C can be estimated with relatively good precision for particle mixes with standardized circularity of 0.81 to 1 (or equivalent sphericity 0.66–1). To increase model precision the circularity of each particle size class C_i can be defined directly from microscopic image analysis. The parameter C^* can be then directly replaced by C_i in regression equation in figure 4 for direct calculation of ϕi. Implementing these results into the calibrated modified 3-parameter packing model, the fourth parameter dealing with ϕ_i as a function of C_i respectively, following results of packing density can be obtained (figure 5).

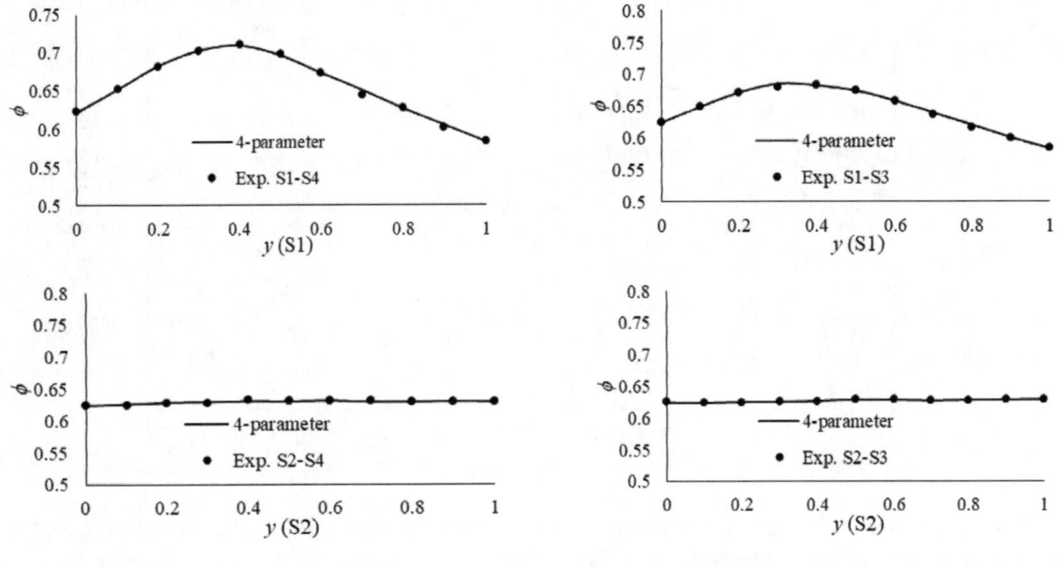

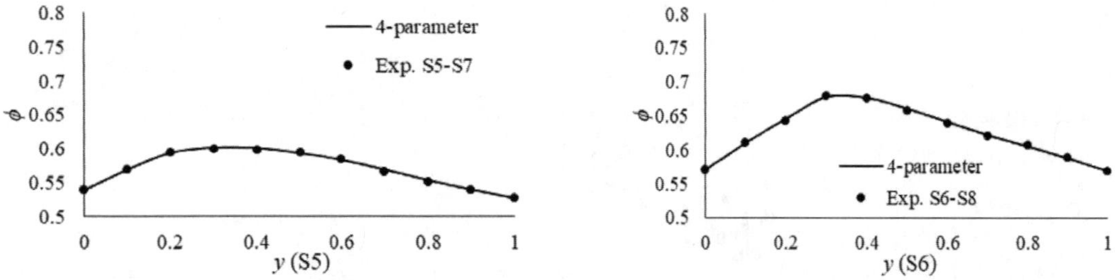

Figure 5. Packing density prediction using 4-parameter packing model.

As can be seen there is a very good agreement between experimentally determined packing densities and those obtained by 4-parameter packing model. Calculation via model is suitable for systems with continual granulometry (S2-S4, S2-S3) as well as for systems with gap in particle size density distribution (S1-S4, S5-S7, S6-S8). The suitability of the 4-parameter model is confirmed for particular systems with particle size from approximately 100 μm to 2000 μm. Therefore it can be a very suitable tool in HPC and UHPC design. It is clearly visible that maximal packing density is reached by sample combination with the largest difference in particle size where combination S1-S4 exceeds the packing density of 0.7 at volume content of S1 0.38. On the other hand it seems to be that continual way of aggregate grading does not lead to significant change of packing density as the mixes S2-S4 and S2-S3 show.

The importance of dense packing with maximal density is demonstrated by the results of mechanical properties. The compressive strength and the flexural strength was monitored by samples of UHPC mortars in dependence on aggregate composition i.e. on aggregate packing density. Figure 6 describes the evolution of mechanical properties; the compressive strength and the flexural strength of test specimens of UHPC mortars with variable composition of aggregate skeleton. The aggregate composition is described via volumetric ratio of fine fraction *y*(K1). The volumetric ratio of fine and coarse aggregate fraction causes the change of aggregate packing density as is described in figure 7.

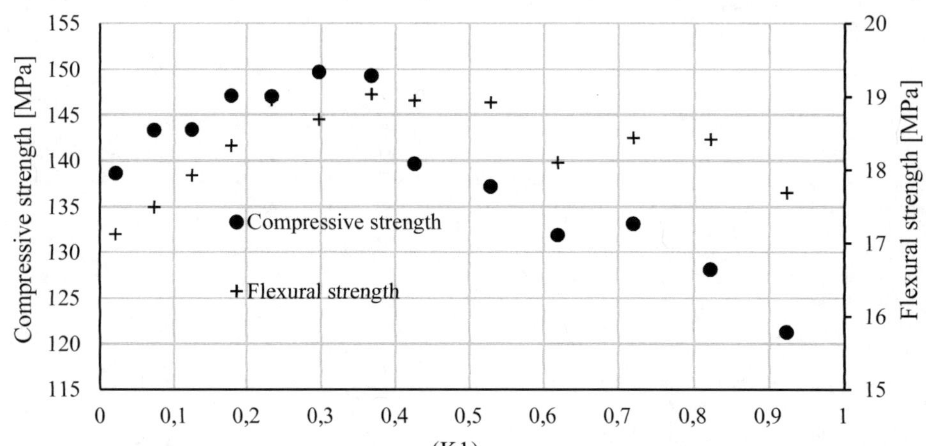

Figure 6. Compressive and flexural strength of prepared samples in dependence on aggregate skeleton composition expressed by volume fraction of fine corundum aggregates *y*(K1).

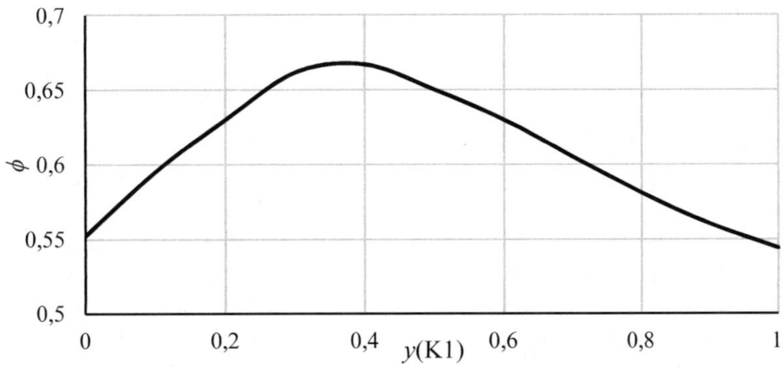

Figure 7. Packing density of particular system K1-K2.

It is evident that maximum of mechanical properties is reached at aggregate composition *y*(K1) and *y* - 1(K2), respectively with the highest packing density (0.66–0.67) of mix K1-K2. With increasing *y*(K1) up to 0.35 the mechanical parameters are increasing as well from value of 138 MPa up to value of 150 MPa for compressive strength and from value of 17 MPa up to value of 19 MPa for flexural strength. It is clearly evident that compressive strength is much more affected by packing density than the flexural strength. Moreover, with further increasing of *y*(K1) the compressive strength drops at the value of 121 MPa when the flexural strength drops at nearly same value as for $y(K1) \approx 0$. The rapid decrease of the compressive strength is caused not only by reduction of packing density but probably by decrease of workability (spread flow) due to increasing amount of fine aggregate fraction as is visible from figure 8.

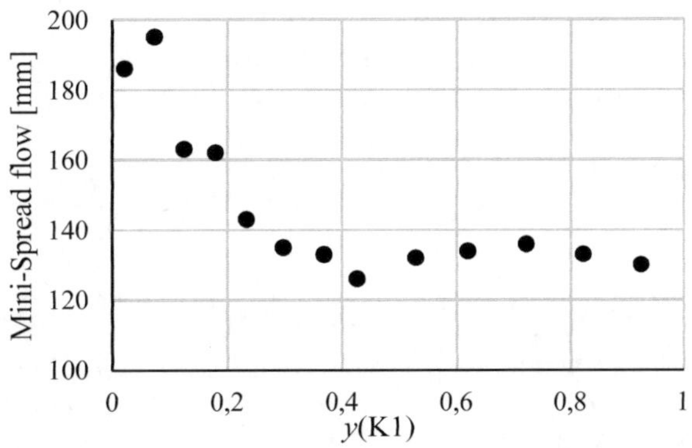

Figure 8. Workability via Mini-spread flow test of fresh samples depending on volumetric fraction of K1 in aggregate skeleton.

Similar results can be found in [13]. The workability decreases with increasing volume fraction of K1 to approximately constant level with the local minimum at maximal packing density of mixture K1 and K2 i.e. *y*(K1) = 0.35. It is believed that low workability for *y*(K1) in range 0.35–1 is partially responsible for decrease of compressive strength. The spread flow is a consequence of the yield stress changes in fresh mortar. Whereas the yield stress is exceeded by vibration during compaction, there has to be an increase in plastic viscosity of mortar as well. The flow behaviour of mortar is driven by this effect after exceeding the yield stress and therefore negatively effects the compaction of material (due to higher value of plastic viscosity).

5. Conclusion

The packing density of particular systems consisting of non-spherical particles was investigated. It was found that there is a significant influence of particle shape on packing density of the system. Packing density of such a "real" system cannot be predicted by packing models based on discrete packing theory which was experimentally and analytically proven. Therefore, another parameter has to be found which puts particle shape in the context to packing density. It was found that particle circularity determined from microscopy image analysis effects the pacing density of individual particle size classes and it can be implemented to the calibrated modified Kwan's 3-parameter packing model for multimodal particular systems derived in author's previous study as a fourth parameter to the form of 4-parameter packing model. This newly developed model shows very good agreement with experimental results of packing density for multimodal particular systems of non-spherical particles. The practical use of developed model as well as importance of packing density point-of-view in ultra-high cementitious composite design was verified. The packing density of aggregate skeleton has a great influence mainly on compressive strength of ultra-high cementitious composites where the highest value is reached by sample with the highest packed aggregates. Moreover, increased packing density of aggregates affects the behaviour of material in fresh state due to reduction of binder volume necessary for system lubrication. It is evident that the packing density represents an approach which is necessary to be used in ultra-high performance concrete and other engineered cementitious composite design.

Acknowledgement
As a doctoral student of the Materials Research Centre at FCH BUT- Sustainability and Development, REG LO1211, with financial support from National Programme for Sustainability I (Ministry of Education, Youth and Sports), collective of authors wants to thank the foundation for its financial support.

References
[1] Dinger D R and Funk J E 1992 Particle Packing II - Review of Packing of Polydisperse Particle Systems. *Interceram* **41**(2) pp 95–97
[2] Ye G, Xin H, Bao-Lin Z, Bao-Gua M and Hong-Bo Z 2008 Method for calculating packing density of powder particles in paste with continuous particle size distribution *Powder Technology* **187**(1) pp 88–93 (doi:10.1016/j.powtec.2008.01.026)
[3] De Larrard F and Sedran T 1994 Optimization of ultra-high-performance concrete by the use of a packing model *Cement and Concrete Research* **24**(6) pp 997–1009 (doi:10.1016/0008-8846(94)90022-1)
[4] Yu R, Spiesz P and Brouwers H J H 2014 Mix design and properties assessment of Ultra-High Performance Fibre Reinforced Concrete (UHPFRC) *Cement and Concrete Research* **56** pp 29–39 (doi:10.1016/j.cemconres.2013.11.002)
[5] De Larrard F 1999 *Concrete mixture proportioning: A scientifil approach* (doi:10.1017/CBO9781107415324.004)
[6] Kwan A K H and Fung W W S 2009 Packing density measurement and modelling of fine aggregate and mortar *Cement and Concrete Composites* **31**(6) pp 349–357 (doi:10.1016/j.cemconcomp.2009.03.006)
[7] Wong V and Kwan A K H 2014 A 3-parameter model for packing density prediction of ternary mixes of spherical particles *Powder Technology* **268** pp 357–367 (doi:10.1016/j.powtec.2014.08.036)
[8] Koutný O, Kratochvíl J, Švec J and Bednárek J 2016 Modelling of Packing Density for Particle Composites Design In *Procedia Engineering* (doi:10.1016/j.proeng.2016.07.386)
[9] Kwan A K H, Wong V and Fung W S W 2015 A 3-parameter packing density model for angular rock aggregate particles *Powder Technology* **274** pp 154–162 (doi:10.1016/j.powtec.2014.12.054
[10] Roquier G 2017 The 4-parameter Compressible Packing Model (CPM) for crushed aggregate particles *Powder Technology* (doi:10.1016/j.powtec.2017.07.028)
[11] Yu A B, Zou R P and Standish N 1996 Modifying the linear packing model for predicting the porosity of nonspherical particle mixtures *Industrial and Engineering Chemistry Research*

(doi:10.1021/ie950616a)
[12] Li L G and Kwan A K H 2014 Packing density of concrete mix under dry and wet conditions *Powder Technology* **253** pp 514–521 (doi:10.1016/j.powtec.2013.12.020)
[13] Sun Y, Wang Z, Gao Q and Liu C 2018 A new mixture design methodology based on the Packing Density Theory for high performance concrete in bridge engineering *Construction and Building Materials* (doi:10.1016/j.conbuildmat.2018.06.062)

IOP Conf. Series: Materials Science and Engineering **583** (2019) 012014 doi:10.1088/1757-899X/583/1/012014

Application of carbon black in conductive fly ash geopolymer mortars

C Mizerová[1], I Kusák[1] and P Rovnaník[1]

[1] Brno University of Technology, Faculty of Civil Engineering, Brno, Czech Republic

Email: mizerova.c@fce.vutbr.cz

Abstract. Materials with enhanced electrical properties are widely studied concerning their future application in smart or self-sensing constructions. Carbon black is known as one of the inexpensive multipurpose admixtures used in polymeric materials. This study is focused on the application of carbon black in fly ash geopolymer mortar to enhance its conductivity and other electrical properties. Geopolymer samples were prepared with the carbon black content from 0.5 to 4.0% to evaluate their performance in selected electrical properties (conductivity, resistance, capacitance). Further investigations included its influence on the mechanical properties (compressive and flexural strength) and microstructure of the binder (mercury intrusion porosimetry, SEM). Despite the considerable improvement in conductivity, the amount of conductive filler over 2% was also associated with increased porosity and reduced compressive strengths.

1. Introduction

Carbon black is an amorphous form of carbon produced by partial combustion or thermal decomposition of hydrocarbon materials. Since it consists of elementary carbon particles of high specific surface, it is widely used in many industrial applications as a pigment or light weight filler with high chemical and thermal stability [1]. Apart from the major application for rubber and plastics reinforcing, it is also incorporated in conductive composites due to its permanent conductivity properties and affordable price [2].

Structural composites with enhanced conductivity can be used in smart structures using their self-sensing properties, i.e. capability to sense strain, internal damages etc. based on the evaluation of composite bulk electrical conductivity/resistivity variation that resulted from the external condition changes. Smart structure is thus designed as a sensor itself and it is suitable to provide easy and cost effective structural health monitoring that does not require installation of embedded or additional sensors [3]. This approach is adopted for number of advanced applications for traffic monitoring solutions (vehicular speed, weighing in motion, traffic density) [4] or thermal and humidity control in structure related to its self-heating properties. For instance, Mingqing [5] successfully incorporated carbon black into mortar for indoor electrical floor heating slab and similar concept is applied in case of possible de-icing solutions for large airport areas or car parks [6]. Moreover, the introduction of carbon conductive admixture in cement matrix increases its electromagnetic shielding ability. Electromagnetic interference shielding of rooms in buildings is needed for protecting electronics, especially those associated with strategic systems such as aircraft, nuclear reactors, transformers, communication and control systems, etc. [7].

The use of carbon black or other conductive fillers has been investigated predominantly in polymeric and cement matrices but this study deals with their use in a fly ash-based geopolymer matrix. Geopolymers are a group of inorganic mineral binders derived from the reactions of aluminosilicate-rich precursor under high alkaline conditions. Their polymeric and highly coordinated three-dimensional structure formed by SiO_4 and AlO_4 units was reported by prof. Davidovits who firstly described the geopolymerization mechanism [8]. Convenient source materials for geopolymer synthesis include any natural or manufactured materials with high Si and Al content in amorphous form, namely low Ca fly ash, metakaolin and calcined kaolin or other clays [9] that are mixed with alkali hydroxide/alkali silicate activator solution [9, 10]. Recently, geopolymers has attracted considerable attention because of their early compressive strength, good chemical resistance and excellent fire resistance behaviour [11]. Depending on the raw materials chosen and curing conditions, alkali activated binders can represent more environment friendly alternative to the ordinary Portland cement [12].

Studies discussing the role of carbon black in geopolymer binders are limited, authors rather used other carbon conductive fillers; graphite, carbon fibres (CF), carbon nanotubes (CNTs) or graphene. Electrical conductivity of fly ash geopolymer binder as such is affected by the activator solution, liquid activator/ash ratio (L/A) and frequency spectrum [13]. The improvement of both electrical and mechanical performance of fly ash geopolymer doped with CNTs was studied by Saafi et al. [14] who states that the optimal concentration is 0.5%. Mackenzie et al. [15] prepared clay-based geopolymers with 0.26% of single wall carbon nanotubes (SWCNTs) and the same concentration of graphite for comparison. As is the case with CNTs, CF is effective conductive filler with the percolation threshold at 0.5% concentration [16]. In another study, partial replacement of CF by carbon black in cement-based matrix maintained the conductivity while lowering the costs and increasing the workability [17]. Excellent results were achieved with application of graphene in conductive composites. Only 0.35% addition resulted in 209% increase of conductivity and lower porosity accompanied by significant increase of strength and modulus of elasticity [18].

2. Materials and methods

Fly ash (FA) from the black coal combustion with chemical composition given in Table 1 was mixed with commercially sold sodium silicate solution (molar ratio $SiO_2/Na_2O = 1.6$, chemical composition in table 2) to prepare the geopolymer binder.

Table 1. Chemical composition of fly ash.

	SiO_2	Al_2O_3	Fe_2O_3	CaO	MgO	S_{total}	Na_2O	K_2O
(%)	49.82	24.67	7.50	3.91	2.68	0.91	0.70	2.78

Table 2. Chemical composition of water glass solution.

	SiO_2	Na_2O	H_2O
(%)	26.43	16.61	56.96

Conductive filler - carbon black Vulcan 7H was added in concentration of 0.5 to 4.0% together with a small amount of water and the dispersing agent, Triton X-100 (2% solution). After proper homogenization, standardized quartz sand with a maximum grain size of 2.5 mm was added as an aggregate together with remaining amount of water. Total amount of water in each mixture varied in order to reach the same workability of the fresh mortars (mortar diameter 180 ±10 mm using the flow table method, EN 1015-3 1999). Mix compositions are shown in table 3.

Table 3. Mix design of fly ash geopolymer mortars.

	REF	S 0.5	S 1.0	S 2.0	S 3.0	S 4.0
Fly ash (g)	350	350	350	350	350	350
Sodium silicate (g)	280	280	280	280	280	280
Carbon black (g)	-	1.75	3.5	7	10.5	14
Sand (g)	1050	1050	1050	1050	1050	1050
Triton X-100 2% (ml)	-	1.75	3.5	7	10.5	14
Water (g)	35	35	35	40	45	50

Fresh mortars were placed into prismatic moulds (40 × 40 × 160 mm) and covered with plastic sealant to protect the binder from moisture loss. After 2 hours in laboratory conditions (22 ±2°C, $\varphi = 45 \pm 5\%$), the samples were thermally treated at 40°C for 24 h. Demoulded specimens were stored in laboratory till the age of testing.

Before the measurements, the prepared prismatic samples were dried at 105°C to reduce the influence of moisture on assessed properties. They were characterized by impedance spectroscopy in the range of 40 Hz to 1 MHz using an Agilent 33220A sinusoidal signal generator and an Agilent 54645A dual-channel oscilloscope. In order to perform impedance analysis, the prismatic specimens were placed between parallel brass electrodes (30 × 100 mm) so that a distance between electrodes was 40 mm. The output voltage of the signal generator was 5.5 V. The input values for the electrical capacity and the resistance of the oscilloscope were 13 pF and 1 MΩ, respectively. These instruments were assembled for fully automated measurement. At higher range of frequencies, 100 MHz up to 3 GHz, R&S ZNC vector analyzer with DAK 12 coaxial probe manufactured by Speag has been used. In this frequency spectrum the electrical conductivity and relative permittivity as a function of frequency have been measured.

Geopolymers were tested for their mechanical properties (compressive and flexural strength). Micromeritics Poresizer 9310 and scanning electron microscope (SEM) Tescan MIRA3 XMU were used to assess the microstructure and porosity of the geopolymers.

3. Results and discussion

3.1. Electrical properties

Electrical resistance of all geopolymers decreased with the carbon black content and with higher frequency applied, as can be seen from Figure 1. At low frequency, samples with carbon black content 2% and higher exhibited considerably lower resistance, initial resistance of the reference sample dropped from 21.3 to less than 3 MΩ. As to frequency from 2 kHz, minor differences among all samples and slower decrease in resistance were observed.

Regarding the influence of frequency, similar trends were noted at changes of electrical capacitance (Figure 2) but with the inverse effect of the filler content. Capacitance of geopolymers S2.0 to S4.0 up to 200 Hz fluctuated and the curves overlapped. At 100 Hz, the electrical capacitance of these geopolymers was twice as high as of the reference sample with the capacitance 55.9 pF. Above 1000 Hz, the curves were smooth and sample S4.0 achieved significantly higher capacitance.

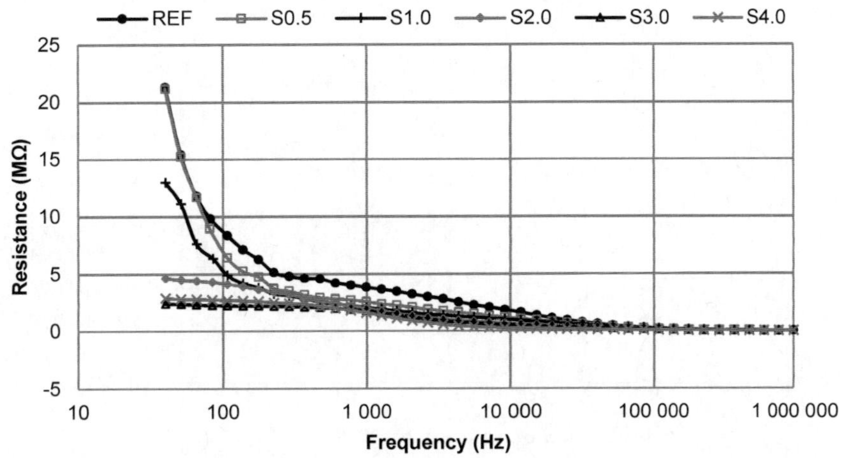

Figure 1. Variation in electrical resistance of FA geopolymer mortars with different carbon black content.

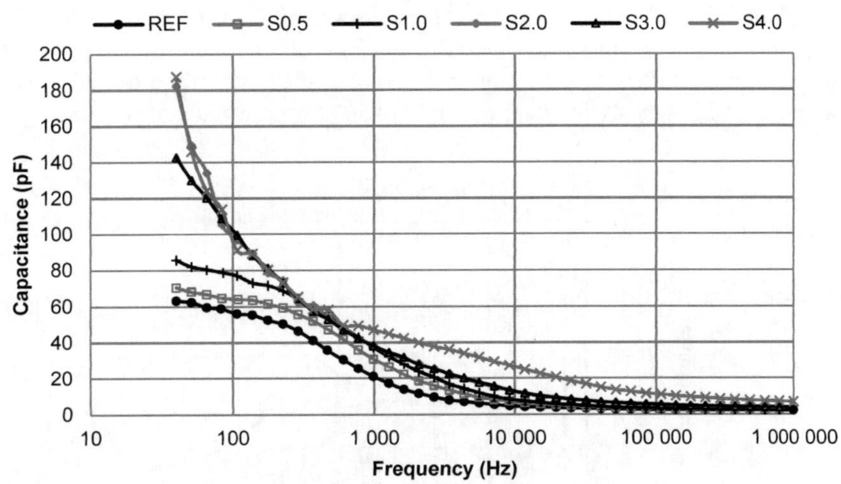

Figure 2. Variation in electrical capacitance of FA geopolymer mortars with different carbon black content.

As expected, higher carbon black content in geopolymers led to improvement in electrical conductivity properties across the frequency spectrum shown in Figure 3. In fact, the resistance of samples S3.0 and S4.0 was nearly the same but S4.0 was characterized by increased capacitance from frequency 1000 Hz and enhanced conductivity at all frequencies. Unlike resistance and capacitance, the electrical conductivity increases at higher frequency.

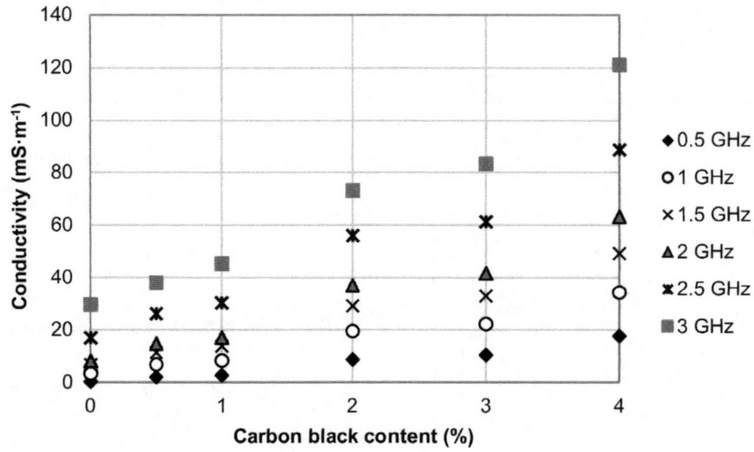

Figure 3. Electrical conductivity of FA geopolymer mortars with different carbon black content at chosen frequencies.

3.2. Mechanical properties

Variations in mechanical performance of geopolymers depending on the carbon black content are displayed in Figure 4. The flexural strength was not negatively affected by the conductive filler and remained constant around 5 MPa. On the other hand, the compressive strength gradually decreased and at carbon black content of 2–4% attained approximately 50% strength in comparison with the reference sample.

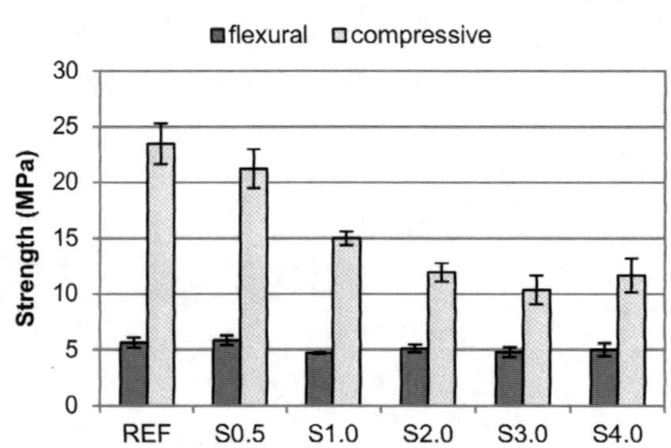

Figure 4. Mechanical properties of FA geopolymer mortars with different carbon black content.

3.3. Microstructure

Excessive amount of water needed during the preparation of samples with higher carbon black content caused an increment in total porosity of the binders (see Figure 5). At low carbon black concentrations, the cumulative intruded volume was about 0.080 $cm^3 \cdot g^{-1}$ but it increased to more than 0.095 $cm^3 \cdot g^{-1}$ at composites with 2–4% carbon black concentration. This fact is in accordance with reduced compressive strength.

SEM imaging of the composites did not reveal the real distribution of carbon black particles within the matrix because of their exceptionally small dimensions (figure 6, picture A). Agglomerates of colloidal size carbon black particles are completely separated during the mixing process, hence no difference in microstructure of REF and S4.0 samples (pictures B, C) is clearly visible. In this case, electron microscopy was not a convenient method to evaluate the quality of their dispergation.

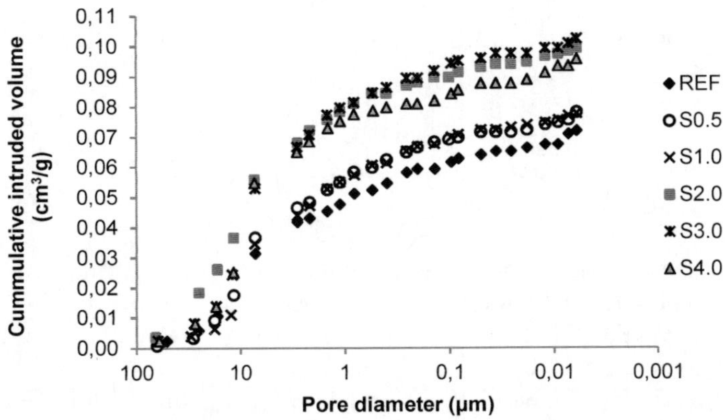

Figure 5. Cummulative intruded volume of FA geopolymer mortars with different carbon black content.

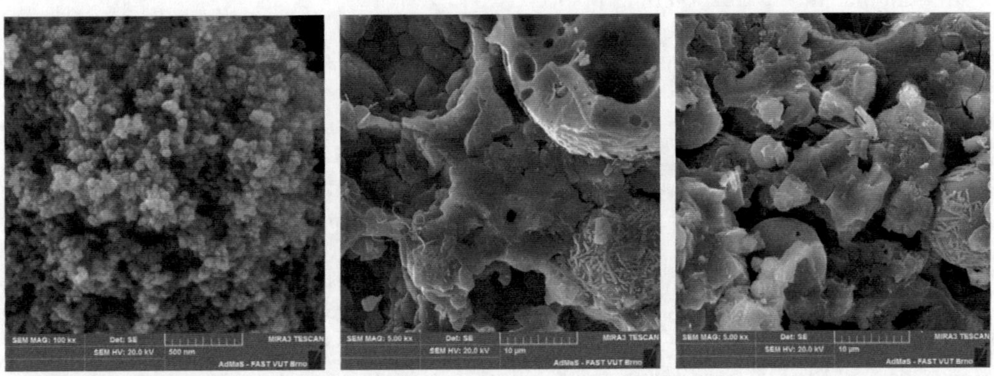

A B C

Figure 6. SEM images, A: carbon black particles (scale: 500 nm), B: sample REF (scale: 10 µm), C: sample S4.0, (scale: 10 µm).

4. Conclusions

This study was performed to evaluate the electrical properties of fly ash geopolymer mortars with carbon black conductive admixture. Geopolymers with carbon black concentration of 0.5–4.0% were analysed by impedance spectroscopy to assess their electrical conductivity, resistance and capacitance, these experiments were followed by testing of the mechanical performance. Carbon black content from 2% resulted in reduced electrical resistance and enhanced capacitance of the mortars, especially at low frequencies up to 100 Hz. At higher frequency range, both resistance and capacitance of all samples showed minor differences. Further increase of carbon black to 4% content resulted in a considerable improvement of electrical conductivity and slight increase in capacitance. Although the nanoparticles of carbon black effectively enhanced the electrical properties, their incorporation into binder structure was accompanied by additional water demand and increased pore volume of hardened specimens determined by the mercury intrusion porosimetry. This led to the deterioration of the compressive strength but the flexural strength of all samples was stable and not negatively affected. As SEM imaging confirmed complete dispergation of carbon particles, carbon black represents a low-cost and effective nanomaterial for conductive composites. Possible combination with another conductive admixture (for instance carbon fibers) may help to eliminate some negative aspects of their application.

Acknowledgements
This study has been financially supported by the Czech Science Foundation under the project No. 19-11516S.

References

[1] Donnet J B ed. 1993 *Carbon black: science and technology* (New York: CRC Press)
[2] Han B, Yu X and Ou J 2014 *Self-sensing concrete in smart structures* (Oxford: Butterworth-Heinemann)
[3] Lu S N, Xie N, Feng L C and Zhong J 2015 Applications of nanostructured carbon materials in constructions: The state of the art *Journal of Nanomaterials* **6**
[4] Han B, Ding S and Yu X 2015 Intrinsic self-sensing concrete and structures: A review *Measurement* **59** pp 110–128
[5] Mingqing S, Xinying M, Xiaoying W, Zuofu H and Zhuoqiu L 2008 Experimental studies on the indoor electrical floor heating system with carbon black mortar slabs *Energy and Buildings* **40**(6) pp 1094–1100
[6] Chung D D L 2004 Self-heating structural materials *Smart materials and structures* **13.3** p 562
[7] Chung D D L 2012 Carbon materials for structural self-sensing, electromagnetic shielding and thermal interfacing *Carbon* **50.9** pp 3342–3353
[8] Davidovits J 1994 Properties of geopolymer cements *First international conference on alkaline cements and concretes* (Kiev: Kiev State Technical University, Scientific Research Institute on Binders and Materials) pp 131–149
[9] Provis L and Van Deventer J S J 2014 *Alkali Activated Materials* (Springer)
[10] Shi C, Roy D and Krivenko P 2003 *Alkali-activated cements and concretes* (CRC press)
[11] Singh B, Ishwarya G, Gupta M and Bhattacharyya S K 2015 Geopolymer concrete: A review of some recent developments *Construction and building materials* **85** pp 78–90
[12] Van Deventer J S J, Provis L, Duxson P and Brice D G 2010 Chemical research and climate change as drivers in the commercial adoption of alkali activated materials *Waste and Biomass Valorization* **1.1** pp 145–155
[13] Hanjitsuwan S, Chindaprasirt P and Pimraksa K 2011 Electrical conductivity and dielectric property of fly ash geopolymer pastes *International Journal of Minerals, Metallurgy, and Materials* **18.1** pp 94–99
[14] Saafi M, Andrew K, Tang P L, McGhon D, Taylor S and Zhou X 2013 Multifunctional properties of carbon nanotube/fly ash geopolymeric nanocomposites *Construction and Building Materials* **49** pp 46–55
[15] Mackenzie K J D, Kenneth J D and Bolton M J 2009 Electrical and mechanical properties of aluminosilicate inorganic polymer composites with carbon nanotubes *Journal of materials science* **44.11** pp 2851–2857
[16] Payakaniti P, Pinitsoontorn S, Thongbai P, Amornkitbamrung V and Chindaprasirt P 2017 Electrical conductivity and compressive strength of carbon fiber reinforced fly ash geopolymeric composites *Construction and Building Materials* **135** pp 164–176
[17] Wen S and Chung D D L 2007 Partial replacement of carbon fiber by carbon black in multifunctional cement–matrix composites *Carbon* **45**(3) pp 505–513
[18] Saafi M, Tang L, Fung J, Rahman M, Sillars F, Liggat J and Zhou X 2014 Graphene/fly ash geopolymeric composites as self-sensing structural materials *Smart materials and structures* **23**(6) 065006

Application of the FRP Rebar in Constructions for Reduction of Thermal Bridges – Compressed elements

J Prokes[1], R Cechmanek[2], F Girgle[3], M Zlamal[3] and P Stepanek[3]

[1] PREFA KOMPOZITY, a.s., Kulkova 4231/10, 61500 Brno, Czech Republic
[2] Research Institute for Building Materials, Hněvkovského 30/65, Brno 617 00, Czech Republic
[3] Brno University of Technology, Veveří 331/95, 60200 Brno, Czech Republic

Email: prokes@prefa.cz

Abstract. The paper describes usage of a composite rebar in applications with reduction of thermal bridges, with emphasis on compression reinforcement. The paper focuses on several areas: The mechanical tests, resulting mechanical properties and temperature behaviour of compressed reinforcement. Compression elements with short fibres of length 6 and 50 mm or long fibres and with or without a rebar were tested. The comparison between materials and results of behaviour of a simple compressed rebar at elevated temperatures is shown. The critical temperature of usage FRP material as a compression element was found.

1. Introduction

1.1. The issue of breaking the thermal bridges in the reinforced concrete constructions

Thermal bridge is a phenomenon of construction which occurs when materials with different thermal conductivity are used for exterior surfaces or parts. It means that the thermal bridges are weaknesses within a building's construction where heat and/or cold is transferred at a significantly higher rate than through the surrounding envelope parts. There are basically two types of this phenomenon: geometric thermal bridges where a part of the structure projects through the building envelope, and material thermal bridges where materials with different conductivity are used in combination. In practice, these effects are often combined. A classic example of this is the balcony slab, where problems occur if the connection is not taken into serious consideration.

1.2. The effect of thermal bridges

The effect of thermal bridges in constructions are:

- *Higher energy consumption.* Due to the thermal loss at the connection, heat is drawn from inside resulting in a significant increase in heating costs and energy consumption.
- *Mould formation.* Interior temperatures of the adjacent rooms can drop well below the dew point. This leads to condensation, deteriorates plaster and paintwork and also mould formation. If there is sustained exposure to condensation, the building is subjected to serious deterioration.
- *Uncomfortable living space.* Cold surface temperatures cause uncomfortable living space for occupants.

1.3. Current methods of solution
Current methods include creating of a gap between main building structure and attached structure, and usage of materials with lower thermal conductivity.

Typically, load-bearing thermal insulation elements made from austenitic steel and thermal insulation are used. These elements form a thermal break between the exterior and the interior construction, while transferring load and maintaining full structural integrity. Also, the higher interior temperature is maintained decreasing condensation, and preventing the formation of mould.

1.4. FRP solution
Application of only FRP (Fiber Reinforced Polymer) rebar in load bearing insulation element is quite new. No other studies dealing with only FRP rebar in load bearing insulation element (it means tensile, shear and compressed reinforcements made only from FRP) are known to the author of this paper. State of the art is described in patents DE 4 102 332 (A1), EP 2 138 641 (A2), DE 1 9711 813 (A1), DE 19508292 (A1), DE 4 040 433 (A1). These patents describe usage of FRP and steel rebar together. At least, shear reinforcement is made of steel. No patents claim usage only FRP rebar. Only our former work is published in the previous article [1] and describe usage of FRP rebar (without steel rebar) in load bearing insulation element.

Reinforcement of most structures with reduction of thermal bridges consist of pulled, compressed and shear elements. The amount of information is published on the pulled FRP reinforcement, but there is a lack information on the compressed FRP rebar.

It is generally not recommended to use FRP rebar as a compressed reinforcement [2], due to lack of information. There is a very low quantity of articles or studies on usage FRP reinforcement as a reinforcing material for compressive loaded concrete constructions. The study [3] describes usage of glass-fiber-reinforced-polymer bars and spirals as reinforcing materials in hollow concrete columns. The crushing strain of the FRP bars embedded in a concrete column was found in study [4]. Another paper [5] describes strength of compression lap-spliced FRP bars in concrete columns with different splice lengths. There are another studies for example [6], which describes usage of FRP rebar in compressed part of beam. But these tests were performed in room temperature and do not utilize insulation material. There is an absence of information on possible solutions, comparison of fiber position in the compressed element and corresponding mechanical properties and temperature resistance. For this reason the article focuses on this topic.

2. Materials and Methods

The shape and positioning of a compressed rebar in structure is shown in Figure 1. Only case where reinforcement is used in element with reduction of thermal bridges (see Figure 1, case B) is solved in this paper.

The cylinder of composite material can be formed at the FRP rebar (see Figure 2). The cylinder may be of different composition. The main component are fibers and polymeric binder (resin). Inorganic fillers can be also added.

The mechanical properties of the cylinder and whole compressed elements depend on

- amount of fibers,
- the length of fibers,
- fiber arrangement,
- resin properties.

For this test the materials were used as follows:

- short fibers (the length of fiber was 6 mm, content of fibers was 25% by weight, position of fibers is random)
- sheet FRP material made of mat (the length of fiber was 50 mm, content of fibers was 50% by weight, position of fibers is random in the plane of the mat),

IOP Conf. Series: Materials Science and Engineering **583** (2019) 012015 doi:10.1088/1757-899X/583/1/012015

- sheet FRP material made of woven roving (plain weave, content of fibers was 60% by weight),
- the rebar in the middle of the cylinder is composed of 80% of unidirectional oriented fibers.

The glass transition temperature (Tg) of resin is 120°C.

Methods of measurements differs for testing in room temperature (chapter 3.1) and in higher temperatures (chapter 3.2).

The cylinders were tested according to the standard [7] in room temperature. The test specimens were placed between the plates of 400 kN test press and then pressure testing was performed with a continuous loading at a constant speed of about 1 mm / min. The obtained working diagram was used to evaluate the compressive strength and modulus of elasticity.

The method for testing the cylinders at the higher temperature has been designed with regard to the standard for fire testing [8]. The test specimens were placed between the load plates and then pressure testing was performed with a constant load and increasing temperature according the standard [8]. The tested samples and pressing equipment were inside the insulation shield to protect them from direct flame, according the considered application (see Figure 3). The load, temperatures of sample and testing chamber were recorded. Specific experimental temperature condition was used because of modeling real conditions during the fire. The temperature rise parameters in the test furnace defined in the standard [8] were used. The temperature was set according to equation:

$$T = T_0 + 345.\log(8t + 1) \tag{1}$$

Where T is temperature in °C in furnace in time t, T_0 is initial temperature in °C in furnace, t is time in minutes from beginning of the test.

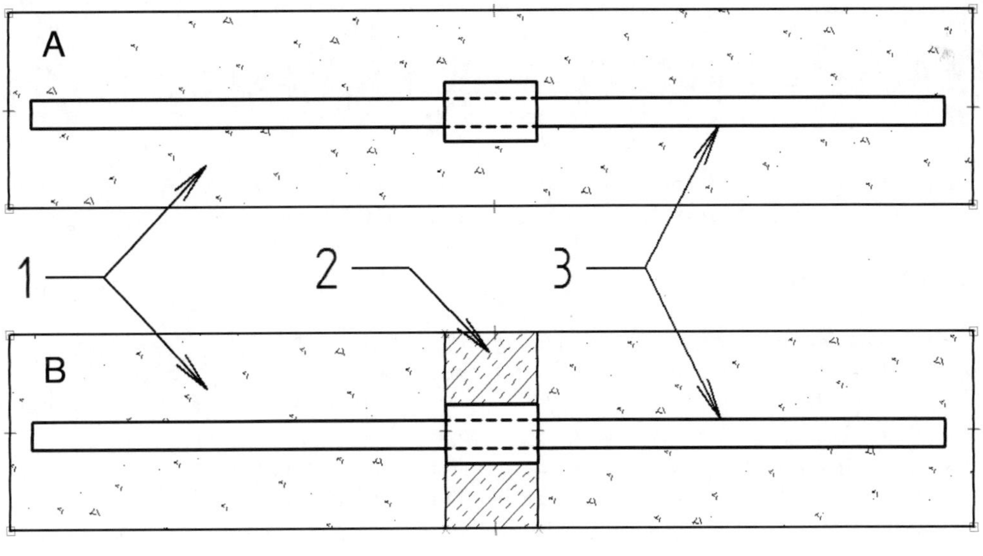

Figure 1. Load-bearing compressed element made from FRP. Symbols: 1 – concrete, 2 – insulation material or gap, 3 – compressed rebar with compression element.

IOP Conf. Series: Materials Science and Engineering **583** (2019) 012015 doi:10.1088/1757-899X/583/1/012015

Figure 2. Load-bearing compressed element made from FRP. FRP rebar diameter 14 mm with molded cylinder diameter 30 mm.

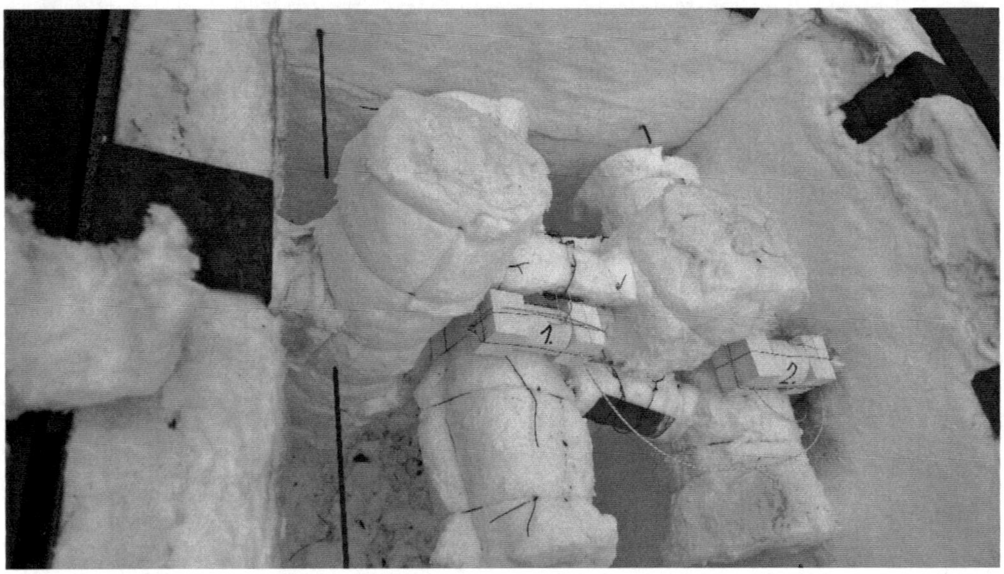

Figure 3. The testing chamber of the furnace.

3. Results and Discussion

3.1. Strength and structure of FRP compressed elements – room temperature testing

Samples of different length and roller diameter ratios were prepared (Figure 4). Also, during the experimental study, the influence of FRP reinforcement on mechanical properties was tested. Only results with short fiber samples are evaluated in this paper.

The compression strength and modulus of the cylinder were measured according to the standard [7], but some parameters (e.g. length to diameter ratio) had to be changed.

The results of mechanical test are shown in Figure 5. Compressive strength and modules are strongly affected by the presence or absence of the rebar. The compressive strengths of the cylinder with FRP rebar of diameter 14 mm is about three to four times higher than strength of the cylinder with FRP rebar of diameter 6 mm or without a rebar. Similarly, the modulus of elasticity is about two times higher for the cylinder with FRP rebar of diameter 14 mm.

The ratio of length and diameter of the cylinders also affects the strength. For example the strength of the cylinder with FRP rebar of diameter 14 mm is decreased about 25% when the length to diameter ratio is 2.00 resp. 2.33.

Figure 4. Load-bearing compressed element made from FRP before pressure testing. White (front row) – without FRP rebar, red (middle row) – with FRP rebar ϕ6 mm, blue (back row) – with FRP rebar ϕ14 mm.

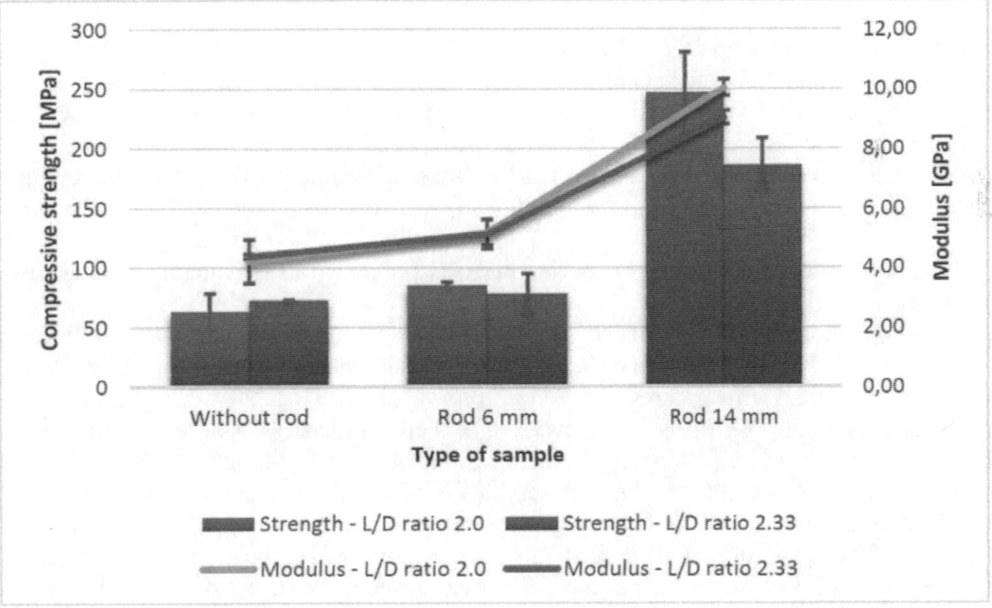

Figure 5. Compressive strength and modulus of elasticity of FRP cylinders. Length to diameter ratio (L/D ratio) 2.0 and 2.33.

3.2. Critical temperature and structure of FRP compressed elements – high temperature testing
Samples made of short fibers, mats, mats and rebar, woven roving and woven roving and rebar were chosen for testing at high temperatures. The goal of tests was to find critical temperatures when the collapse of a loaded sample occurs. The load was 20, 30 and 40 MPa.

Experimental load was applied and then the heating of the furnace started. There were measured temperatures at the cylinder surface and inside the cylinder, temperature in the furnace and compression load. Example of results for load 40 MPa and the cylinder made of woven roving and rebar of diameter 14 mm is shown in Figure 6.

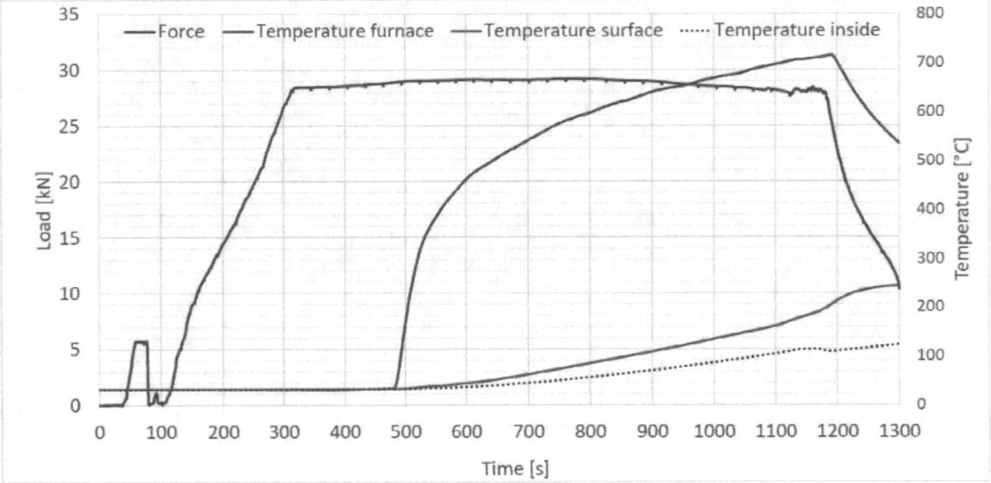

Figure 6. Test of load bearing capacity of FRP cylinders made of woven roving and rebar of diameter 14 mm. Test of critical temperature for compressive force 40 MPa. [9]

The aim of the experiment was to determine critical temperatures, when the cylinder collapses. Overview of these temperatures for compressive force 30 MPa is shown in Figure 7. It can be seen that critical temperatures for these materials is in a narrow area 210 – 220 °C (except samples with short fibers). It is because the failure of samples with long fibers is controlled by material of binder. Although the cylinder made of woven roving and rebar has the lower critical temperature than the cylinder made of mat, the reliability was very good and the residual strength after reaching the critical temperature was best.

Unfortunately, very low amount of experiments was published in the literature and these results cannot be compared with similar research.

The tests of dependency of load intensity and critical temperature were performed only for samples with woven roving and rebar (Figure 8). It is clear that the load intensity in the range from 20 to 40 MPa does not affect the critical temperatures in this solution.

Against expectations, the press elements retain their properties even at significantly higher temperatures than the glass transition temperature (Tg) of the resin. Although the Tg is 120°C only, the critical temperature is above 200°C. Even the samples with sort fibers reach critical temperature about 170°C. This temperature is 50°C above the Tg.

The results of these measurements describe the performance of composite materials under pressure at both room temperature and fire conditions. Such experiments have not been published yet. Some authors [3], [4] or [5] conducted interesting pressure tests at room temperature but their experiments did not contain contribution of fire conditions. The used method of fire testing is very interesting with regards to the finding the parameters for design the members of reinforced concrete. The method is relatively cheap and very fast and provides important results describing the behavior of FRP material in fire conditions.

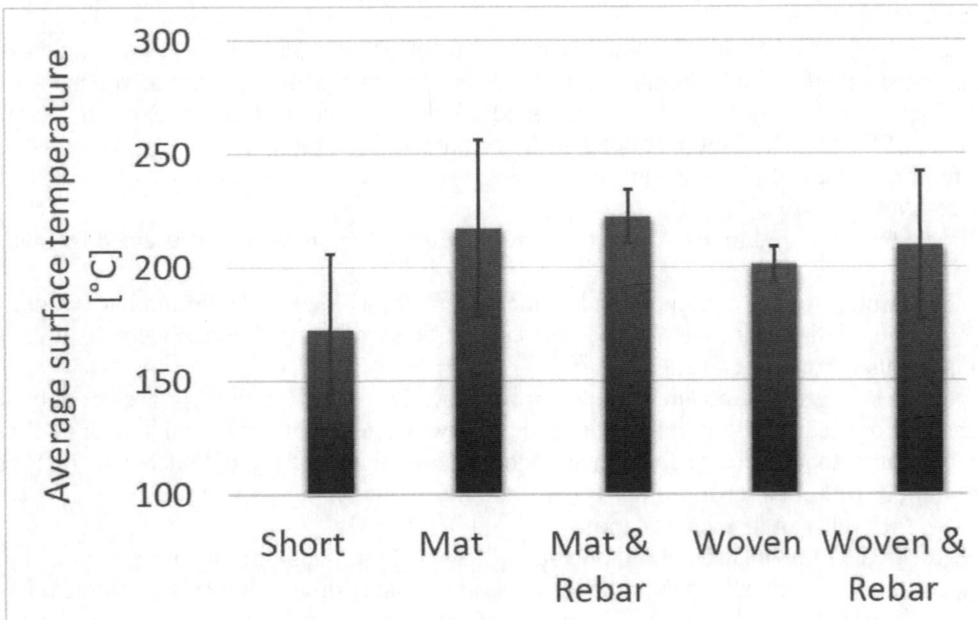

Figure 7. Test of load bearing capacity of FRP cylinders made of short fibres, mat, mat and rebar of diameter 14 mm, woven roving and woven roving and rebar of diameter 14 mm. Test of critical temperature for compressive force 30 MPa. [9]

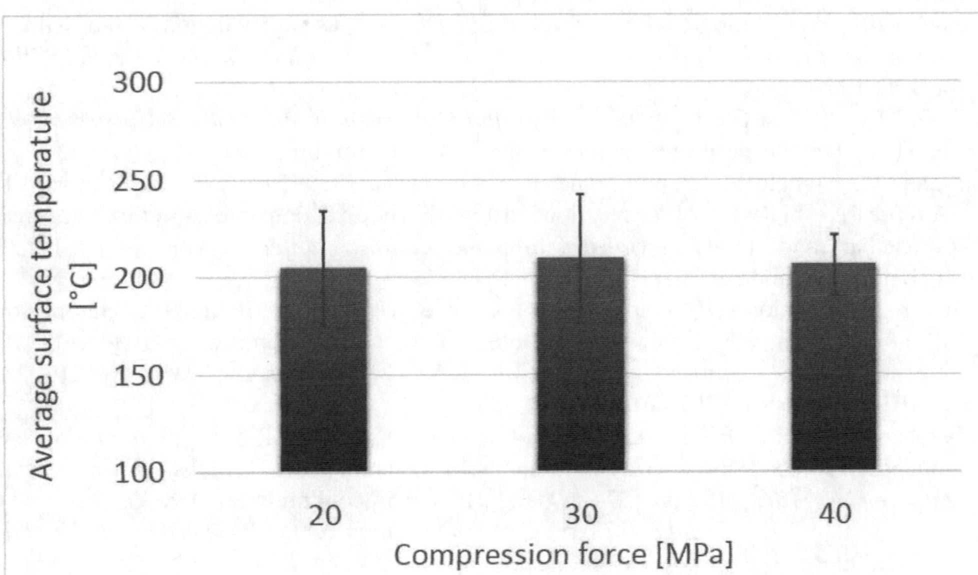

Figure 8. Test of load bearing capacity of FRP cylinders made of woven roving and rebar of diameter 14 mm. Test of critical temperature for compressive force 20, 30 and 40 MPa. [9]

4. Conclusions

Loading capacity of the element reinforced with FRP rebar is determined by load-bearing capacity of the implemented compression reinforcement. To optimize the technical solution, we need to know the mechanical characteristics of the compressed elements. The results of experiments showed the properties of FRP reinforcement applied as the compressed reinforcement.

It was found that the critical temperature is above 200°C. Except material with short fibers. Its critical temperature is about 170°C.

The critical temperature samples from woven roving and 14mm rebar is not affected by load 20, 30 or 40 MPa

The load bearing capacity of compressed element from short fibers and 14mm rebar is significantly affected by length to diameter ratio. The compression strength of samples is reduced about 25% when this ratio is changed from 2 : 1 to 2,33 : 1.

Compressive strength and modules samples from short fibers and 14mm rebar are strongly affected by the presence or absence of the rebar. The compressive strengths of the cylinder with FRP rebar of diameter 14 mm is about three to four times higher than strength of the cylinder with FRP rebar of diameter 6 mm or without a rebar. Similarly, the modulus of elasticity is about two times higher for the cylinder with FRP rebar of diameter 14 mm.

Future research will be focused on finalization of the long-term compression tests of all materials in room temperature and finding the maximum load bearing capacity of the compression element for higher temperatures, especially for temperatures close to critical temperature.

Acknowledgments

This work was published with support of Technology Agency of the Czech Republic. The project number TH04020431.

References

[1] Prokes J 2018 Application of the FRP rebar in constructions for reduction of thermal bridges – new insights In *IOP Conference Series: Materials Science and Engineering*, vol 379 012026 pp 180–190

[2] ACI 440.1R-06:2006 *Guide for the Design and Construction of Structural Concrete Reinforced with FRP Bars* (American Concrete Institute, ACI Committee 440)

[3] Ajarmeh O, Manalo A, Benmokrane B, Karunasena W, Mendis P and Nguyen K T Q Compressive behavior of axially loaded circular hollow concrete columns reinforced with GFRP bars and spirals In *Construction and Building Materials* vol 194 pp 12–23 (Doi: 10.1016/j.conbuildmat.2018.11.016)

[4] Ajarmeh O, Manalo A, Benmokrane B, Karunasena W and Mendis P Axial performance of hollow concrete columns reinforced with GFRP composite bars with different reinforcement ratios *In Composite Structures* vol 213 (Doi: 10.1016/j.compstruct.2019.01.096)

[5] Tabatabaei Kashani A, Eslami A, Mohamed H and Benmokrane B Strength of compression lap-spliced GFRP bars in concrete columns with different splice lengths, In *Construction and Building Materials* v182 pp 657–669 (Doi: 10.1016/j.conbuildmat.2018.06.154)

[6] ASTM D695–15:2015 *Standard Test Method for Compressive Properties of Rigid Plastics* (ASTM International, West Conshohocken, PA)

[7] ČSN EN 13501-2:2004 *Fire classification of construction products and building elements - Part 2: Classification using data from fire resistance tests, excluding ventilation services* (Czech office for standards, metrology and testing)

[8] Zlamal M and Čairović D 2018 *The study of the influence of temperature on the behavior of compressed composite cyllinders, not published results*

Development of lightweight composites based on foam glass aggregate

J Zach[1], M. Sedlmajer[1], J Bubenik[1], M Drdlova[2]

[1] Brno University of Technology, Faculty of Civil Engineering, Centre AdMaS, Purkyňova 139, 612 00 Brno, Czech Republic
[2] Research Institute for Building Materials, Hněvkovského 30/65, 617 00 Brno, Czech Republic

Email: zach.j@fce.vutbr.cz

Abstract. Building composites based on various types of lightweight aggregate have been widely used in the last decades for production of precast elements (e.g. panels) or whole structures with a good thermal insulation/mechanical properties ratio. The presented research investigated the feasibility of using foam glass aggregate, prepared from waste glass powder, in lightweight insulating composites for building industry. The effect of various binders (both organic and inorganic), binder/filler ratios and 3 types of fibres addition (PET, cellulose and mineral wool waste) on the physico-mechanical properties (bulk density, compressive strength) and thermal conductivity of the composite was evaluated. The results indicate, that applicable composites with sufficient mechanical parameters and low thermal conductivity (coefficient of thermal conductivity about 0,09 W/(m·K) can be prepared using both types of binders.

1. Introduction

Thermal protection of buildings (in the Czech Republic according to ČSN 730540) is currently one of the key requirements placed on building structures (in the Czech Republic according to Act 406/2000 Coll., as amended, and Decrees No. 268/2009 Coll. and 78/2013 Coll., as amended). The increasing demands on thermal protection of buildings respond to the global need to reduce CO_2 emissions and to reduction of consumption of non-renewable resources. Simultaneously, there is the pressure to improve the situation in the field of waste management and increase the share of waste recycling and the use of secondary raw materials in industry and construction.

Lightweight aggregate based on foam glass is a relatively new and environmentally friendly product that can be used alone in unbound applications or as a part of composite materials. This aggregate is made from waste glass powder, which is not suitable to be reused in the glass production, in terms of granulometry and chemical composition [1]. Glass powder can also be used directly in cement-based composites as an admixture. The final properties of foam glass aggregate depend on the production technology. Round grained aggregate is produced by packing and firing in a rotary kiln, sharp-edged aggregate is produced by the disintegration and crushing of larger foam glass plates produced continually in tunnel kilns. In both cases, different aggregate fractions are obtained (either directly, or subsequently by crushing and sorting) namely the broad or narrow aggregate fractions. This is very important for optimal design of composites, in particular of optimal design of granulometric curve [2].

The main advantage of the foam glass aggregate is its low bulk density, which varies (depending on the fraction) in the range of 90–430 kg/m^3, low thermal conductivity in the range of 0.06–0.11 W/(m.K) and high value of crush resistance (compression resistance) in the range of 0.3–2.0 N/mm^2.

2. Composites based on foam glass aggregates

Up to now, the foam glass aggregate was mainly investigated as partial substitute of natural aggregate in lightweight concrete. The properties at fresh and hardened state were evaluated, the results showed the potential for the use of foam glass aggregates in such concretes. [1] Foam glass aggregate can be also bonded to plates or shaped products by suitable binders. From the point of view of possible use in the building industry, composite materials based on foam glass can be used as insulating materials for applications with higher demands on the mechanical properties of the insulator. In this case, the insulator transmits a higher permanent load (e.g. roof and floor construction) or must withstand a higher random load in the structure (e.g. skirting board construction in thermal insulation systems, etc.). Such insulators are used primarily to eliminate or at least partially reduce thermal bridges [3]. Sometimes they can also be used in terms of positive acoustic properties (especially sound absorption) [4]. Studies regarding the utilization of foam glass aggregate in gap-graded type of composite with very low density intended to be used in thermal insulating applications are scarce. The presented research investigates this posibility with emphasis on evaluation of the effect of various binders, binder/filler ratios and fibre addition on the physico-mechanical properties and thermal conductivity of the proposed composite.

3. Materials and methods

Sharp-edged lightweight aggregate (fractions 4/16 and 16/32 mm) from crushed foam glass was chosen for experimental works. This aggregate is produced by Refaglass company from waste packaging glass, which is not suitable to be reused in glass industry. Cement 42.5 R Mokra was used as an inorganic binder, the polyurethane resin Leeson D3149/20 and polyester resin Havelpol 2 were used as organic binders. The amount of binder was designed as to achieve the optimum ratio of mechanical parameters and thermal insulation properties. In the case of cement-based composites, only the aggregate of fraction 4/16 was used and the cement dose was uniformly chosen to be 100 kg per 1 m^3 (based on the results of previous research [2]). In order to improve the properties (primarily mechanical) of the composite, recycled fibres based on cellulose, garnetted mineral wool and garnetted PET bottles were added, each at 5% of the weight of the cement. A total of 4 formulae were designed that differed in the used fibre type (one formula was reference without fibres). The difference was still in the amount of water due to the different type of fiber, the design was made for the same consistency. A superplasticizer was used to reduce the water-cement ratio. The individual formulae used are described in table 1.

Table 1. Composition of individual formulae of lightweight concrete with open structure.

Entry component			Recipe 1 m^3			
			CEM 1	CEM 2	CEM 3	CEM 4
Cement CEM I 42.5 R Mokrá		[kg]	100	100	100	100
Aggregate 4–16 mm REFAGLASS		[l]	1000	1000	1000	1000
Water		[l]	67	88	82	61
Plasticizing admixture MURAPLAST FK 19		0.8% m_c [kg]	0.80	0.80	0.80	0.80
Fibers	PET	5% from m_c [kg]	-	-	5	-
	cellulose		-	5	-	-
	Mineral wool		5	-	-	-
Water cement ratio		[-]	0.67	0.88	0.82	0.61

In the case of polymer binder composites, both fractions 4/16 and 16/32 were used with two binder doses, 25 and 30 wt. %. Those doses were selected based on the previous research as to achieve uniform coating of the aggregate grains at the lowest possible binder amount. The excessive amount of binder increases production costs and also reduces the thermal insulation properties of the resulting composite. A total of 4 formulae were designed, which are listed in table 2.

Table 2. Composition of optimized samples with polymer binder.

Sample identification/component	Binder	Binder weight ratio [%]
PU 16/32	Polyurethane	25
PU 4/16	Polyurethane	30
PES 16/32	Polyester	25
PES 4/16	Polyester	30

Based on the above-mentioned formulae, 150 mm cube-shaped samples (see example in figure 1) were prepared from both types of composites to determine physico-mechanical properties, as well as 300 x 300 x 50 mm plate-shaped samples to determine thermal insulation properties. The mixtures were placed into the moulds treated with the releasing agent and compacted with light 1kN pressure and vibration. Determination of following properties was carried out:

1. Density in fresh/hardened state (EN 12350-6, ČSN EN 12390-7), [9, 10]
2. Compressive strength (EN 12390-3), [11]
3. Thermal insulation properties (EN 12667, ISO 8301). [6, 7]

Figure 1. Example of composite samples with polymer binder.

4. Measurement results

Physico-mechanical and thermal insulation properties were determined on the test samples of both types of composite materials. The results of the determination of physico-mechanical properties (average values of at least 5 test samples) are given in tables 3 and 4:

Table 3. Physico-mechanical properties of cement-based samples.

Sample identification /parameter	Bulk density [kg/m^3]	Compressive strength [MPa]
CEM 1	490	1.10
CEM 2	490	0.69
CEM 3	500	0.78
CEM 4	530	1.48

Table 4. Physico-mechanical properties of polymer-based samples.

Sample identification /parameter	Bulk density [kg/m^3]	Compressive strength [MPa]
PU 16/32	324	0.67
PU 4/16	363	0.65
PES 16/32	306	0.28
PES 4/16	365	0.41

As can be seen from the values given in tables 3 and 4, the composites based on cement binder showed better mechanical properties and higher bulk density than the organic polymer binder composites. Comparing the PU and PES resin binders, better mechanical properties were achieved in case of PU binder, the compressive strength was about twice higher. This is partly connected with the mechanical properties of the PU resin itself (higher toughness and flexural strength compared to PES resin), but can be also attributed to higher viscosity of PU resin, which is more suitable for bonding the absorptive foam glass aggregate from the technological point of view. The porous aggregate tends to absorb low viscosity PES resin extensively, which reduces the amount of resin for binding and thus final strength of the composite.

In compliance with expectations, the compressive strength raises with increasing amount of binders in both cases.

Regarding the specimens bonded with cement-based matrix, it was observed, that addition of any waste fibres does not improve compressive strength of the composite. Presumed reason is overall low amount of cement matrix and the gap-graded structure of the whole composite, in which the fibres cannot be strongly anchored and evenly dispersed to bring proper load bearing capacity.

Determination of thermal insulation properties of samples in dried state was performed. The thermal conductivity was determined as $\lambda_{10,dry}$ [W/(m.K)]. The results are shown in tables 5 and 6.

Table 5. Thermal technical properties of cement-based samples.

Sample identification /parameter	Bulk density [kg/m^3]	Coefficient of thermal conductivity [W/(m·K]
CEM 1	420	0.1036
CEM 2	430	0.0975
CEM 3	450	0.1002
CEM 4	430	0.1019

Table 6. Thermal technical properties of polymer-based samples.

Sample identification /parameter	Bulk density [kg/m^3]	Coefficient of thermal conductivity [W/(m·K)]
PU 16/32	324	0.0927
PU 4/16	363	0.0913
PES 16/32	306	0.0877
PES 4/16	365	0.0867

As can be seen from the results, the samples bonded with organic binder show slightly lower thermal conductivity, which is given both by the organic binder with lower thermal conductivity than cement and also by the lower density of the samples. It is also apparent that samples with a 4/16 aggregate fraction with a higher density compared to samples with 16/32 fraction showed better results of thermal insulation properties. This is due to the structure of the air pores and their thermal properties in the individual types of composite. The 4/16 fraction composite contains smaller air pores and cavities compared to the 16/32 fraction, which is beneficial for the final insulating properties of the composite. In the case of cement bonded composites, relatively small differences in thermal conductivity have been found among investigated mixes. The best values were found for formula no. 2 with cellulose based fibres, but the difference can be considered as insignificant. Addition of any tested fibre into the designed gap graded composite does not affect the thermal insulating properties.

In general, it can be stated that both types of composites exhibited comparable mechanical properties at a given density, as did lightweight concretes described in scientific papers [1, 3]. Regarding the observed dependencies of thermal conductivity coefficient to density, the results correspond to the conclusions of other scientific papers dealing with similar issue of lightweight composites with comparable density and porosity [5].

5. Conclusion

The objective of the presented research was to develop gap-grading composite with the optimal ratio of thermal insulating and mechanical properties and to evaluate the effect of various binders (both organic and inorganic), binder/filler ratios and 3 types of fibres addition (PET, cellulose and mineral wool waste) on the physico-mechanical properties (bulk density, compressive strength) and thermal conductivity of the composite.

The following conclusions can be drawn:

- The usage of the foam glass aggregate in gap-grading composites for thermal insulation applications in building industry is feasible with both cement and organic resin binders. All prepared specimens showed low thermal conductivity (around 0.09 W/(m.K)) and sufficient mechanical parameters to be implemented in applications with higher demands on the mechanical properties, such as skirting board construction in thermal insulation systems, etc.
- Samples bonded with organic binder show slightly lower thermal conductivity, which is given both by the organic binder with lower thermal conductivity (compared to cement) and also by the lower density of the samples.
- Samples with a 4/16 aggregate fraction with a higher density compared to 16/32 fraction showed lower thermal conductivity, which is mainly attributed to the structure of the air pores of the 4/16 aggregate and the whole composite.
- The addition of waste fibres does not positively affect the mechanical properties of the composite and the influence of the fibres presence on the thermal conductivity is minimal. On the other hand, the fibres positively influenced compactness and homogeneity of samples.

Acknowledgements
This paper was written under the project No. LO1408 "AdMaS UP - Advanced Materials, Structures and Technologies", supported by the Ministry of Education, Youth and Sports under the "National Sustainability Programme I" and project FV20086 "Development of light new-age building materials with utilization of light-weight aggregate based on waste glass powder" supported by the Ministry of Industry and Trade.

References
[1] Limbachiya M, Meddah M S and Fotiadou S 2012 Performance of granulated foam glass concrete *Construction and Building Materials* **28** pp 759–768
[2] Zach J, Sedlmajer M, Novák V and Bubeník J 2018 Utilization of lightweight aggregate based on foam glass in lightweight concrete with extremely low density *In 20. Internationale Baustofftagung - IBAUSIL 2018* **1** pp 295–301
[3] Real S, Gomes M G, Rodrigues A M and Bogas J A 2016 Contribution of structural lightweight aggregate concrete to the reduction of thermal bridging effect in buildings *Construction and Building Materials* **121** pp 460–470
[4] Park S B, Seo D S and Lee J 2005 Studies on the sound absorption characteristics of porous concrete based on the content of recycled aggregate and target void ratio *Cement and Concrete Research* vol 35 issue 9 pp 1846–1854 (doi:10.1016/j.cemconres.2004.12.009)
[5] She W, Zhang Y and Jones M R 2014 Three-dimensional numerical modeling and simulation of the thermal properties of foamed concrete *Construction and Building Materials* **50** pp 421–431
[6] ISO 8301:1991 *Thermal insulation - Determination of steady-state thermal resistance and related properties - Heat flow meter apparatus*
[7] EN 12667:2001 *Thermal performance of building materials and products. Determination of thermal resistance by means of guarded hot plate and heat flow meter methods. Products of high and medium thermal resistance* (Geneva, Switzerland: STANDARD by International Organization for Standardization)
[8] EN 12350-2:2009 *Testing fresh concrete - Part 2: Slump-test*
[9] EN 12350-6:2009 *Testing fresh concrete - Part 6: Density*
[10] EN 12390-7:2009 *Testing hardened concrete - Part 7: Density of hardened concrete*
[11] EN 12390-3:2009 *Testing hardened concrete - Part 3: Compressive strength of test specimens*

IOP Conf. Series: Materials Science and Engineering **583** (2019) 012017 doi:10.1088/1757-899X/583/1/012017

Effect of curing environment on length changes of alkali-activated slag/cement kiln by-pass dust mixtures

V Bilek[1, 2], L Kalina[1] and H Simonova[2]

[1] University of Technology, Faculty of Chemistry, Institute of Materials Science, Purkyňova 118, 61200 Brno, Czech Republic
[2] University of Technology, Faculty of Civil Engineering, Institute of Structural Mechanics, Veveří 331/95, 602 00 Brno, Czech Republic

Email: bilek@fch.vut.cz

Abstract. This paper investigates the combination of commonly used blast furnace slag (BFS) and quite non-traditional cement kiln by-pass dust (BD) for the preparation of alkali-activated binder. More specifically, the possibilities of alkali-activated slag (AAS) shrinkage compensation and cracking mitigation using BD were studied. For this purpose, mixtures with ratios of BFS/BD ranging from 100/0 to 60/40 were prepared and cured in different environment, namely autogenous conditions, water curing, air curing and their combinations. The results showed that curing regime is a determinative parameter for performance of the mixtures: BD containing specimens showed a certain expansion during the first days of water curing leading to AAS shrinkage compensation, while in absence of water curing increasing dose of BD increased drying shrinkage significantly.

1. Introduction

Nowadays, it is a great challenge to develop and use some binders with lower ecological impact compared to traditional Portland cement (PC). The most common way is partial PC replacement by various industrial by-products like granulated blast furnace slag and fly ash. Nevertheless, these aluminosilicate precursors can also be used to produce non-cement binders thanks to the development of hydration products after mixing with alkaline solutions. Such alkali-activated binders can reach excellent mechanical properties, durability, resistance to elevated temperatures and relatively low heat of hydration [1, 2]. On the other hand, particularly the binders based on alkali-activated slag (AAS) suffer from severe shrinkage and cracking.

Several possibilities of AAS shrinkage reduction have been studied. Very common is, for example, alkaline activation of slag/fly ash blends, which is advantageous not only from the viewpoint of shrinkage, but also setting time. However, this is usually connected with the decrease of strengths, particularly when cured at room temperatures [3]. Moreover, introducing of nitrogen oxides reductions technologies to power plants result in the presence of ammonium sulfate and bisulfate salts [4] which are converted to ammonia under alkaline pH or can be leaching issues and thus such fly ashes are not usable even in common PC-based concretes.

Other industrial by-product with potential to reduce or rather compensate AAS shrinkage is cement kiln by-pass dust (BD). The role of rotary kiln by-pass is to take away part of the kiln gases rich in alkali chlorides and sulfates and thus prevent formation of coatings on kiln walls that can even lead to its blockage as well as to reduce chloride content in clinker. Therefore, BD is rich in chlorides and sulfates, but also in particles of partially burned raw meal, particularly free CaO. This means that BD would act as an expansive agent (EA).

For PC-based concretes, the role of EAs is well understood [5, 6], while only a few papers focused on the use of EAs in AAS have been published. Authors in the paper [7] used two forms MgO differing in their reactivity and observed that their effect depended on their nature, The more reactive (faster) MgO reduced drying shrinkage up to 25% during the first days but its effectiveness decreases with increasing time, while the less reactive MgO started having effect on shrinkage after 15 days. Nevertheless, maximum shrinkage reduction was only 13%. Shrinkage reduction was attributed to the formation of hydrotalcite, while in another study [8] beneficial effect of MgO related to the brucite formation. Also EA containing quick lime, periclase and anhydrite was investigated [9]. Shrinkage was reduced by up to 75% for high doses of EA, but increasing dose of EA simultaneously led to significant acceleration of setting time and workability decrease.

It can be seen that some more or less successful attempts to reduce AAS shrinkage using EAs have been done. This paper pioneers the possibilities of the use of BD as EA which would be beneficial from the viewpoint of both AAS shrinkage reduction and wider utilization of BD.

2. Experimental part

2.1. Materials and mixtures composition

Ground granulated blast furnace slag from Czechia (Kotouč, spol. s r. o.) with Blaine fineness of 400 m^2/kg was used as a main part of a binder. As determined using XRD with Rietveld evaluation method, it contained about 90% of amorphous phase and some merwinite and akermanite. Slag was partially replaced with BD produced in Slovakia (cement plant Horné Srnie, Cemmac, a.s.) which also consisted mainly of amorphous phase (~44%), free CaO (~16%), KCl (~20%), larnite (~10%) and arcanite (~9%) and traces of quartz and alunite. Chemical composition of both slag and BD as determined using X-ray fluorescence (XRF) is given in table 1.

Table 1. Chemical composition of slag and BD as determined using XRF.

	CaO	SiO_2	Al_2O_3	Fe_2O_3	MgO	Na_2O	K_2O	SO_3	Mn_2O_3	TiO_2	P_2O_5	Cl
slag	39.6	38.9	8.9	0.27	8.8	0.39	0.40	1.4	0.83	0.32	0.01	–
BD	25.9	6.5	2.5	1.6	0.48	0.44	17.1	5.2	–	0.14	0.08	10.3

Waterglass with silicate modulus of 2.0 was used to alkali-activate slag/BD blends with ratios ranging from 100/0 to 60/40 (or even 0/100 in the case of setting time). The dose of waterglass was adjusted to 8% with respect to binder (slag + BD) weight. Siliceous sand with maximum grain size of 2 mm was used as a fine aggregate. Its weight was three times higher than the weight of the binder. Water to binder ratio including water from waterglass was 0.35 and 0.46 for pastes and mortars, respectively.

2.2. Specimens preparation and testing

2.2.1. Mixing. Mixing procedure was based on that standardized one for testing of PC in EN 196-1. First of all, waterglass and water were homogenized, then the binder was added and after 30 s addition of sand in the case of mortars followed. Total mixing time was four minutes.

2.2.2. Workability and setting time. Just after the mixing of mortar, workability was tested using a jolting table and expressed as a spread diameter after lifting the truncated cone with lower base of 100 mm as well as after 15 jolts of the table. Unlike workability, initial and final setting time was determined on pastes using Vicat needle apparatus.

2.2.3. Strength development. After the mixing, the mixtures were cast into the steel molds to prepare prisms with dimensions of 40 × 40 × 160 mm further used for flexural and compressive strength testing after 24 hours, 7 days and 28 days. After 24 hours, specimens were demolded and sealed with polyethylene foil to ensure curing in autogenous conditions, i.e. to prevent the mass exchange between specimens and ambient laboratory climate (23°C, 60% relative humidity). Two specimens of each mixture at each age were tested, i.e. two values of flexural and four values of compressive strength were obtained and averaged.

2.2.4. Length changes. The same procedure was used to prepare prisms with dimensions of 25 × 25 × 285 mm for length changes determination, but specimens were demolded earlier, after 5 hours, to start the length changes measurement as soon as possible. At the age of 24 hours, some of the specimens were unsealed and immersed in water or left to dry at ambient laboratory conditions (23°C, 60% relative humidity). Also combination of water and air curing was tested starting with water curing for 1, 3 and 7 days followed by air curing. Length changes were determined using ASTM C490 apparatus.

3. Results and discussion

Effect of partial slag replacement by BD on setting time and workability is given in figure 1. It can be seen that while workability is not significantly affected by various slag/BD ratios up to 60/40, both initial and final setting time slightly decrease at low replacement levels (up to 10% BD), then they somewhat increase with increasing replacement levels up to 30% of BD and after this threshold value, they rapidly decrease. The use of EAs in AAS is often accompanied with significant acceleration of the setting time and early loss of workability [9] and therefore no or slight setting time acceleration with BD doses below 30 or 40% is desirable. Although the described behavior appears strange, the same trends were observed during the repeating of this experiment so it is not a random error. It is related to the chemistry of BD since, according to our experience, also rapid setting even for low doses of BD can be observed. It depends on the given BD batch composition which can be highly variable even for the same source of BD. Acceleration of setting time in presence of BD was expected due to presence of high contents of free CaO in BD which suggest consequent formation of portlandite. Accelerated setting time related to portlandite and ettringite formation has already been reported for slag/BD systems [10].

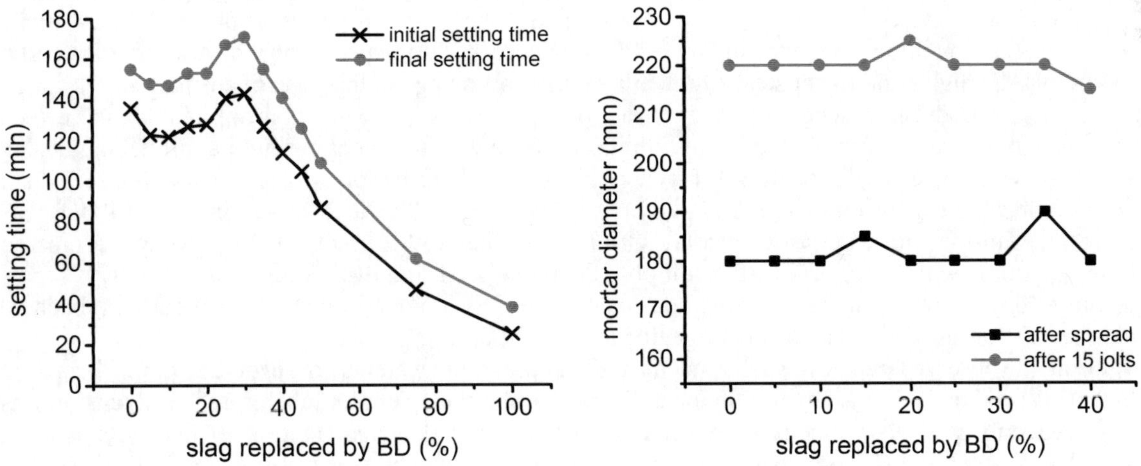

Figure 1. Effect of slag replacement by BD on and workability of AAS-based mixtures.

Gradual decrease in compressive strength with increasing amount of incorporated BD is obvious from figure 2. Quite similar trend can also be seen for flexural strength, but lack of statistics and naturally scattered values limits further discussion. The main reason for decrease in compressive strengths with increase of BD is simultaneously decreasing slag amount and thus amount of amorphous aluminosilicate phase needed for C-S-H formation. Furthermore, formation of other hydration products without strong binding properties is possible, e.g. chloro-aluminate-hydrates as was reported for PC/BD mixtures [11].

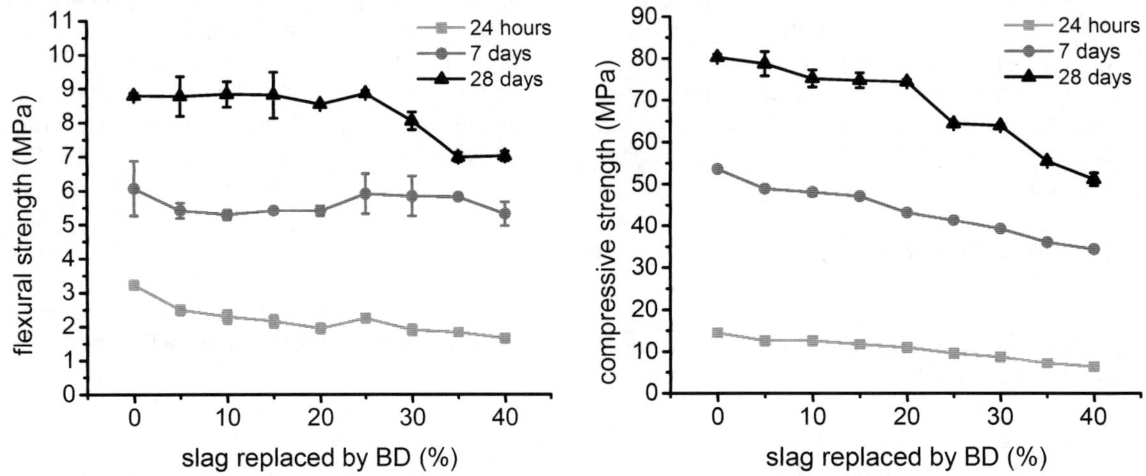

Figure 2. Effect of slag replacement by BD on strength development of AAS-based mixtures.

Length changes of AAS/BD mixtures depending on the curing environment are given in figure 3. It should be noted that due to high number of mixtures and curing regimes only one specimen of each mixture at each curing environment was used. Also, in some cases, some specimens were broken during the demolding which results in missing data.

Nevertheless, the obtained results show that curing environment plays a crucial role in dimensional stability of the prepared mixtures. In autogenous conditions only, some shrinkage reduction for BD incorporating mixtures was recorded, but no expansion. If air curing is applied after 24 hours of autogenous conditions, presence of BD resulted in increased drying shrinkage of AAS mortars and it seems that the higher the BD dose, the higher the drying shrinkage in this case of curing.

Furthermore, different length of water curing before the start of drying was applied. It can be seen that one day of water curing led to results similar to those without water curing, i.e. increasing drying shrinkage with increasing dose of BD. However, if water curing period was prolonged to three or six days, noticeable expansion of mortars, particularly those with 30% and higher amounts of BD, was achieved. This led to a decrease in total shrinkage during the subsequent drying stage compared to the mixtures with lower or no BD addition which tend to shrink even immersed in water. It seems that three days of water curing are enough to finish the expansion but prolonged water curing (six days) is beneficial during following drying stage due to reduced drying shrinkage.

From the observed trends, it is clear that the amount of mixing water is not sufficient for the formation hydration products responsible for volume expansion of the specimens and therefore the expansion was observed only if the specimens are surrounded by additional water. This is in line with general knowledge of EA in PC-based materials [5, 6]. Increased drying shrinkage of AAS by BD observed in the cases of insufficient curing before the start of drying would be related to the changes in the hydration process leading to a decrease of C-S-H phase content and thus reduced mechanical properties, particularly elastic modulus, as well as possible changes in porosity that play key role in shrinkage processes. These aspects will be the object of further study. Now, it can be noted that it was reported that the application of EAs in AAS can lead to portlandite or ettringite formation [10, 12], but additionally to these phases also monosulfoaluminate and Friedel's salt were found in slag/BD systems [13].

IOP Conf. Series: Materials Science and Engineering **583** (2019) 012017 doi:10.1088/1757-899X/583/1/012017

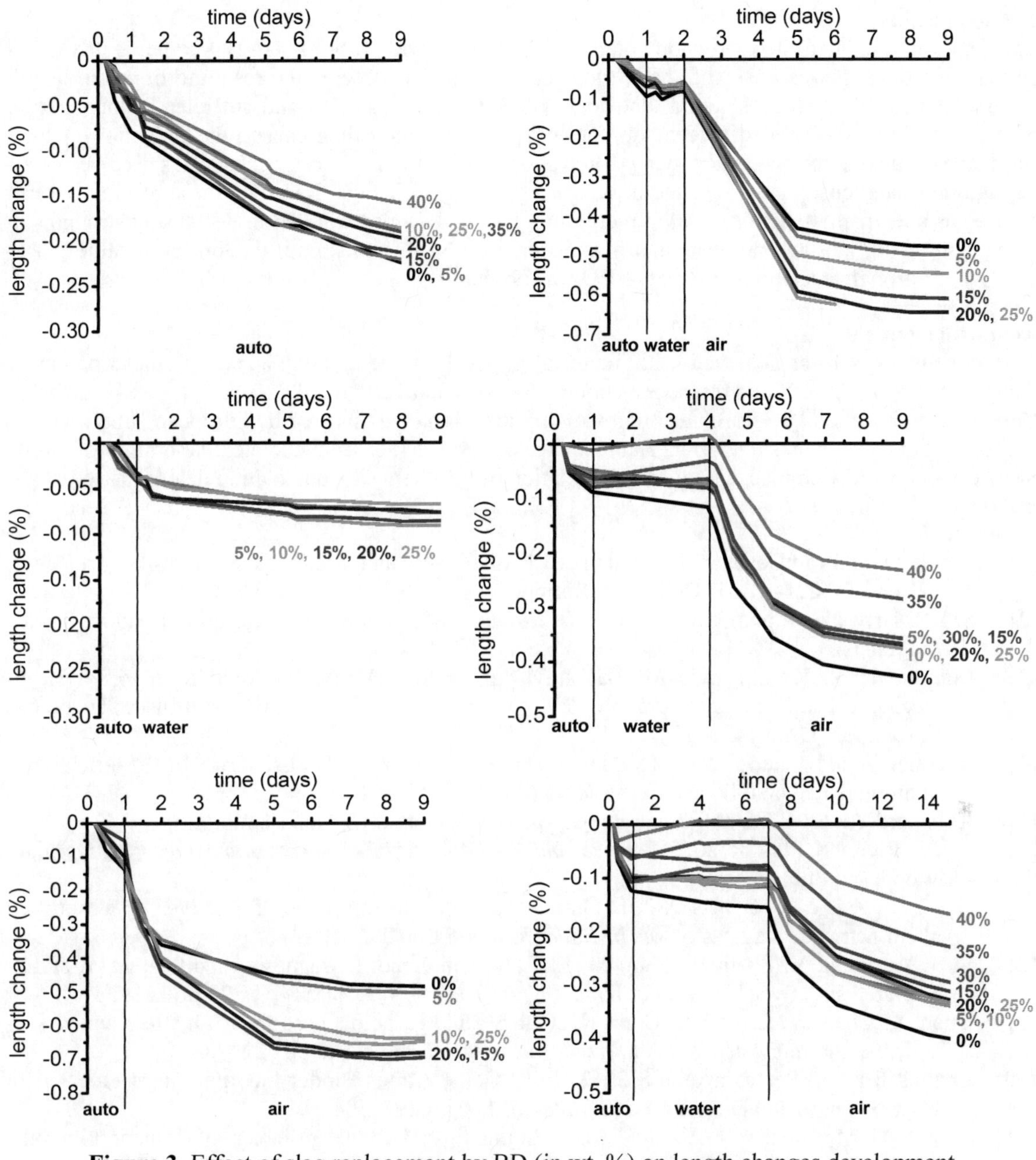

Figure 3. Effect of slag replacement by BD (in wt. %) on length changes development of AAS-based mixtures at various curing regimes.

4. Conclusions

This paper investigated the expansion potential of mixtures with various slag/BD contents in various curing conditions. It was observed that curing regime strongly affected the resulting trends in length changes development. It was found that doses of BD exceeding 25% and sufficiently long period of water curing were needed to develop some expansion which then caused the reduction in total shrinkage during the following air drying. If too short (24 hours) or no water curing was used before air curing, increasing dose of BD resulted in increasing drying shrinkage. Based on these results, BD can be used as an expansive agent in AAS. Increasing dose of BD also resulted in decreasing strengths in autogenous conditions and not very significant changes in setting time until the threshold value (30%) after which strongly accelerating effect of BD was observed.

Acknowledgement

This outcome has been achieved with financial support of the Czech Science Foundation under the project No 18-12289Y and by the project no. 734833 "GeoDust" of H2020 call Marie Sklodowska-Curie Research Actions: Research and Innovation Staff Exchange financed by European Union funds.

The authors also thanks to Zuzana Hantaková and Ivan Rachler from chemical laboratory of Horné Srnie cement plant (Cemmac, a.s.) for the facilitation of the chemical composition determination.

References

[1] Provis J L and van Deventer J S J (eds.) 2014 Alkali Activated Materials: State-of-the-Art Report RILEM TC 224-AAM (Dordrecht: Springer)

[2] Shi C, Krivenko P and Roy D 2006 *Alkali-activated cements and concretes* (London: Taylor & Francis)

[3] Marjanović N, Komljenović M, Baščarević Z, Nikolić V and Petrović R 2015 Physical–mechanical and microstructural properties of alkali-activated fly ash–blast furnace slag blends *Ceram. Int.* **41** pp 1421–35

[4] Roessler J and Cheng W Z Evaluation of the leaching risk posed by the beneficial use of ammoniated coal fly ash *Fuel* **184** pp 613–19

[5] Collepardi M 2006 *The new concrete* (Castrette di Villorba: Grafiche Tintoretto)

[6] Aïtcin P-C and Flatt R 2015 *Science and Technology of Concrete Admixtures* (Cambridge: Woodhead Publishing)

[7] Jin F, Gu K and Al-Tabbaa A 2014 Strength and drying shrinkage of reactive MgO modified alkali-activated slag paste *Constr. Build. Mater.* **51** pp 395–404

[8] Shen W, Wang Y, Zhang T, Zhou M, Li J and Cui X 2011 Magnesia modification of alkali-activated slag fly ash cement *J. Whuan Univ. Technol.-Mater. Sci. Ed.* **26** pp 121–25

[9] Yuan X, Chen W, Lu Z and Chen H 2014 Shrinkage compensation of alkali-activated slag concrete and microstructural analysis *Constr. Build. Mater.* **66** pp 422–28

[10] Chaunsali P and Peethamparan S 2013 Novel Cementitious Binder Incorporating Cement Kiln Dust: Strength and Durability *ACI Mater. J.* **110** pp 297–304

[11] Shoaib M M, Balaha M M and Abdel-Rahman A G 2000 Influence of cement kiln dust substitution on the mechanical properties of concrete *Cem. Concr. Res.* **30** pp 371–77

[12] Bakharev T, Sanjayan J G and Cheng Y-B 2000 Effect of admixtures on properties of alkali-activated slag concrete *Cem. Concr. Res.* **30** pp 1367–74

[13] Konsta-Gdoutos M S and Shah S P 2003 Hydration and properties of novel blended cements based on cement kiln dust and blast furnace slag *Cem. Concr. Res.* **33** pp 1269–76

IOP Conf. Series: Materials Science and Engineering **583** (2019) 012018 doi:10.1088/1757-899X/583/1/012018

Development of acoustic insulating materials from recycled textile fiber

J Zach[1], J Bubenik[1], V Novak[1] and P Trisko[2]

[1] Brno University of Technology, Faculty of Civil Engineering, Centre AdMaS, Purkyňova 139, 612 00 Brno, Czech Republic
[2] RETEX a.s., U nádraží 894, 672 01 Moravský Krumlov, Czech Republic

Email: bubenik.j@fce.vutbr.cz

Abstract. Nowadays, there is an increasing demand for recycling materials and hence for greater environmental friendliness. The issue of waste management and recycling is a very actual topic in the society of most developed countries. One area where the vast amount of waste is generated every year is the textile industry. Textile waste (old clothing, fashion accessories, car upholstery, home textiles and other textile products) is most often made up of a combination of different types of natural fibers (cotton, silk) and man-made fibers (polyester, viscose). Thanks to the large amount of this waste and recycling technologies available, the secondary textile fibers are an interesting secondary raw material that can be used for the production of thermal and acoustic insulators. The article describes research results in the development of medium and heavy insulating mats usable in vertical constructions and floating floor structures.

1. Introduction

In order to assure protection against the adverse effects of noise and vibrations, a quality design of building construction and the use of high quality acoustic insulating materials is needed during the execution of new constructions and reconstruction (especially in civil housing construction). Acoustic insulating materials help to reduce unwanted noise spreading through dividing constructions (horizontal partition structures, internal partitions, building envelope), or provide sound absorption from surrounding areas for technology and other devices with increased noise levels.

Problems with the spread of noise in the area of horizontal and vertical partition structures are covered by the standard ČSN 730532 „Acoustics - Protection against noise in buildings and evaluation of acoustic properties of building elements - Requirements". The last valid amendment to this Z3 standard occurred in 2017, where conditions for hygienic use of buildings were changed. When assessing acoustic insulating properties of structures, impact insulation and airborne sound insulation are assessed in insulators. From the material point of view, the low dynamic value of the used insulator is important to achieve high impact sound insulation. For airborne sound insulation, it is important to have a high sound absorption value of the insulator in the frequency spectrum of the sound insulation band [1–6].

Textile recycling is a very complicated operation, but due to the constant growth of textile waste, it is essential to make this technology more efficient and use it more effectively. After careful sorting and defibering of textile waste it is possible to obtain a high quality secondary raw material, which can be used, among other things, for the production of acoustic and thermal insulation materials with very good utility properties. Such prepared insulators are widely used and offer an alternative to traditional insulating materials. The article describes insulators made of purely synthetic PES fibers or a combination of polyester and cotton fibers that have been thermally bonded with bicomponent fibers.

2. Acoustic insulation materials from textile fibers

The joint research by Retex and Brno University of Technology is attempting to find use of different types of textile waste, which is very difficult to use in other applications within the textile industry, for the production of thermal and acoustic insulation in the construction industry. In the case of acoustic insulation, the most important insulators are for insulating internal partitions, as well as insulators for the construction of floating floors.

Two types of artificial synthetic fibers (recycled textiles) were selected for the production of test samples. These were crushed carpet edges and torn filters. The microscopic analysis and determination of fiber thicknesses and lengths were performed on samples of recycled fibers using the VHX-950F digital microscope. At the crushed ends, the thickness of the individual fibers ranged from 25 to 40 µm and the length corresponded to the chopping at 15 mm. The fibers were smooth, dark in color and slightly curved. The detail of individual fibers can be seen in figures 1 and 2 [7–9].

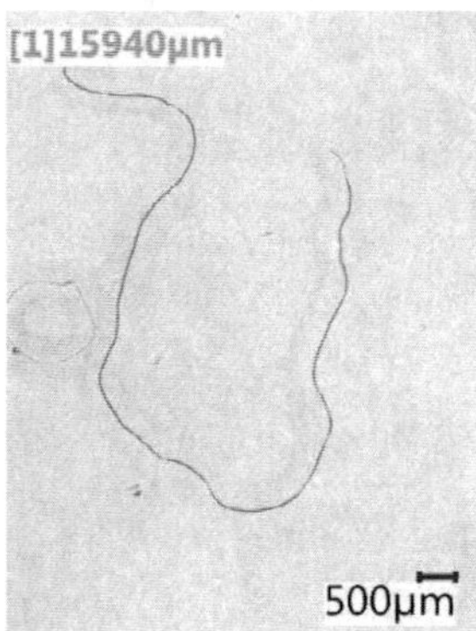

Figure 1. Length of the crushed edge fibers.

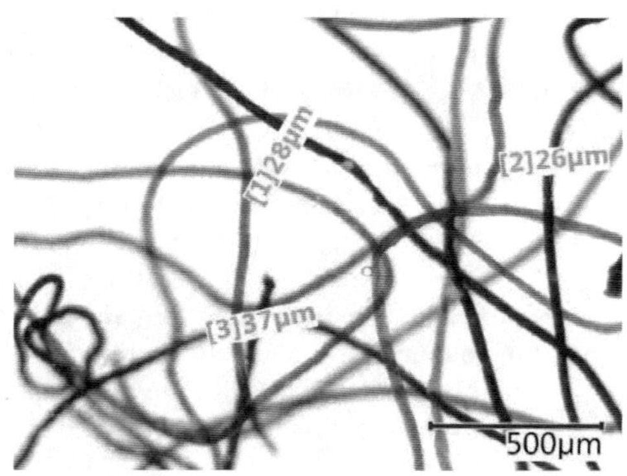

Figure 2. Thickness of the crushed edge fibers.

For torn filters, the fiber thickness ranged from 10 to 15 µm. The length of these significantly thinner fibers was in the range of 15 to 25 mm. The fibers had a rough surface, a light color, and were more curved than the crushed edges. The fibers can be seen in figures 3 and 4.

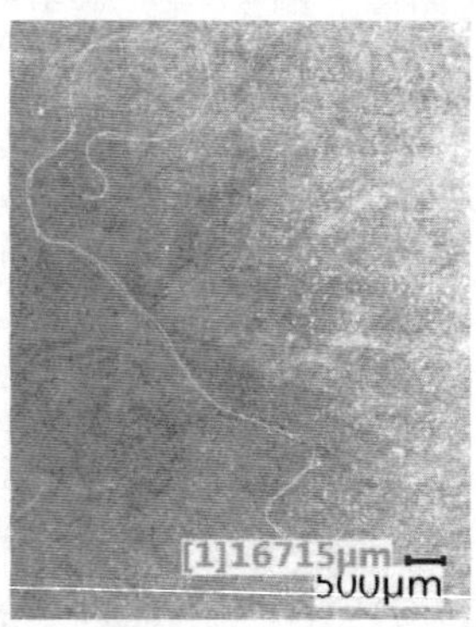

Figure 3. Length of the torn filters fiber.

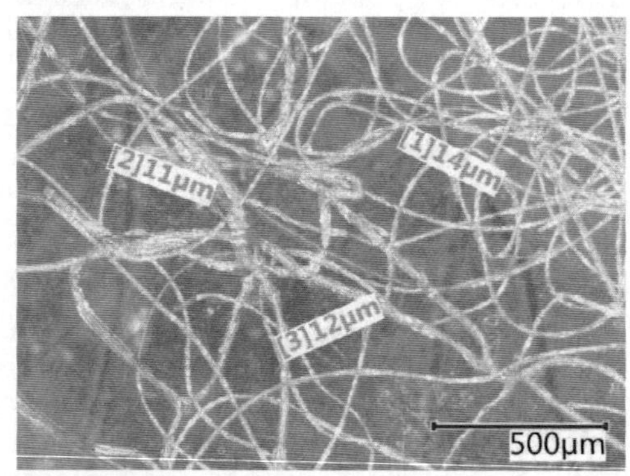

Figure 4. Thickness of the torn filters fiber.

Homogenization of these two types of fibers resulted in the formation of the mixture that forms the main skeleton of the samples described in this article, see figure 5.

Figure 5. A mixture of crushed edges and chopped filters.

3. Preparations of test specimens

Five insulators were proposed in the experimental work and they were divided into two groups:

1. The first group was designed for vertical constructions and the samples were of medium to higher bulk weight. The main objective was to determine the influence of the bulk density on the sound absorption of the individual types of samples.
2. The second group of insulators was designed for the construction of floating floors, these being high density bulk insulators. In this case, they were high density bulk samples, which differed from each other mainly by their thickness.

All samples had the same raw material composition, namely 80% of the fibers (crushed edges, chopped filters) and 20% bicomponent binder fibers. Sample production was performed by the thermal bonding method and the design composition of the individual insulators is shown in table 1 below.

Table 1. Recipes and design features of individual test specimens.

	Sample number	Basic weight [g/m²]	Production density [kg/m³]	Thickness [mm]	Composition [%] BiCo	15 mm crushed edges	Chopped filters
1. Group (vertical structures)	1	5000	145	35	20	40	40
	2	5000	100	50	20	40	40
	3	5000	50	100	20	40	40
2. Group (use in floor constructions)	4	2000	200	10	20	40	40
	5	4000	200	20	20	40	40

After thorough homogenization, the fibers were deposited on a conveyor, on which a fiber mat was formed by the layering of material. The fibrous insulating mat was transported via the conveyor to the baking furnace, where the material is densified at the desired bulk density at elevated temperatures. After cooling, the final phase of production was reached, namely the formatting of the insulator to the desired size (most often 1000 x 500 mm).

Samples of the required dimensions were formatted from the produced materials to determine the required properties.

1) Square samples 300 x 300 mm (determination of thermal insulation properties),
2) Square samples 200 x 200 mm (determination of mechanical, thermal, physical-mechanical and acoustic properties),
3) Circular samples Ø 100 mm and Ø 30 mm (determination of acoustic properties).

IOP Conf. Series: Materials Science and Engineering **583** (2019) 012018 doi:10.1088/1757-899X/583/1/012018

4. Methods

A number of laboratory measurements were performed on the above samples to determine the key properties that are essential for the use of acoustic insulation products in building practice. The following properties were determined on the samples:

1) Acoustic properties (ISO 10534-2 and EN ISO 11654),
2) Dynamic stiffness (ISO 9052-1),
3) Thermal insulation properties (EN 12667, (ISO 8301)),
4) Compressibility (EN 12431),
5) Determination of thickness (EN 12431 and EN 823).

5. Results and discussion

5.1. Physical and mechanical properties

The key properties for insulating materials include the determination of thickness (at standard load 50 /samples 1–3/ or 250 Pa /samples 4 and 5/), bulk density and compressibility. The bulk density was given by production, but for the sake of clarity the actual values according to EN 1602 were measured. The thickness of the sample was determined according to EN 12431 and EN 823 and compressibility according to EN 12431. All tests were carried out on samples measuring 200 x 200 mm. The measurement results are shown in the following table [10].

Table 2. Overview of measured thicknesses, bulk density and compressibility in test samples of acoustic insulation from waste textile.

Sample identification	Thickness [mm]	Bulk density [kg/m^3]	Compressibility [%]
Sample 1	42.54	112.7	13.8
Sample 2	52.80	98.3	18.8
Sample 3	84.47	56.3	40.9
Sample 4	15.12	218.4	8.8
Sample 5	23.41	175.6	6.6

The highest bulk density was achieved in sample 4 (218.37 kg/m^3) thanks to the highest compaction level and the lowest resulting sample thickness. On the other hand, the lowest bulk density had sample 3 (56.31 kg/m^3) because of the greatest thickness and therefore low compaction during production. The compressibility corresponds to bulk densities and the lowest was in sample 5 (6.6%) and highest in sample 3 (40.9%).

5.2. Acoustic properties

For insulating materials used in vertical dividing structures (partitions, load-bearing walls), airborne sound insulation and sound absorption in the sound insulation band of 100 to 5000 Hz are an important feature. The single digit value of the sound absorption factor was determined in accordance with ISO 10534-2 in the third octave bands on circular samples with a diameter of 100 and 30 mm. For better clarity and comparison of the individual insulators we determined the weighted sound absorption coefficient according to EN ISO 11654. For materials suitable for use in floor structures, the main criterion is the impact sound insulation of the floor as a whole. In this respect, the dynamic stiffness of the insulating materials according to ISO 9052-1 was determined by the resonance method on samples of 200 x 200 mm [11–13].

Table 3. Overview of measured values of acoustic properties for test specimens of acoustic insulation.

Sample identification	Dynamic stiffness [MPa/m]	Weighted sound absorption factor [-]
Sample 1	6.76	0.55
Sample 2	5.12	0.55
Sample 3	2.92	0.65
Sample 4	19.40	0.30
Sample 5	14.70	0.35

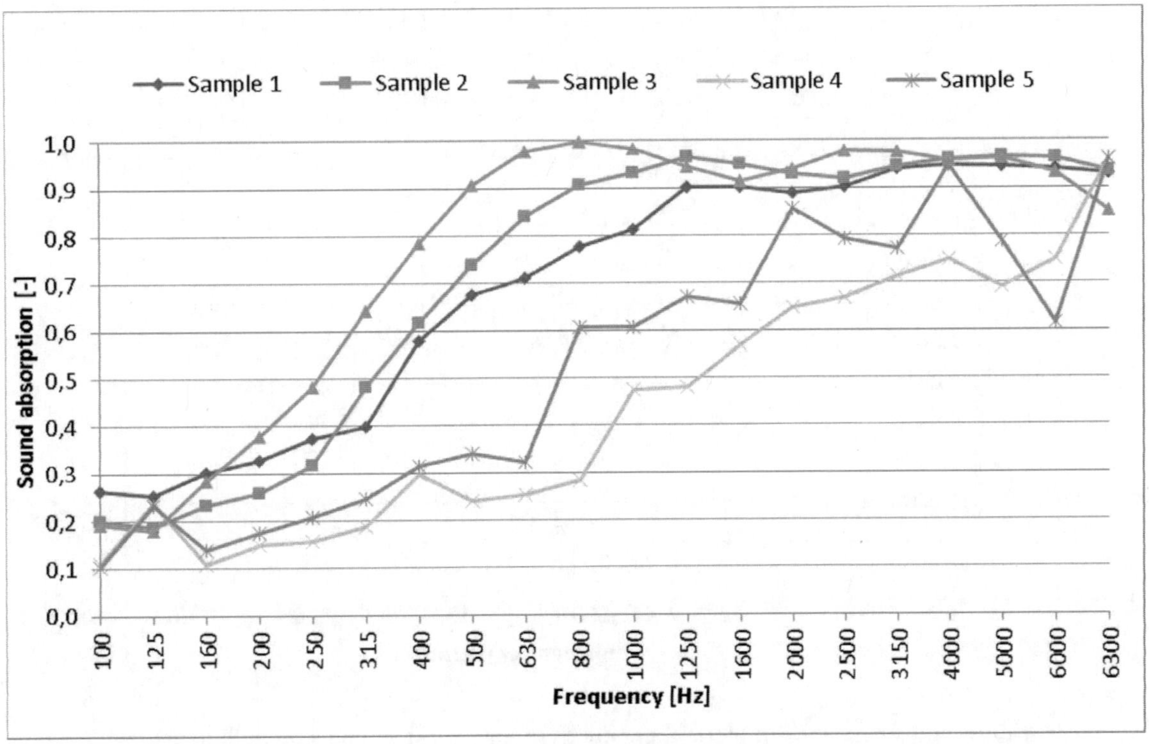

Figure 6. The course of the sound absorption factor for each test specimen in relation to the frequency.

The measured values show the dependence of the acoustic properties on the bulk density of the samples. Samples 4 and 5 with high bulk density reached high dynamic stiffness values (highest sample 4 at 19.4 MPa/m). For samples 1 and 3, the dynamic stiffness values were noticeably lower depending on the bulk density (lowest sample 3 at 2.92 MPa/m). In the area of sound absorption, samples with a lower bulk density (1 to 3) achieved significantly higher values of the weighted sound absorption factor (highest sample 3 with a value of 0.65) than samples 4 and 5 with high density (the lowest sample 4 with a value of 0.3).

5.3. Thermal insulation properties

The thermal insulation properties are very important for both vertical and horizontal structures. This is in particular a structure designed to separate two spaces with a different climatic environment inside. In the case of floor constructions, it is most often a matter of insulating the space from the ground or unheated space. It is similar in vertical constructions with the issue of insulating the external cladding against the external environment or insulating the space with a different internal environment. Determination of the thermal conductivity coefficient was performed by a stationary plate method on instruments Lambda 2300 and Fox 200 at a temperature gradient of 10 K. The samples for determination of the thermal conductivity coefficient had dimensions 300 x 300 mm or 200 x 200 mm [14–16].

Table 4. Overview of the measured values of the thermal conductivity coefficient.

Sample identification	Thickness [mm]	Coefficient of thermal conductivity [W/(m.K)]
Sample 1	42.54	0.0396
Sample 2	52.80	0.0405
Sample 3	84.47	0.0434
Sample 4	15.12	0.0419
Sample 5	23.41	0.0421

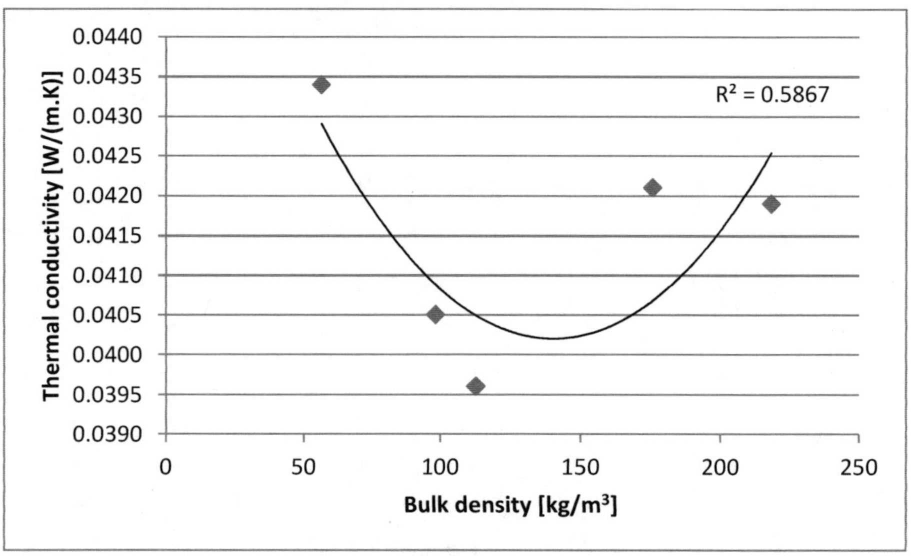

Figure 7. Dependence of thermal conductivity of all test samples on bulk density.

The measured thermal conductivity values showed minimal differences between the individual samples. The thermal conductivity values ranged from 0.0396 to 0.0434 W/(m.K). The most favorable thermal conductivity value was measured for sample 1 (0.0396 W/(m.K)). Is possible to see, that optimal interval of density (from point of view of lower thermal conductivity) is between 100–50 kg/m^3 (see in figure 7).

When evaluating individual properties, samples 1 to 3 show very good acoustic properties from point of view of airborne sound insulation (high sound absorption coefficient – see in figure 6). These samples seem to be very suitable for use in partitioning vertical constructions to ensure the airborne sound insulation of the structure. For samples 4 and 5, the acoustic properties are slightly worse (but it depends on their low thickness and higher density), but the advantage is minimal compressibility and hence their use in more load-bearing structures (float floor structures). For these insulators, a relatively low value of the thermal conductivity is important to ensure the thermal insulation capacity of acoustic insulation materials. There is very interesting dependence of thermal conductivity on bulk density – see in figure 7. Optimal interval of bulk density was found for test samples higher as usual [3, 5, 7]. It is caused by composition of test samples and content of crushed edges and chopped filters with lower degree of pulping of textile particles.

Conclusion

As part of the development work, the possibility of using textile waste that is difficult to process was verified. These were used filters and edges from the carpet production. By recycling this waste, a secondary raw material was obtained, which was used for the production of medium, higher and high density acoustic insulators for vertical and horizontal construction.

Test samples were used to determine the basic physical-mechanical, acoustic and thermal insulation properties. Based on the determined values it can be said that these materials produced from waste textile (polyester) fibers exhibit very good results in all the identified key characteristics and are comparable to commonly used materials on the construction market.

The greatest advantage of these materials is the processing of waste which is difficult to dispose of, namely textile waste. By using this waste as a primary raw material there is a saving of non-renewable resources and hence the overall reduction of mining and extraction of mineral resources.

Acknowledgements

This paper was elaborated with the financial support of the project MPO FV20127 "Development and research advanced thermal and acoustic insulation based on waste textile and natural fibers" and project No. LO1408 "AdMaS UP - Advanced Materials, Structures and Technologies", supported by Ministry of Education, Youth and Sports under the „National Sustainability Programme I".

References

[1] Latif E, Lawrence R M H, Shea A D and Walker P 2018 An experimental investigation into the comparative hygrothermal performance of wall panels incorporating wood fibre, mineral wool and hemp-lime *Energy and Buildings* (Switzerland) vol 165 pp 76–91

[2] Battegazzore D, Alongi J, Duraccio D and Frache A 2018 All Natural High-Density Fiber- and Particleboards from Hemp Fibers or Rice Husk Particles *Journal of Polymers and the Environment* (USA) vol 26 issue 4 pp 1652–1660

[3] Lau K T, Hung P Y, Zhu M H and Hui D 2018 Properties of natural fibre composites for structural engineering applications *Composites Part B-Engineering* (England) vol 136 pp 222–233

[4] Lacoste C, El Hage R, Bergeret A, Corn S and Lacroix P 2018 Sodium alginate adhesives as binders in wood fibers/textile waste fibers biocomposites for building insulation *Carbohydrate polymers* (England) vol 184 pp 1–8

[5] Zheng C, Li D F, Ottenhall A and Ek M 2017 Cellulose fiber based fungal and water resistant insulation materials *Holzforschung* (Germany) vol 71 issue 7–8 pp 633–639

[6] ČSN 730532:2010 *Acoustics - Protection against noise in buildings and evaluation of acoustic properties of building elements - Requirements* (Prague, Czech Republic: The Office for Standardization, Metrology and State Testing)

[7] Zach J, Hroudova J and Korjenic A 2016 Environmentally efficient thermal and acoustic insulation based on natural and waste fibers *Journal of Chemical Technology and Biotechnology* (USA) vol 91 issue 8 pp 2156–2161

[8] Ali M E and Alabdulkarem A 2017 On thermal characteristics and microstructure of a new insulation material extracted from date palm trees surface fibers *Construction and Building Materials* (England) vol 138 pp 276–284

[9] Dang M N, Grillet A C, Diep T M H, Bui Q B and Woloszyn M 2018 Influence of thermo-pressing conditions on insulation materials from bamboo fibers and proteins based bone glue *Industrial crops and products* (Netherlands) vol 111 pp 834–845

[10] EN 823:2013 *Thermal insulating products for building applications. Determination of thickness* (Geneva, Switzerland: STANDARD by International Organization for Standardization)

[11] EN ISO 11654:1997 *Acoustics -- Sound absorbers for use in buildings -- Rating of sound*

IOP Conf. Series: Materials Science and Engineering **583** (2019) 012018 doi:10.1088/1757-899X/583/1/012018

absorption (Geneva, Switzerland: STANDARD by International Organization for Standardization)
[12] ISO 10534-2:1998 *Acoustics — Determination of sound absorption coefficient and impedance in impedance tubes — Part 2: Transfer-function method* (Geneva, Switzerland: STANDARD by International Organization for Standardization)
[13] ISO 9052-1:1989 *Acoustics -- Determination of dynamic stiffness -- Part 1: Materials used under floating floors in dwellings* (Geneva, Switzerland: STANDARD by International Organization for Standardization)
[14] EN 12667:2001 *Thermal performance of building materials and products. Determination of thermal resistance by means of guarded hot plate and heat flow meter methods. Products of high and medium thermal resistance* (Geneva, Switzerland. STANDARD by International Organization for Standardization)
[15] EN 12431:2013 *Thermal insulating products for building applications. Determination of thickness for floating floor insulating products* (Geneva, Switzerland: STANDARD by International Organization for Standardization)
[16] ISO 8301:1991 *Thermal insulation -- Determination of steady-state thermal resistance and related properties -- Heat flow meter apparatus* (Geneva, Switzerland: STANDARD by International Organization for Standardization)

IOP Conf. Series: Materials Science and Engineering **583** (2019) 012019 doi:10.1088/1757-899X/583/1/012019

Performance of ternary gypsum-based mortars in wet environment

M Dolezelova[1] **and A Vimmrova**[1]

[1] Department of Materials Engineering and Chemistry, Faculty of Civil Engineering, Czech Technical University in Prague, Thákurova 7, 166 29 Prague, Czech Republic

Email: magdalena.dolezelova@fsv.cvut.cz

Abstract. Gypsum is one of the most environmentally friendly materials, but it could be used only in the dry environment. When gypsum is wet, its mechanical properties worsen substantially and namely its strength decreases significantly. Ternary gypsum-based binders, composed from gypsum, lime and pozzolan evince better resistance against the moisture, because water-resistant phases are created as products of pozzolanic reaction. The behaviour and properties of several gypsum composites with different types of pozzolans, exposed to the moisture, were investigated. Samples of ternary composites were stored in the water for 360 days and their properties were tested during the period and compared with the properties of gypsum mortar. While gypsum mortar degraded significantly, the ternary mortars retained its properties and the strength of some of them even increased.

1. Introduction

The gypsum is one of the most environmentally friendly materials and this is the reason why the gypsum is nowadays extensively studied. Wider use of gypsum in the construction process is restricted to its placement only in the interiors without the presence of moisture. The loss of strength caused by the wet environment is generally known [1]. One possibility how to improve the moisture resistance of gypsum is to combine the gypsum with pozzolan and ensure the activation of pozzolan in gypsum. Predominantly the ternary gypsum-based materials combining gypsum, lime or cement and different types of pozzolan are tested [2]. Fraire-Luna et al. [3] compared pure gypsum material with materials containing blast furnace slag and metakaolin, which have been stored under the water for 360 days and found out that the resistance of composed materials is better.

The solution to the problem of the loss of mechanical properties in the wet environment is thus the creation of ternary gypsum pastes. While the ternary gypsum-based pastes were studied extensively, the long-term durability of gypsum mortar has not yet been researched sufficiently. In general, we can say that the gypsum composites with any filler are described in the literature only rarely. Herrero et al. [4] focused on recycling by using the mechanical grinding of rubber particles from end-of-life tires as a filler in their article. The behaviour of the gypsum paste with a combination of the different types of sand was studied by authors in the previous article [4]. The next logical step is the description of the behaviour of ternary binders in combination with the filler. Therefore, the research of the ternary gypsum mortars, especially of their mechanical properties and also moisture resistance, is an important task.

2. Materials

Four gypsum-based composites and one gypsum mortar as a reference were tested. Gypsum, lime, three pozzolans were used as dry components of the mixtures. The specification of the used materials is in the table 1.

Table 1. Materials specification.

	Specifications	Producer
Gypsum	Gypsum binder G2 BII	Gypstrend s.r.o.
Lime	CL 90 S	Vápenka Čertovy schody a.s.
Crushed ceramic	-	HELUZ cihlářský průmysl v.o.s.
Silica fume	Stachesil S	STACHEMA CZ s.r.o.
GBS	Granulated blast furnace slag (SMŠ 400)	KOTOUČ ŠTRAMBERK, spol. s r.o.
Aggregates	Standardized sand (CEN, ČSN EN 196-1)	Filtrační písky spol. s.r.o. (Chlum)

The final compositions of tested mortars are in the table 2. The composites were composed of ternary or binary binder and sand. The gypsum mortar (G) was made as comparative material. This mortar contents only gypsum and sand. The binder to sand ratio was 1:2. The dosage of pozzolan in the ternary binders was calculated according to the amount of the amorphous phase (determined by XRD analysis; amorphous phase of crushed ceramic is 45.3%; silica fume - 90%; GBS - 87.7%) in order to achieve the same amount of amorphous phase in each mixture and therefore the results could be compared. The amount of lime was the same for all mixtures. The water/binder ratio was determined by the flow test for diameter 185 ±5 mm.

Table 2. Material compositions.

Designation of the mixtures	Pozzolan		Gypsum	Lime	Sand	Water/binder ratio
	[%] by mass					[-]
KC	Crushed ceramic	5.6	24.4	3.3	66.6	0.78
KM	Silica fume	2.8	27.2	3.3	66.6	0.82
KS	GBS	2.9	27.1	3.3	66.6	0.78
K	-		30.0	3.3	66.6	0.81
G	-		33.3	-	66.6	0.90

3. Methods

3.1. Basic properties

The basic properties were measured on the samples at the age of 28 days. The density ρ of tested materials was measured by helium pycnometry (Pycnomatic ATC, Thermo Fisher Scientific). The gravimetric method was used for determination of bulk density of samples. The porous structure was studied by mercury intrusion porosimeter Pascal 140 and 440 (Thermo Fisher Scientific).

The microstructure of all samples was studied by scanning electron microscopy (SEM) using a ZEISS Merlin electron microscope with Gemini II column at an acceleration voltage of 7-10 kV, probe current of 46–71 pA and working distance of 8–13 mm. The samples were dried and then covered with a carbon and they were not polished.

3.2. Compressive and tensile strength

Production of samples for strength measuring was managed according to EN 13279-2 [6]. Two sets of test specimens were made for each period of measuring. The mechanical properties were measured after 7, 28, 90, 180 and 360 days. Each test set contained three prisms 160 × 40 × 40 mm. Samples were stored in laboratory condition (marking A) and also under the water (marking W). Samples were not dried before the testing. Bending strength f_f [MPa] was determined by three-point flexural test by mechanical press FP 100 (VEB Industriewerk Ravenstein) and compressive strength f_c [MPa] was tested after the bending test. The strength tests were also performed according to standard EN 13279-2. Moisture by mass was determined immediately after the compressive strength test.

4. Results and discussion

The values of bulk density, density and porosity are shown in table 3. The density of samples stored in the laboratory condition is the same as density of samples stored under the water, Density of all materials is about 2550 kg·m^{-3}. The storage condition, however, had an effect on the bulk density and porosity of the samples.

The porosity of all materials was lower when the samples were stored under the water (W). The decrease of porosity of ternary composites (KC, KM, KS), stored under water is caused by the formation of CSH phases and ettringite, because wet environment is more favourable for pozzolanic reaction and therefore the faster transformation of the structure. The presence of these phases has been proven by scanning electron microscopy. The structure of ternary composites stored under the water (W) is distinctly finer, as depicted in figure 1, where the structure of composites with crushed ceramic after 360 days is shown. The presence of ettringite and CSH phases was observed e.g. by Fraire-Luna et al. [3] and Escalante-Garcia et al. [7][8].

The gypsum mortar (G) has the highest values of porosity. The slightly lower porosity of samples stored under the water is caused by the dissolution of gypsum.

Table 3. Bulk density, density and porosity at the age of 28 days.

Designation of the mixtures	**A** - laboratory environment			**W** - stored under the water		
	ρ_V [kg·m^{-3}]	ρ [kg·m^{-3}]	p [%]	ρ_V [kg·m^{-3}]	ρ [kg·m^{-3}]	p [%]
KC	1627	2552	36.2	1709	2576	33.7
KM	1576	2520	37.5	1698	2551	33.5
KS	1629	2574	36.7	1747	2549	31.5
K	1529	2553	40.1	1716	2540	32.5
G	1459	2536	42.5	1511	2551	40.8

Figure 1. Structure of composite with crushed ceramic (*KC*): left - storage in the laboratory, right - storage in water.

The values of the compressive strength were evaluated by Dean-Dixon test [9]. The results of compressive strength are in figures 2 and 3. The moisture by mass (right axis) is added because the samples were not dried before the test. The values of moisture in the individual time period are connected by the line for better understanding.

The moisture by mass of composites stored in the laboratory environment (A) was in the range from 0.1 to 0.4%. The gypsum mortar (G) showed to hygroscopic behaviour, which is typical for the gypsum and therefore its moisture by mass is higher than the moisture of gypsum composites. The values are between 2.3 and 3.1%. The compressive strength of the materials stored in the laboratory environment (A) and its development in time are shown in figure 3. There is not any substantial loss of compressive strength for each material over time. That was expected because the gypsum was not exposed to increased moisture load.

Really interesting is the slight decrease in compressive strength at the age of 180 days. The decrease in samples with pozzolan was more pronounced than the decrease in samples without pozzolan. The decrease is caused by crystallization changes in the structure with the result in loss of strength. Similar decrease of compressive strength was observed by Fraire-Luna et al. [3] and Magallanes-Rivera and Escalante –Garcia [11]. The compressive strength at the age of 360 days increased again. In a further study of these systems, it would be appropriate to perform scanning electron microscopy supplemented by EDS analysis which could help to clarify these changes of compressive strength at the 180 days.

The more fundamental are results of samples stored under the water (W) (figure 3). The compressive strength of gypsum mortar decreased during the time, as expected [1]. The strength of gypsum mortar is significantly lower than the strength of samples, stored in the laboratory condition (A). The moisture by mass was higher, between 17.1 and 22.3%. The maximum moisture by mass of all composites was 14.6%. The decrease of strength and higher value of moisture by mass were accompanied by an increase of volume, see figure 4. The composite without pozzolan (K) had lower values of compressive strength than the other composites. The values were constant in the time.

The compressive strength of composites with pozzolan (KC, KM, KS) was growing over the time. The compressive strength of composite with crushed ceramic (KC) was increasing smoothly gradually. Compressive strength at the age of 360 days was three and a half times higher than the strength at the age of 7 days. The development of strength of samples with GBS (KS) was similar but the increase of strength was less than twofold. The increasing of strength of the materials with pozzolan was proven by other authors [3][7][8][12], while other authors the use of ternary gypsum based materials in wet environment do not recommend [13], [14].

The composites with silica fume (KM) achieved the highest value of compressive strength at all times. Its strength was 6.04 MPa at the age of 360 days and it was more than threefold than the value at the age of 7 days. The most significant increase took place up to 90 days. If we compare the composite with silica fume stored in the laboratory environment (KM – AU) with the same composite stored under the water (KM – WU), we can see that the compressive strength of samples stored in the laboratory was relatively constant at all time. On the contrary, the strength of samples stored under the water increased gradually. The compressive strength of samples at the age of 7 days was more than three times lower when the samples were stored under the water.

The compressive strength of composites with silica fume was the same for both types of storage (around 6 MPa) after 360 days. Use of silica fume as the pozzolan addition in the ternary gypsum-based composites seems therefore very appropriate.

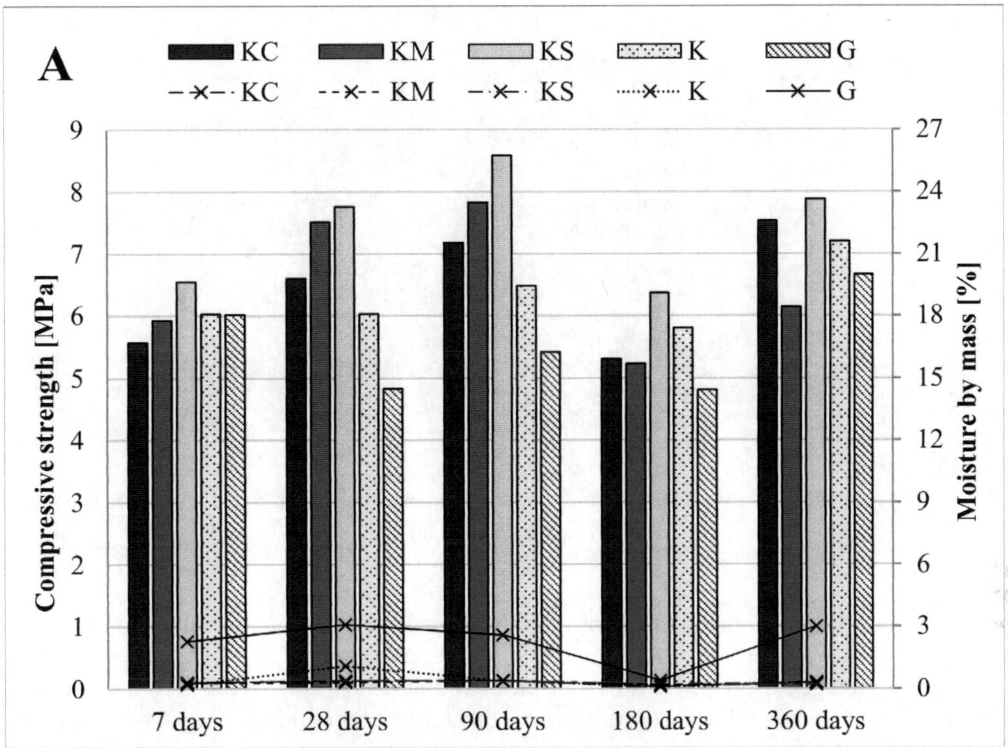

Figure 2. Mechanical properties and moisture by mass of samples stored in the laboratory (A).

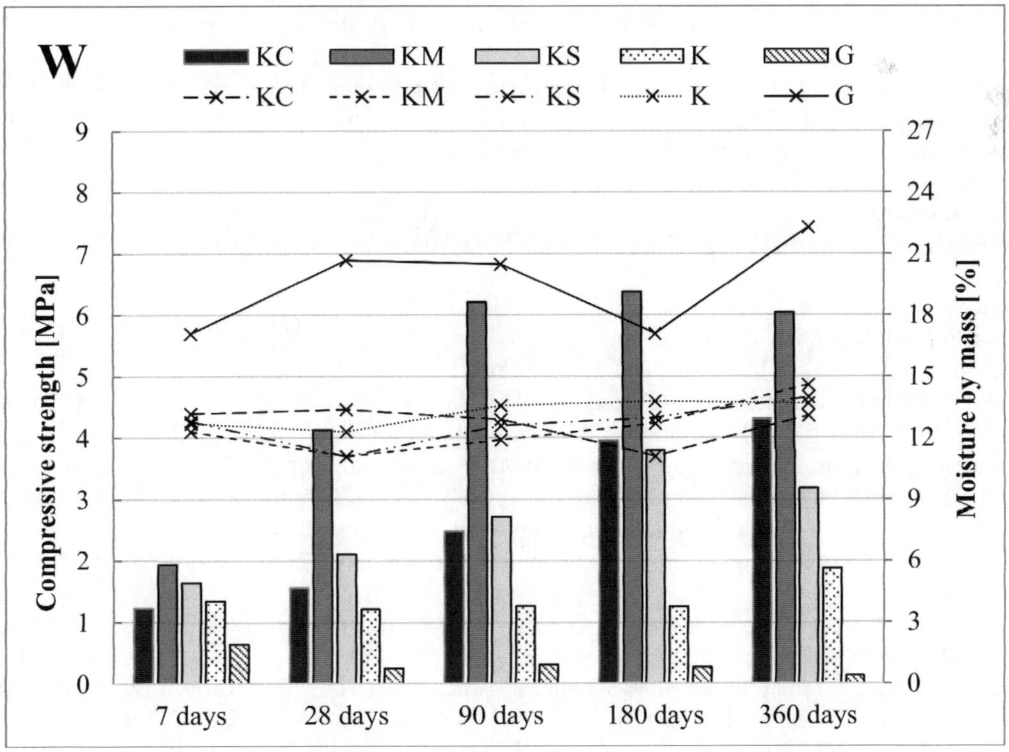

Figure 3. Mechanical properties and moisture by mass of samples stored in the water (W).

Figure 4. Photo of the gypsum-based composites and gypsum mortar stored under the water (W), 360 days old; from the left: crushed ceramic-KC, Silica fume-KM, GBS-KS, without pozzolan-K, gypsum mortar-G.

5. Conclusions

The influence of the storage on the mechanical properties of gypsum composites and gypsum mortar was investigated. The samples were stored in two environments, in the laboratory and under the water. The compressive strength of all materials was measured at 7, 28, 90, 180 and 360 days. The basic properties, i.e. bulk density and porosity depend on the storage conditions. The values of compressive strength of all materials stored in the laboratory conditions were over 4.8 MPa, regardless of the age. The storage under the water had a more fundamental impact on the values of compressive strength. The gypsum mortar had a negligible compressive strength at the age of 360 days. The binary composite did not lose strength during the storage under the water, the values were up to 1.87 MPa. The compressive strength of ternary composites was increasing during the time. The ternary composite with silica fume proved the best results. Its compressive strength at the age of 360 days was the same for both types of storage. This combination is therefore considered as a good solution if we want to improve the resistance of gypsum against wet environment.

Acknowledgements

This research was supported by the Czech Science Foundation, Project No. 19-08605S.

References

[1] Karni J and Karni E 1995 *Mater. Struct.* **28** 9
[2] Camarini G and De Milito J A 2011 *Constr. Build. Mater.* **25** pp 4121–5
[3] Fraire-Luna P, Escalante-Garcia J and Gorokhovsky A 2006 *Cem. Concr. Res.* **36** pp 1048–55
[4] Herrero S, Mayor P and Hernández-Olivares F 2013 *Mater. Des.* **47** pp 633–42
[5] Krejsova J, Dolezelova M, Pernicova R, Svora P and Vimmrova A 2018 *Cem. Concr. Compos.* **92** pp 188–97
[6] EN 13279-2:2014 *Gypsum binders and gypsum plasters. Test methods*
[7] Escalante-García J, Magallanes-Rivera R and Gorokhovsky A 2009 *Constr. Build. Mater.* **23** pp 2851–5
[8] Escalante-García J, Rios-Escobar M, Gorokhovsky A and Fuentes A 2008 *Cem. Concr. Compos.* **30** pp 88–96
[9] Dixon's Q-test: Detection of a single outlier 2018 *EDUCATIONAL APPLETS* [accessed 2018-12-10] (*Available at*: http://195.134.76.37/applets/AppletQtest/Appl_Qtest2.html)
[10] Svoboda L 2007 *D Dtest* [accesed 2019-01-10] (*Available at*: http://people.fsv.cvut.cz /~svobodal/DDtest/)

[11] Magallanes-Rivera R and Escalante-García J 2014 *Constr. Build. Mater.* **71** pp 317–26
[12] Martinez-Aguilar O, Castro-Borges P and Escalante-García J 2010 *Constr. Build. Mater.* **24** pp 631–9
[13] Marinkovic S and Kostic-Pulek A 2007 *J. Phys. Chem. Solids* **68** pp 1121–5
[14] O'Rourke B, McNally C and Richardson M 2009 *Constr. Build. Mater.* **23** pp 340–6

IOP Conf. Series: Materials Science and Engineering **583** (2019) 012020 doi:10.1088/1757-899X/583/1/012020

Stabilization of the ammonia in SNCR fly ash - the influence of tannins presence on the preparation of an autoclaved aerated concrete

L Galvankova[1], M Ledl[2], D Beranova[1] and T Opravil[1]

[1] Brno University of Technology, Faculty of Chemistry, Institute of Material Chemistry, Purkyňova 118, 612 00, Brno, Czech Republic
[2] Brno University of Technology, Faculty of Civil Engineering, Institute of Technology of Building Materials and Components, Veveří 331/95, 602 00, Brno, Czech Republic

Email: xcgalvankova@fch.vut.cz

Abstract. Due to the legislative regulations on NO_x emissions, a selective non-catalytic reduction (SNCR) technology had been introduced to a coal combustion process in power plants. The valuable by-product, fly ash, contains ammonia in the form of soluble salts, e.g. NH_4HSO_4 and $(NH_4)_2SO_4$. After mixing SNCR fly ash with cement, thanks to the rise of pH, the toxic ammonia releases and contaminates the working area, so the presence of these salts is undesirable in an autoclaved aerated concrete (AAC) manufacturing process. A possible solution is the conversion of soluble ammonium salts to an insoluble form. Tannins are a class of polyphenolic biomolecules, which react with the ammonium ions to form insoluble compounds. The AAC samples were mixed using SNCR fly ash and two different ammonium binding additives - pure tannic acid and the cost-effective animal food supplement containing chestnut tannins. The influence of additives on the hydration process of the starting mixture was studied by isoperibolic calorimetry. The results suggest that the cost-effective source of tannins retards the hydration. The presence of insoluble compounds was studied by the infrared spectroscopy.

1. Introduction

The strict regulations of nitrogen oxides (NO_x) emissions induce the introduction of NO_x control technologies to all industry levels. A significant amount of NO_x emissions is produced in power plants during the combustion process, so the introduction of air pollution control technologies is inevitable. There are several technologies concerning NO_x removal, but thanks to its efficiency and low capital expense, the selective non-catalytic reduction process (SNCR) plays an important role in European countries [1]. SNCR is a post-combustion technology for reducing NO_x by injecting ammonia (NH_3) or urea or other ammonia type reactant into the furnace at the determined location. The reactants selectively react with NO_x at the temperature ranging from 875°C to 1150°C to produce nitrogen and water. Besides other factors, e.g. the NO_x concentration in flue gas, reagent injection rate, etc., the temperature of the reaction is the most important parameter which affects the efficiency of the process [2]. By increasing the temperature above 1150°C, the reducing ammonia based reactant is oxidized results in additional NO_x production. Below the 800°C, the temperature is too low to the reaction could proceed so the injected NH_3 remains unreacted. Even if the temperature rises above 870°C, a part of NH_3 does not react. The unreacted NH_3 is so-called "ammonia slip" [3].

Besides the ammonia slip, the flue gas contains SO_3 and water. These compounds react to form soluble ammonium salts, mainly ammonium bisulfate (NH_4HSO_4) and ammonium sulfate ($(NH_4)_2SO_4$) [4]. During the cooling process, these salts are absorbed on the surface of fly ash. Fly ash from coal-fired power plants is a useful secondary raw material partially re-used in building industry as a pozzolan material [5]. It is commonly used as a source of SiO_2 instead of silica sand in lightweight building material - autoclaved aerated concrete (AAC) [6]. AAC has outstanding properties such as fire resistance, thermal and sound insulation and excellent manipulability. In past decades the fly ash AAC attracts attention thanks to the global interest in re-using of energetic by-products. Ammonia captured in the fly ash in the form of soluble salts could be released as an ammonium gas or could be leached to water. Under a highly alkaline environment, which is created when water is added to a system containing cement and lime (CaO), the ammonium salts react with water and liberate ammonium gas as shown in the equation (1) [3]. The release of toxic gas during the production deteriorates the working place and could cause serious injuries. Dilmore and Neufeld [7] used the ammoniated fly ash after SCR (selective catalytic reduction) for the production of AAC. They conclude that using ammoniated fly ash as a source of SiO_2 does not affect the mechanical properties of prepared AAC samples.

$$NH_4^+ + OH^- \leftrightarrow NH_3 \uparrow + H_2O \qquad (1)$$

Tannins are a class of natural polyphenolic biomolecules that bind to and precipitate proteins and other organic compounds [8]. They are found in a variety of plants, mainly in wood and fruits. Tannin compounds could be divided into two main groups differing in chemical composition. Hydrolysable tannins are a mixture of simple phenols (gallic and ellagic acid) and could be hydrolysed by weak acids or bases to carbohydrates and phenolic acids. Condensed tannins are polymers formed by the condensation of flavans. The industrially produced tannins are mainly extracted from plans by hot water or other solvents. Tannins react with NH_3 in water to form insoluble compounds. Hashida et al. [9] concluded that the amination of gallic acid and related polyphenols with pyrogallol nucleus is regioselective and proceeds only in the presence of O_2 (figure 1). Amination of condensed tannins was studied by Braghiroli et al. [10]. The reaction with ammonia leads to multi-amination of phenolic hydroxy groups and cross-linking through the nitrogen bridges, which causes the gelation of the tannin/ammonia mixture in water. The possible reaction products are shown in figure 2. The amination of organic compounds in wood is used for a process called ammonia fuming, during which the colour of the wood changes. Miklečić et al. [11] concluded that the amount of water extractable tannin decreases rapidly after the ammonia fuming, while the amount of bonded nitrogen increases.

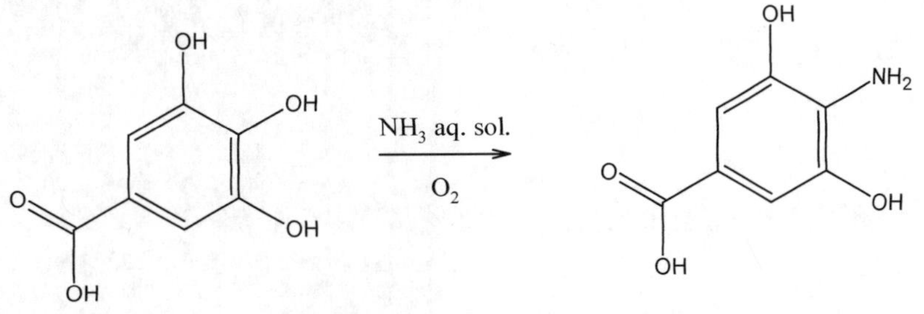

Figure 1. Reaction scheme of amination of gallic acid [9].

As was mentioned above, during the preparation of AAC the ammonium gas is released from the SNCR fly ash which deteriorates the working space. The possibility of binding the ammonia into the insoluble compound using natural tannins is examined. For this purpose, the extract from chestnut wood and pure tannic acid was used. The influence of these additives on the production of AAC was observed.

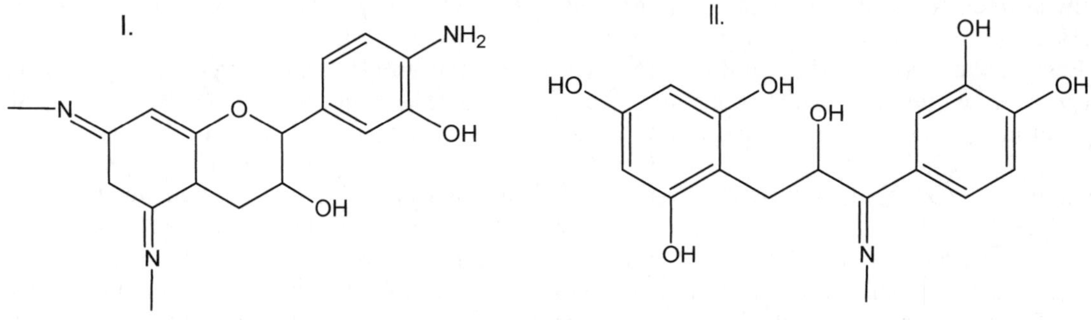

Figure 2. Two possible products of amination of condensed tannins [10].

2. Experimental

2.1. Starting materials

The coal fly ash from Počerady power plant was used as a source of SiO_2. The phase composition was determined by X-ray diffractometry analysis; the diffractogram is shown in figure 3. The main detected crystalline phases were mullite ($Al_6Si_2O_{13}$), quartz (SiO_2), hematite (Fe_2O_3) and magnetite (Fe_3O_4). The morphology of fly ash particles was observed by scanning electron microscope (figure 4). The fly ash consists of spherical particles with diameter around 10 μm. The total ammonium nitrogen in a raw fly ash was determined to 48 mg of NH_3/kg of fly ash.

As tannin additives, two sources were used. The first was the commercial dietary product for animals called Farmatan (F). It is a natural chestnut wood extract and the producer declares that there are 74 grams of chestnut tannin in 100 grams of Farmatan, the rest is a sweetener. As the second tannin source, the tannic acid (T, pharmaceutical purity, Lach-Ner) was used. Tannins were dissolved in water and mixed with the dry fly ash. After 24 hours the mixture was dried in a laboratory dryer and stored. The amount of added tannins differs from 0 (refer as reference) to 2 wt. % to the amount of fly ash. Except for the tannin enriched fly ash, the starting mixture was mixed from Portland cement (CEM I), quicklime (CL 90), gypsum ($CaSO_4{\cdot}2H_2O$) and aluminium powder. Additionally, the tap water at a temperature of around 50°C was used.

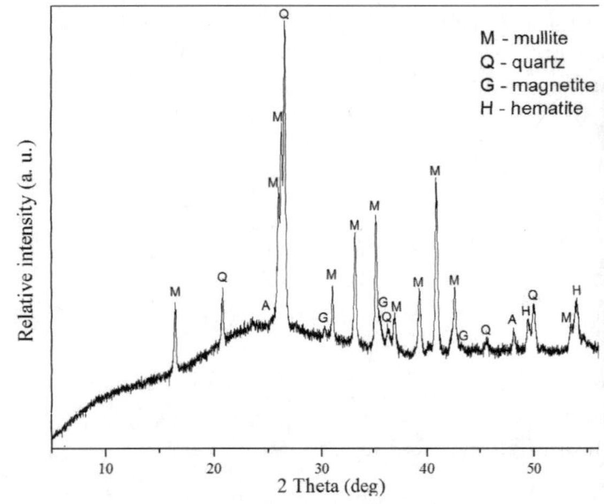

Figure 3. XRD pattern of raw fly ash.

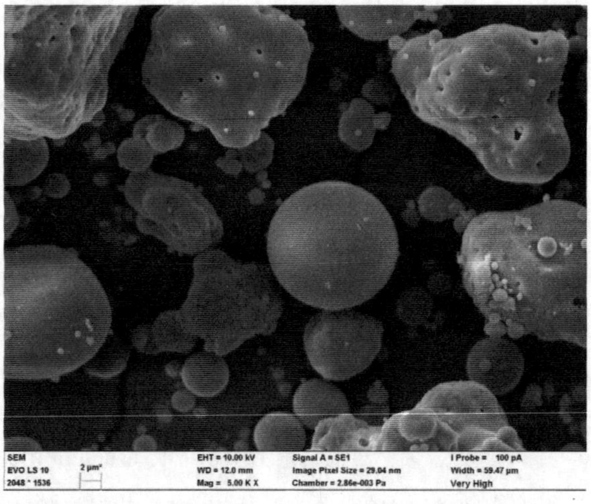

Figure 4. SEM image of raw fly ash.

2.2. Isoperibolic calorimetry

The influence of tannin additives on the hydration process was studied by isoperibolic calorimeter constructed at Faculty of Chemistry, Brno University of Technology. The construction of the calorimeter is described elsewhere [12]. The mixture for isoperibolic calorimetry was made from 64.6 wt. % fly ash, 25.5 wt. % quicklime, 9.9 wt. % Portland cement. The w/b (water-binder) ratio was set to 0.55. Immediately after the mixing, the mixture was placed to the sealed container and the measurement was started. The samples are labelled according to the type and amount of tannin additive added to fly ash, e.g. sample C_F0.5 corresponds to a sample prepared from fly ash with 0.5 wt. % of Farmatan.

2.3. AAC manufacturing and characterization of products

The mixture for manufacturing of AAC blocks was made from 57.3 wt. % fly ash, 25.4 wt. % quicklime, 11.6 wt. % Portland cement, 5.7 wt. % gypsum and a trace of aluminium powder acting as foaming agent. The w/b ratio was set to 0.6. The prepared mixture was poured into iron moulds, where it expanded to its final volume. The hardened samples were removed from moulds after 24 hours and dried at 40°C for 24 hours. The prepared blocks were cured in a laboratory autoclave to obtain their final properties. The conditions of the hydrothermal reaction were set to 190°C and 1.2 MPa. The reaction was stopped after 8 hours and the prepared AAC blocks were stored and examined. The samples are labelled according to the same principle as the samples for calorimetry, but instead of C_F0.5, it is A_F0.5 as an autoclaved sample.

Phase composition of autoclaved samples were examined by X-ray powder diffraction (XRD) using X-ray diffractometer Empyrean (Pananalytical) with Cu Kα radiation from 5 to 90° 2 theta and the step 0,013°. FTIR spectroscopic measurements were conducted on chosen samples using FTIR-ATR spectrometer iS50 from Thermo scientific. The total ammonium nitrogen (TAN) concentration in all autoclaved samples was determined according to an international standard ISO 7150-1 [13].

3. Results and discussion

The influence of the tannin additives to a cement-lime-fly ash system was examined by isoperibolic calorimetry. This method is suitable for a study of the evolution of heat during the hydration of silicate binders. The results are presented as a dependence of the temperature of the testing mixture placed in the sealed container on time (figure 5–6 a). The course of heat development during the hydration process was determined by integration of temperature curve and is shown in figure 5–6 b.

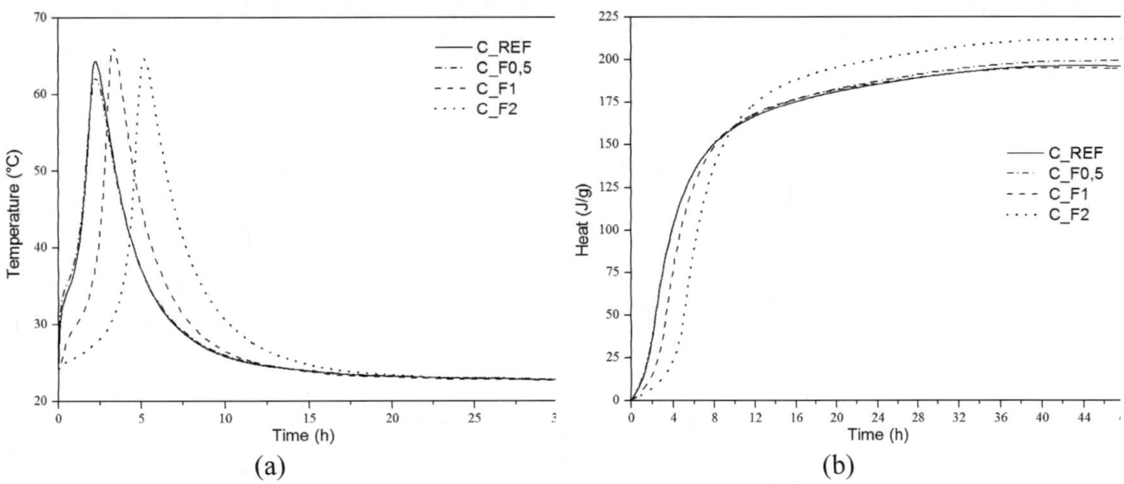

Figure 5. The results of isoperibolic calorimetry for samples prepared using Farmatan: (a) differential curves (b) integral curves.

IOP Conf. Series: Materials Science and Engineering **583** (2019) 012020 doi:10.1088/1757-899X/583/1/012020

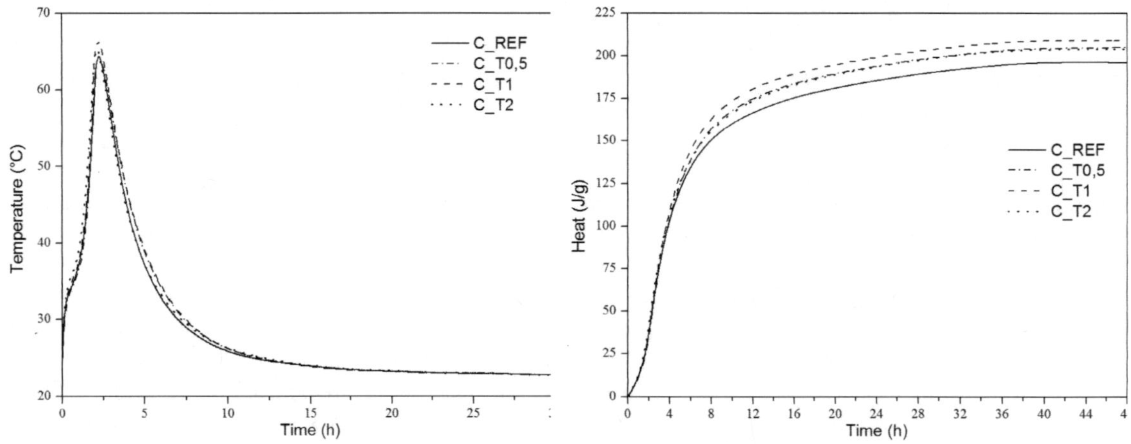

(a) (b)

Figure 6. The results of isoperibolic calorimetry for samples prepared using Tannic acid: (a) differential curves (b) integral curves.

An important advantage of calorimetry is the possibility to estimate the start of setting of the mixture. The results indicate that the addition of Farmatan in more than 1 wt. % delays the start of the setting. As was mentioned earlier, Farmatan consists of chestnut tannin and sugar. It is known, that sugars retard cement hydration [14]. After the addition of water, the main clinker minerals (C_3S and C_3A), started to dissolve and after the saturation of the system with Ca^{2+} and SiO_4^{4-} ions, the C–S–H phase and portlandite are formed. In the presence of sugars, the Ca^{2+} ions form calcium complexes with sugars, which are bound to the C_3S surface and thereby slow down the nucleation of C–S–H phases [15]. In the sample containing 2 wt. % of Farmatan, the setting was delayed for about 3 hours. The amount of evolved heat in Farmatan samples is comparable with the reference. Contrary, when using pure tannic acid as a tannin additive, the hydration is not influenced in all studied samples. Because the chestnut tannin in Farmatan belongs to a group of hydrolysable tannins [16] same as the tannic acid, it can be stated that the delay in hydration in Farmatan samples is caused only by sugars. Pehanich et al. [17] suggested that tannins could act as cement hydration retardants, but this problem was not sufficiently examined until now.

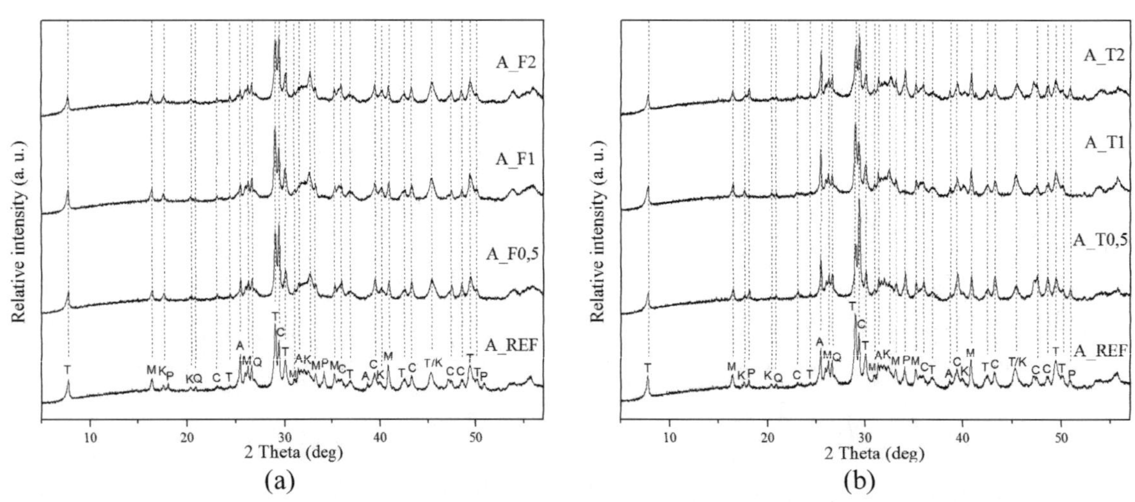

(a) (b)

Figure 7. XRD patterns of final AAC products with different tannin additive:(a) Farmatan (b) Tannic acid (T = Tobermorite, M = Mullite, K = Katoite, P = Portlandite, Q = Quartz, C = Calcite, A = Anhydrite).

The phase composition of autoclaved samples was determined by XRD analysis. The diffractograms of all samples are shown in figure 7. During the hydrothermal reaction mineral tobermorite ($Ca_5Si_6O_{17}\cdot 5H_2O$) is formed acting as a main binding phase in the AAC [6]. Besides tobermorite, katoite ($Ca_3Al_2(SiO_4)_{3-x}(OH)_{4x}$, x = 1.5-3.0) was formed. There is a trace of unreacted CaO in a form of portlandite ($Ca(OH)_2$) and calcite ($CaCO_3$). The unreacted phases from fly ash, mullite ($Al_6Si_2O_{13}$) and quartz (SiO_2), were detected in all samples. The presence of unreacted gypsum in the form of anhydrite ($CaSO_4$) was observed in all samples. Generally, this phase composition well agrees with the reported ones [6, 18]. There is no significant influence of the presence of tannin-based compounds to the final phase composition of prepared AAC samples.

Because the ammonium in fly ash is in the form of water-soluble salts, the amount of ammonium nitrogen in prepared AAC samples was measured. The results are presented in table 1. It is evident that there is no water extractable ammonium nitrogen in all samples, even in the reference sample without the tannin additives. This is probably caused by the fact that during the mixing phase of the starting mixture, the pH rise and the ammonium is released in the gas form, see equation (1). This problem was studied earlier [19], and the results are with good agreement with that obtained in this study. From the obtained results there is no possibility how to prove that the ammonium nitrogen reacts with tannin additives. It could be stated, that if the reaction proceeds the reaction products are not soluble in the water.

Table 1. The concentration of total ammonium nitrogen in autoclaved samples.

Sample	A_REF	A_F0.5	A_F1	A_F2	A_T0.5	A_T1	A_T2
C_{TAN} [mg of NH_3/kg of fly ash]	0.66	0.28	0.27	0.35	0.39	0.52	0.49

The presence of different functional groups could be studied by Infrared microscopy (IR). Five samples were chosen for IR analysis, two samples after the calorimetry measurement and two samples of final AAC, all containing 2 wt. % of tannin additive and a calorimetry reference. The results of the IR analysis are shown in figure 8. The stretching vibrations of O–H groups in H_2O or possibly hydroxyls are characterized as broadband centered at around 3400 cm^{-1} mainly in samples after calorimetry. The band at 1630 cm^{-1} corresponds to the bending vibration of molecular water, probably in C–(A)–S–H structures. The sharp band at 3640 cm^{-1} can be attributed to less strongly hydrogen-bonded water molecules in an interlayer of $Ca(OH)_2$, which is consistent with the results obtained by Yu et al. [20]. The intensity of this band decreases in autoclaved samples because during the hydrothermal reaction the excess portlandite reacts with SiO_4^{4-} to form tobermorite and katoite. The bands at 1400–1500 cm^{-1} and 870 cm^{-1} can be attributed to asymmetric stretching of CO_3^{2-}. The presence of carbonates in autoclaved samples was detected by XRD, while in the samples after calorimetry, the unreacted portlandite probably undergo contamination of air CO_2. The most intense bands occurring in the range 1100–900 cm^{-1} are attributed to asymmetrical stretching vibrations of silicate tetrahedra (SiO_4). Yu et al. [20] assigned the different stretching bands of Si–O in calcium silicate hydrates. The band at 960 cm^{-1} corresponds to Si–O stretching of Q^2 sites. The intensity increase with the ordering of structure of C–S–H compounds. The broadening of this band towards higher frequency is due to the presence of Q^3 and Q^4 sites in SiO_2 gel. The IR results of autoclaved samples are in good agreement with results obtained by Wu et al. [21], who performed an IR spectroscopy on AAC samples. The broadband centered at 1090 cm^{-1} could be also attributed to stretching vibrations of C–N as $R\text{-}NH_2$, but it could not be differentiated to which kind of stretching this band corresponds.

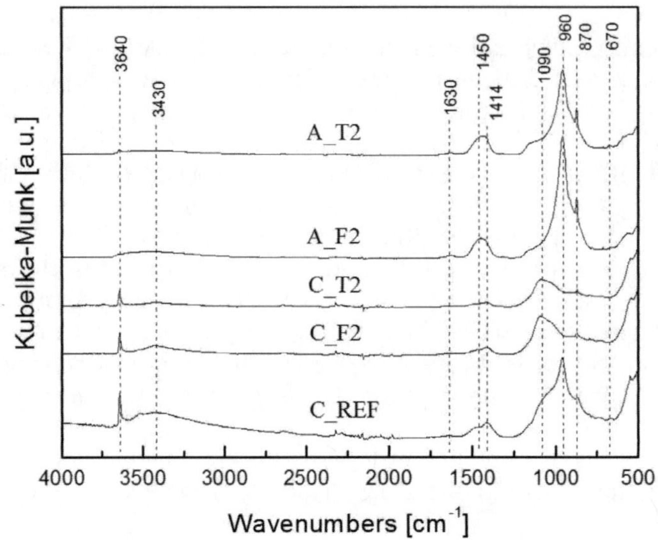

Figure 8. IR spectra of chosen samples.

4. Conclusions

The influence of the addition of tannin-based additives to cement-lime-SNCR fly ash system was studied. Tannins react with ammonium ions in water to form insoluble compounds. Fly ash after SNCR is enriched by NH_3 in a form of water-soluble salts (NH_4HSO_4 and $(NH_4)_2SO_4$). During the mixing phase, when the pH of the mixture rises, ammonium is released in a form of gas and deteriorates the working space.

Tannin additives were added to a starting mixture in a range from 0.5 to 2 wt. % to the amount of ammoniated fly ash. Two tannin additives were used; the cost-effective dietary product for animals containing chestnut tannin and sugar called Farmatan and pure tannic acid. Both additives belong to a group of hydrolysable tannins. The presence of Farmatan in a cement mixture causes delays in the start of the setting, probably thanks to obtained sugars in an additive. Sugars act as retarding agents. The pure tannic acid does not affect the hydration of the cement-lime-fly ash system. From this reason even if Farmatan is a low-cost material, it is not appropriate to use it as an ammonium binding additive in a cement-lime system. The effect of tannin additives to a final phase composition of the autoclaved aerated concrete using fly ash as a SiO_2 source was not observed. The final composition of all prepared AAC samples was much alike. Unfortunately, the creation of insoluble ammonium compounds with tannins was not confirmed, because no suitable analytical method to prove the presence of $R\text{-}NH_x$ was found.

Acknowledgements

This study was supported by the project of specific research FAST/FCH-J-18-5411 with the financial support of Brno University of Technology.

References

[1] Javed M T, Irfan N and Gibbs B M 2007 *J. Environ. Manage.* **83**(3) pp 251–89

[2] Randall D D, Fry C R and Schaffner K S 2016 Selective Noncatalytic Reduction *The EPA Air Pollution Control Cost Manual* (Research Triangle Park: U.S. Environmental Protection Agency) pp 3–70

[3] Laskawiec K, Gebarowski P and Kramek-Romanowska K 2016 *Acta Energetica* **29**(4) pp 58–63

[4] Wang H, Ban H, Golden D and Ladwig K 2002 *Prepr. Pap. Am. Soc. Div. Fuel Chem.* **47** pp 837–8

[5] Ahmaruzzaman M 2010 *Prog. Energy Combust. Sci.* **36** pp 327–63
[6] Qu X and Zhao X 2017 *Constr. Build. Mater.* **135** pp 505–16
[7] Dilmore R M and Neufeld R D 2001 *J. Energy Eng.* **127** pp 37–50
[8] Kalia S and Avérous L 2016 *Biodegradable and Biobased Polymers for Environmental and Biomedical applications* (Hoboken, New Jersey: John Wiley & Sons; Salem, Massachusetts: Scrivener Publishing) pp 96–149
[9] Hashida K, Makino R and Ohara S 2009 *Holzforschung* **63**(3) pp 319–26
[10] Braghiroli F, Fierro V, Pizzi A, Rode K, Radke W, Delmotte L, Parmentier J and Celzard A 2013 *Ind. Crops Prod.* **44** pp 330–5
[11] Mikleĉić J, Ŝpanić N and Jirous-Rajkovic V 2012 *Bioresources* **7**(3) pp 3767–78
[12] Brandštetr J, Polcer J, Krátký J, Holešinský R and Havlica J 2001 *Cem. Concr. Res.* **31**(6) pp 941–7
[13] ISO 7150-1:1984*Water quality – determination of ammonium – Part 1: Manual spectrometric method* (International Organization for Standardization)
[14] Thomas N L and Birchall J D 1983 *Cem. Concr. Res.* **13** pp 830–42
[15] Bishop M and Barron A G 2006 *Ind. Eng. Chem. Res.* **45** pp 7042–9
[16] Tang H R, Hancock R A and Covington A D 1992 *Basic Life Sci.* **59** pp 221–43
[17] Pehanich J L, Blankehorn and Silsbee M R 2004 *Cem. Concr. Res.* **34**(1) pp 59–65
[18] Matsui K, Kikuma J, Tsunashima M, Ishikawa T, Matsuno S, Ogawa A and Sato M 2011 *Cem. Concr. Res.* **41** pp 510–9
[19] Galvánková L, Lédl M, Opravil T, Beranová D and Tkacz J 2018 *Solid State Phenomena* **276** pp 167–72
[20] Yu P, Kirkpatrick R J, Poe B, McMillam P F and Cong X 1999 *J. Am. Ceram. Soc.* **82** pp 742–8
[21] Wu Q, Guang J, Li S, Wang S, Zou X, Zhu H and Zhang Ch 2018 *J. Mater. Civ. Eng.* **30**(7) 04018134

Substitution of aeration additive with recycled rubber and its effect on compressive strength and modulus of elasticity

R Chylik[1], T Trtik[1], J Fladr[1] and K Seps[1]

[1] Faculty of Civil Engineering CTU Prague, Thákurova 7, 166 29 Praha 6, Czech Republic

Email: roman.chylik@fsv.cvut.cz

Abstract. The article follows on previous research by authors dealing with the properties of concrete with the addition of recycled rubber. This paper describes the substitution of aeration additives with rubber recycled material and the effect of this substitution on strength and modulus of elasticity. For concrete mixes containing 40 kg/m^3 of recycled rubber, the air content of the fresh concrete was measured. Concrete with the same amount of fresh air was made using an aeration additive. Compressive strength and modulus of elasticity were compared.

1. Introduction

Humans consistently exhaust the planet's natural resources and produce a large amount of waste that can not be recycled. The constantly growing number of manufactured tires (1 billion in 2016, expected 1.2 billion in 2030) and permanently increasing number of worn tires in landfills created an issue of reuse of the worn tires. Fortunately, used tires have found several ways of reuses.

Recycling is carried out in several ways. Worn tires can be incinerated at high temperatures or crushed to rubber powder. The difference between shredding at normal temperature and shredding at temperature below zero (cryogenic shredding) is in quality of rubber powder. Cryogenic shredding leads to higher grinding fineness, but it is more expensive.

One of the ways how to use a shredded tire is to add rubber powder to concrete. On this subject, several studies were created. The results of the tests showed that an additive of rubber powder could have a positive influence on deformability [1] and durability [2]. Frankowski [3] studied the influence of crumb rubber on reinforced concrete for structures. The results of his research showed that the addition of rubber powder in reinforced concrete improved the resistance to cracking, shock wave absorption and acoustic wave damping and lowered the weight and thermal conductivity of the material.

This paper is focused on the substitution of an aeration additive by recycled crumb rubber. Different concrete mixes with defined amount of air in fresh concrete were performed. The values of concrete with an additive of rubber powder (crumb rubber concrete, abbreviated as CRC) were compared with concrete mixture with an aeration additive (control mix). The effect of substitution of an aeration additive by recycled rubber powder on compressive strength and modulus of elasticity has been investigated.

Protecting the environment, using waste material and conserve natural resources. This is a task for our and future generation. Especially, in civil engineering, there are many opportunities to save natural resources and reuse of waste material of other industries. The use of secondary raw materials can also positively affect the mechanical properties of concrete.

This article explores the possibility of using recycled tires as a replacement for aeration additives and the effect of this substitution on the strength o and the modulus of elasticity of concrete.

2. Mixture composition

According to previous research by the authors, the addition of rubber powder has a positive effect on freezing and thawing durability [4].

At first, the content of the air in fresh concrete was measured for mixes contain 40 kg/m^3 of recycled rubber powder. Normal strength concrete with the defined amount of air in fresh concrete as FCRC was made. Values of air content are presented in table 1.

The tests were carried out in accordance with a valid European Committee for Standardization (CEN) 2009 Testing fresh concrete; Air content – Pressure methods Part 7 (EN 12350-7) [7]. The final compounds of concrete are shown in table 2.

Table 1. Air content.

Concrete mix	Air content (%)
CCRC	2.7
FCRC	7.2
AA	7.2

FCRC – crumb rubber concrete with finely (0-1 mm) crushed rubber powder
CCRC – crumb rubber concrete with coarsely (1-3 mm) crushed rubber powder
AA – Normal strength concrete with an aeration additive

Table 2. Mix design.

	Quantity in FCRC (kg/m^3)	Quantity in AA (kg/m^3)
Cement CEM I 42,5 R	410.0	410.0
Water	200.0	200.0
Fine aggregate 0-4 mm	840.0	840.0
Coarse aggregate 4-8 mm	340.0	340.0
Coarse aggregate 8-16 mm	620.0	620.0
Shredded rubber	40.0	0.0
Water reducer	0.8	0.8
Aeration additives	0.0	2.9

3. Results

3.1. Compressive strength

The compressive strength of concrete specimens was determined according to European Committee for Standardization (CEN) 2009 Testing hardened concrete; Compressive strength of test specimens Part 3 (EN 12390) [8] after 28 days. Measurements were performed on cylindrical specimens with 150 mm diameter and 300 mm height. Measured values are shown in table 3.

IOP Conf. Series: Materials Science and Engineering **583** (2019) 012021 doi:10.1088/1757-899X/583/1/012021

Table 3. Compressive strength (MPa).

	Density (kg/m^3)	Average force (kN)	Average strength (MPa)
FCRC 1	2270	617.8	34.3
FCRC 2	2262	595.3	33.7
FCRC 3	2258	595.0	33.6
CCRC 1	2210	601.3	33.2
CCRC 2	2267	594.7	33.7
CCRC 3	2245	607.7	34.1
AA 1	2208	552.7	31.3
AA 2	2241	563.8	31.9
AA 3	2226	547.5	31.0

Figure 1. The test configuration of compressive strength.

Figure 2. Selected specimen after test of compressive strength.

3.2. Modulus of elasticity

The modulus of elasticity is the ratio of stress and deformation. Materials with a higher modulus of elasticity require higher stresses to achieve the same deformation. In conventional construction, a higher modulus of elasticity is usually advantageous, but in particular cases material with a lower modulus of elasticity may be advantageous. These specific cases include load by shock-impacted or explosive. The explosive load of construction is treacherous, except that the pressure waves are dangerous as well as flying pieces of concrete released from the structure.

There is a relationship between the strength and the modulus of elasticity of concrete. If we produce a structural element that is capable of withstanding of load by explosion from high-strength concrete with a higher modulus of elasticity, the overall damage rate will be smaller, but the speed of flying parts of concrete from the structure will be greater. This could be dangerous as well as the explosion.

The solution to this problem could be to use a material with sufficient strength but a lower value of the modulus of elasticity. The modulus of elasticity of the concrete depends on the composition of the recipe. If we want to reduce the modulus of elasticity of concrete, we need to add a compound with low modulus of elasticity to the concrete mixture.

The modulus of elasticity of concrete specimens was determined according to [9]. Measurements were performed on cylindrical specimens with 150 mm diameter and 300 mm height. Measured values are shown in tables 4.

Table 4. Modulus of elasticity (GPa).

	Modulus of elasticity (GPa)
FCRC 1	26.6
FCRC 2	36.8
FCRC 3	30.1
CCRC 1	33.0
CCRC 2	32.8
CCRC 3	30.4
AA 1	30.1
AA 2	27.6
AA3	28.5

4. Discussion

Nowadays, there are quite a number of different studies dealing with the addition of recycled rubber in various forms to the concrete mix.

For example, Zheng et al. [5] investigated the effect of replacing coarse aggregate with crushed rubber. In their study, among other things, they confirmed the indirect relationship between the amount of rubber added and the compressive strength and value of the modulus of elasticity.

Gerges et al. [6] observed the mechanical and dynamic properties of concrete with the addition of a variable amount of crushed rubber. They confirmed the assumption of a negative effect of rubber recycled material on strength, but also a favorable effect of rubber recycled material on the resistance of the material against dynamic loading.

The findings of the above and other studies are consistent with the author's review [4]. The aim of this paper was to understand whether the negative effect of adding rubber recycled material on strength and modulus of elasticity is primarily due to the recycled rubber and its low strength, or is due to increased air content in fresh concrete.

In the case of concrete containing coarsely crushed rubber recycled material (CCRC), the reduction in strength and modulus of elasticity is exclusively attributable to rubber recycled as the difference in fresh air content from the reference mixture is 0.6% and coarse rubber recycled can be considered as partial aggregate replacement.

Concrete containing finely ground rubber recycled material had comparable strength and modulus of elasticity to concrete with coarse rubber recycled material, but the air content was significantly higher than CCRC and REF. The reduction in strength and modulus of elasticity can be attributed to the increased of air content of fresh concrete since on the surface of rubber powder is more air and pores are situated around rubber powder.

The AA mixture tests which contained the same amount of fresh air as the FCRC mixture confirmed the relative compliance of the compressive strength. The modulus of elasticity was slightly lower than the FCRC and CCRC mixture.

5. Conclusion

Several tests of concrete with substitution of aerating additive by crushed rubber powder were performed. At first, concretes with the admixture of 40 kg/m^3 of finely (FCRC) and coarsely (CCRC) crushed rubber were performed and the air content in fresh concrete was measured. Thereafter, concrete with the same air content as FCRC was made using aeration additive (AA). On these concretes compressive strength and modulus of elasticity were investigated. The results of tests showed the following:

- Finely crushed rubber powder is more efficient than coarsely crushed rubber powder as aerating additives.
- Using finely crushed rubber powder as substitution of aerating additive has a low positive influence on compressive strength and modulus of elasticity.
- Concrete made by aerating additive is more homogenous compared with finely and coarsely crumb rubber concrete.
- Shredded rubber powder can be used as a full replacement aeration additive without a negative influence on the strength and modulus of elasticity.

Acknowledgements

This paper was prepared thanks to the support of the Center for Effective and Sustainable Transport Infrastructure within the Center for Competence of the Technology Agency of the Czech Republic (TAČR), project number TE01020168 and financial support of SGS18 / 115 / OHK1 / 2T / 11.

References

[1] Topçu U B 1995 The Properties of Rubberized Concrete *Cement and Concrete Research* **25** p 304

[2] Benazzouk A and Queneudec M 2002 Durability of Cement- Rubber Composites under Freeze Thaw Cycles *Pro- ceedings of International congress of Sustainable Concrete Construction* **5** p 355

[3] Frankowski R. 1995 *Rubber-Crumb-Reinforced Cement Concrete* U.S. Patent 5391226

[4] Chylík R, Trtík T, Fládr J and Bílý P 2017 Mechanical properties and durability of crumb rubber concrete *IOP Conference Series*: *Materials Science and Engineering* **236** p 236

[5] Zheng L, Huo X and Yuan Y 2008 Strength, Modulus of Elasticity, and Brittleness Index of Rubberized Concrete *Journal of Materials in Civil Engineering* **20** p 692

[6] Gerges N, Camille A and Fawaz S 2018 Rubber concrete: Mechanical and dynamical properties *Case Studies in Construction Materials* **9**

[7] EN 12350-7:2009 *Testing fresh concrete, Part 7 - Air content - Pressure methods* (European Committee for Standardization)

[8] EN 12390:2009 2009 *Testing hardened concrete, Part 3 - Compressive strength of test specimens* (European Committee for Standardization)

[9] ISO 1920-10 *Part 10 - Determination of static modulus of elasticity in compression*

IOP Conf. Series: Materials Science and Engineering **583** (2019) 012022 doi:10.1088/1757-899X/583/1/012022

Improving the dispersion of silica fume in cement pastes and mortars

M Janca[1], P Siler[1], T Opravil[1] and J Kotrla[1]

[1] Brno University of Technology, Faculty of Chemistry, Materials Research Centre, Purkyňova 464/118, Brno 612 00, Czech Republic

Email: xcjancam@fch.vut.cz

Abstract. There are many possibilities how to improve concrete properties such as compressive strength, flexural strength, durability, and others. The change in the composition of ordinary Portland concrete (OPC) is one possible way. This change may result in the formation of concrete with added properties known as high-performance concrete (HPC). Traditional methods used to receive high-performance concrete design and manufacture emphasize the removal of coarse aggregates. This is achieved by using the specially sorted fine aggregates at relatively low doses along with super-plasticizers and silica fume. Problem with fine reactive aggregates is their agglomeration which causes reduction of the reactive surface. This work is focused on evaluation of the ultrasound technology for the dispersion of silica fume. For comparison, different cement mixtures were prepared to study the effects of the treatment of silica fume prior to its use in HPC applications.

1. Introduction

Silica fume (microsilica) is well known supplementary cementitious material which is used in cement mixtures to affect microstructure with focus to interfacial zone between binder – aggregate. Silica fume is the byproduct of producing silicon metal or ferrosilicon alloy. Silicon vapor released from the furnace oxidizes and forms small spherical particles, which are collected in silos. Besides silica fume which is an industrial by-product, natural Pozzolans are known. These natural Pozzolans are mostly of volcanic origin.

A most common form of silica fume used in current concrete practice is in dry powder form, the alternative is slurried silica fume. Dry powder usually consists of agglomerates which can be a size up to several millimeters and they are generally not dispersable into individual silica fume spheres under conventional concrete mixing. Substantial content of agglomerates almost always remain in the concrete. The size of undispersed agglomerates remaining in the cement matrix after mixing can be bigger than the particles of used cement. This is very limiting for the potential benefits regarding fine particle filler effect [1].

Silica fume is a high surface material. The surface is usually measured by absorption of nitrogen and results show that the surface area of silica fume is around 20 000 m^2/kg. For example values of the surface area of Portland cement measured by the absorption of nitrogen is around 1 500 m^2/kg. Together with the high values of the surface area, the amorphous character of SiO_2 in the silica fume which gives high potential to react in the Pozzolanic reactions is very advantageous property. Surface area is very important and can be increased by precise deagglomeration of the raw microsilica powder [2].

The reaction between the particles of amorphous SiO_2, portlandite ($Ca(OH)_2$) and water create products similar to hydration of Portland cement, well known as the calcium – silicate – hydrate (C – S – H) gel. Crystals of portlandite are consumed by this reaction. Residual portlandite can have a negative influence on the mechanical properties due to its low compressive strength. Pozzolan reaction is shown in equation (1). Its rate is slower then hydration of cement minerals and secondary C – S – H* gel makes structure much more compact [3].

At early stages (within a few hours of hydration), the extra surface of the pozzolan results in acceleration of the overall hydration kinetics by enhanced heterogeneous nucleation of C – S – H gel [4]. This acceleration effect is typically termed as the filler effect, in the filler term material is described as chemically inert material [5],[6]. Formation of secondary pozzolanic C – S – H* gel, results in improvement of solid to solid state connectivity in the microstructure of the matrices, lead to improvement in materials properties, such as higher compressive strength and improved durability [7].

$$\text{puzzolan} + CH + H \rightarrow C\text{–}S\text{–}H^{*} \tag{1}$$

* secondary C-S-H gel

Elimination of agglomerates can be accomplished by many ways such as grinding or mixing. Both of these methods have physical limitations of deagglomeration and can add impurities because the sample is in direct contact with the mill. On the other hand, ultrasonication of liquid suspensions of powders can provide different possibilities for agglomeration reduction. Because there is no direct contact between the sample particles and ultrasonic source, contamination of the product is minimal after the ultrasonication. Mechanism of deagglomeration by using sonication is described by the interaction of agglomerates with collapsing cavities or with other particles [8].

During the ultrasonication, agglomerates interact with cavities which are caused by the propagation of mechanical waves in the liquid. The sound waves that propagate into the liquid media result in alternating high-pressure (compression) and low-pressure (rarefaction) cycles. The individual particles can be held together by different attraction forces on physical or chemical nature. Ultrasonic cavitation in liquids causes high-speed liquid jets. This jets press liquid at high pressure between the particles and separate them from each other [9].

There is very strong interest in the reasonable use of secondary materials. According to this fact, it is necessary to utilize the highest potential properties of each component in the concrete. Deagglomeration and increasing silica fume surface area among possible ways how to improve the mechanical properties of the concrete. For this reason ultrasonication technology was chosen, which can be easily introduced into industry and result in better function of the silica fume in concrete production.

2. Materials and methods

Silica fume ELKEM Microsilica 920E and Portland cement CEM I 42,5 R plant Mokrá were used for the mixtures. For improving rheological properties superplasticizer MasterGlenium ACE 446 which allowed the low water-binder ratio was used. Two different aggregates for mortars, normalized sand PG II and calcinated bauxite were used. To achieve a similar particle size distribution of aggregates bauxite was sieved on 0.75 mm and 1.25 mm mesh sieves. The fraction between the sieves was analyzed by laser analysis of particle size (HELOS KR Sympatec) and compared with the PG II (middle sand according to ČSN EN 196–1). Figure 1 and Figure 2 show the particle size distribution of the aggregates. Figure 3 and Figure 4 show the particle size distribution of Portland cement and undispersed silica fume.

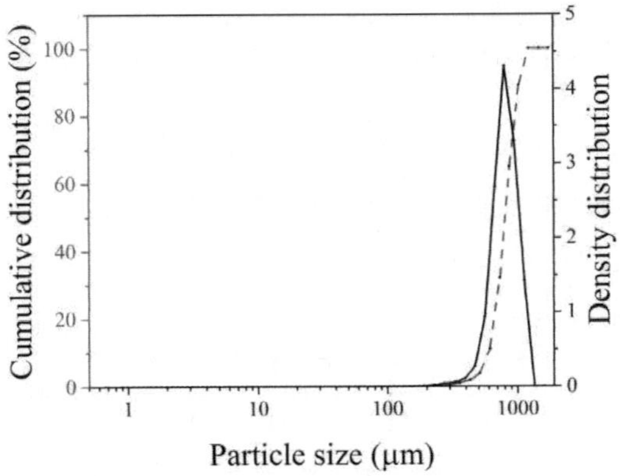

Figure 1. The particle size of PG II.

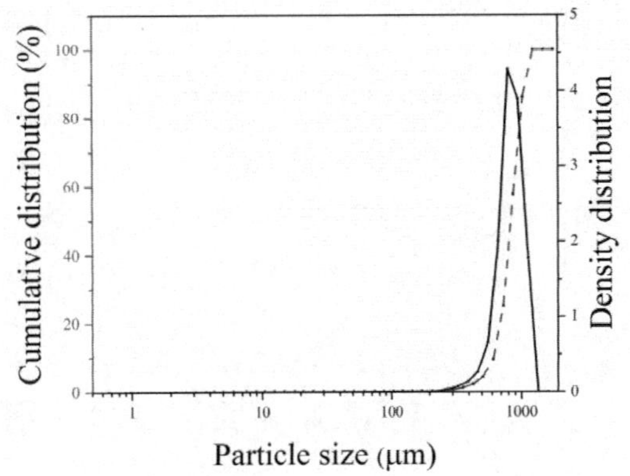

Figure 2. The particle size of bauxite.

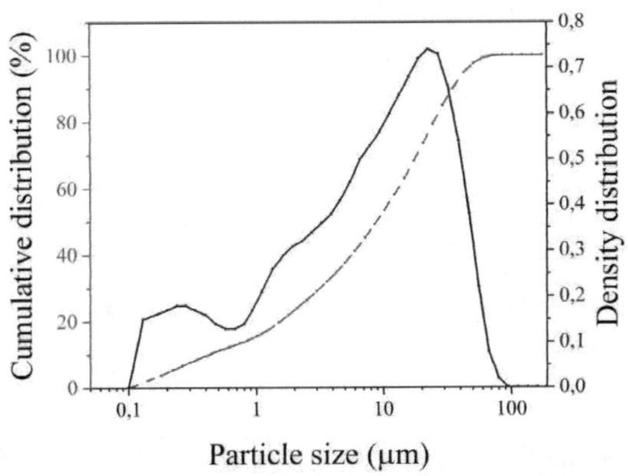

Figure 3. The particle size of CEM I 42.5 R.

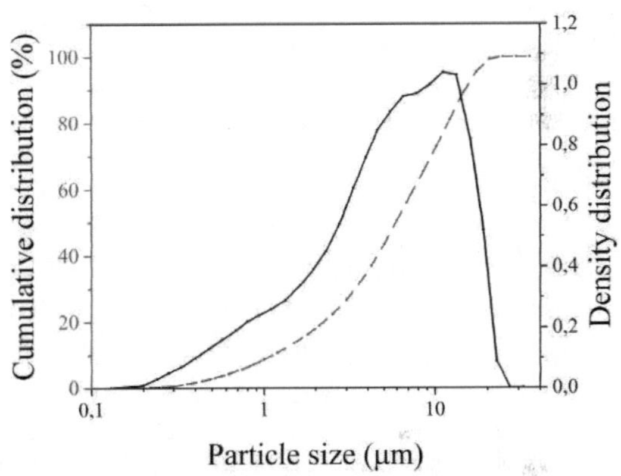

Figure 4. The particle size of raw silica fume.

Table 1. Particle size distribution.

Sample (μm)	D_{50} (μm)	D_{90} (μm)	Modus (μm)
CEM I 42,5 R	8.91	35.18	22.51
Silica fume	5.92	14.95	11.7
PG II	806.58	1047.06	796.93
Bauxite	837.05	1084.26	796.93

Table 1 shows the characteristical values of the measured materials. Modus value of the aggregates is the same and D_{50} and D_{90} are very similar. The particle size of the binder materials (silica fume and cement) are different. Silica fume before using ultrasonication contains two times smaller particles and it can provide the effect of the filler in the samples.

The composition of prepared samples is shown in Table 2. All samples had a similar W/B (water/binder) ratio 0.3 and dose of the plasticizer was counted as 1.2% of the binder components (cement + puzzolan). Cement pastes with number 1 contain no aggregates other samples contain 50% aggregates counted to all other components. All mixtures were prepared in three different ultrasonication times 0, 5 and 35 minutes. Mixing of the components was performed according to valid standard ČSN EN 196-1.

For the mixtures, without ultrasonication, the superplasticizer was dissolved in water and added to the cement and silica fume in the mixer. Producing samples of mortars the aggregates were added to this mixture after 30 seconds of mixing. For samples where the sonication was used, the silica fume was mixed with the water and superplasticizer. This suspension was then ultrasonified for 5 and 35 minutes. After the ultrasonication, it was added to the mixer with cement and in the case of mortars, the aggregates were added later on, after 30 seconds od mixing.

For structure observation the scanning electron microscope Zeiss EVO LS 10 was used. Samples were covered with golden/ platinum layer for better conduction of electron. Conditions of the measurement accelerating voltage 15 kV and current to 100 pA were set.

Table 2. The composition of the prepared samples.

Sample	1	1U	1UU	2_Kp	2_KpU	2_KpUU	3_Kb	3_KbU	3_KbUU
CEM I 42.5 R [g]	3672	3672	3672	1836	1836	1836	1836	1836	1836
Silica fume [g]	324	324	324	324	324	324	324	324	324
PG II [g]	0	0	0	2160	2160	2160	0	0	0
Bauxite [g]	0	0	0	0	0	0	2160	2160	2160
H_2O [g]	1296	1296	1296	648	648	648	648	648	648
W/B	0.3	0.3	0.3	0.3	0.3	0.3	0.3	0.3	0.3
Superplasticizer [g]	51.8	51.8	51.8	25.9	25.9	25.9	25.9	25.9	25.9
Sonification [min]	0	5	35	0	5	35	0	5	35

3. Results and discussion

For dispergation by sonification ultrasonic bath Bandelin DT 514 H was used. Table 3 shows the properties of the ultrasonic bath.

Table 3. Properties of Bandelin DT 514 H ultrasonic bath.

Internal dimensions [mm]	Container volume [l]	Ultrasonic power $P_{eff.}$ [W]
325 x 300 x 150	13.5	215

Consumption of the energy by the ultrasonic bath for the dispergation of silica fume was calculated according to equation (2) and (3). Time of dispergation was 0, 5 or 35 minutes and the values energy consumpted are 64.5 kJ for 5 minutes of sonification and 451.5 kJ for 35 minutes. When *W* is work, *P* stands for ultrasonic power and *t* for time of ultrasonication.

$$W = P \cdot t = 215 \cdot 300 = 64\,500 \text{ J} = \underline{\underline{64.5}} \text{ kJ} \tag{2}$$

$$W = P \cdot t = 215 \cdot 2\,100 = 451\,500 \text{ J} = \underline{\underline{451.5}} \text{ kJ} \tag{3}$$

Mechanical properties were tested on three 40 x 40 x 160 mm test specimens for each sample and day of storage. Data were measured by equipment for testing of building materials (DESTTEST, BETONSYSTEM). The average values of compressive strength are shown in Figure 5.

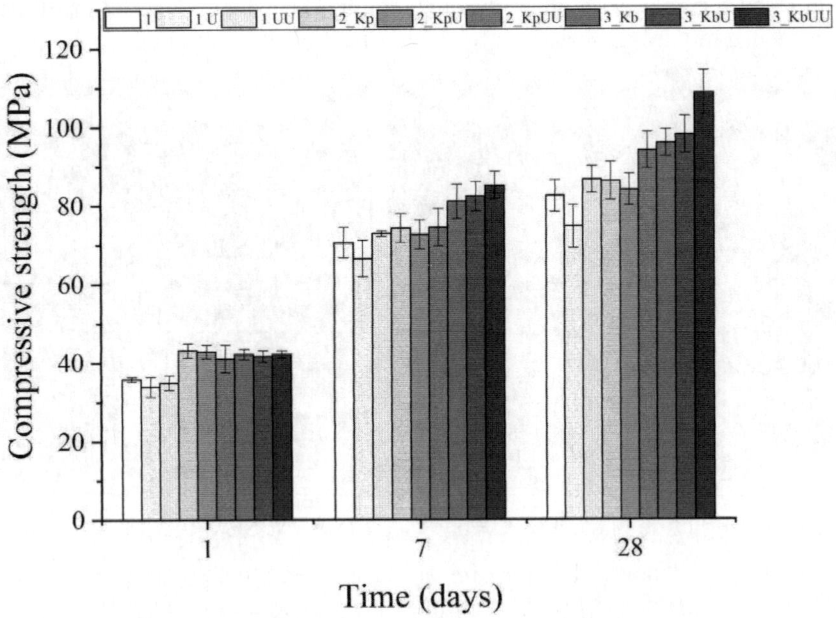

Figure 5. The compressive strength of prepared samples.

Generally, the compressive strength can be seen as the trend of gradual increase of strength in time. Each sample was prepared in three different ways with the same chemical composition. These three different ways of mixture preparation showed different values of the strength. In sample 1 and 2, the lowest compressive strength shows the mixtures with a short time of sonification. Highest compressive strength shows mixtures with 35 minutes of sonification and the lowest compressive strength shows mixtures with no sonification.

The mixture number three with calcinated bauxite as an aggregate showed different results. It was assumed that using aggregate with higher compressive strength will result in higher compressive strength of the mortar but the trend will be preserved. Sample 3 which contains calcinated bauxite showed different mechanical properties.

Sample batch 1 (paste) and 2 (mortar with PG II) exhibit the same trend in compressive strength increase. Lowest compressive strength was observed in the samples with 5 minutes of sonification. Samples with no sonification shown middle values of compressive strength and highest compressive strength were observed with 35 min of sonification. This trend is caused by different stages of sonification. In the first stage, the silica fume particles are hitting each other and making bigger agglomerates [10]. During this time it was necessary to mix the silica fume suspense and exhibit the sedimentation of silica fume agglomerates. After approximately 15 minutes of sonification, there is no need for the mixing of the suspension because silica fume particles are small enough, that buoyancy force is bigger than the gravity force. This is the reason why the 5 minutes of sonification has a negative influence on the compressive strength of the samples 1 and 2.

In sample batch 3, which contain bauxite as the aggregate it was observed, that with increasing time of dispergation, also the compressive strength was increased. This is probably caused by the different shape of the bauxite. Bauxite is agglomerate with higher mechanical properties than the quartz. Grinding of the bauxite is performed by dynamical forces during the crushing which results in the different shape of the bauxite particles [11].

This particle shape has a very big influence on the dispergation of the silica fume agglomerates because it acts as a sharp blade which can cut the agglomerates compared to the quartz spherical particles. Literature shows that there is a clear relationship between shape, texture, and grinding of aggregates and the voids content of aggregates. This fact supports the theory about the shape influence on the mixing and deagglomeration of the silica fume. The deagglomeration of the mixing, in case of bauxite, exceeded the deagglomeration by the ultrasonic disperser [12].

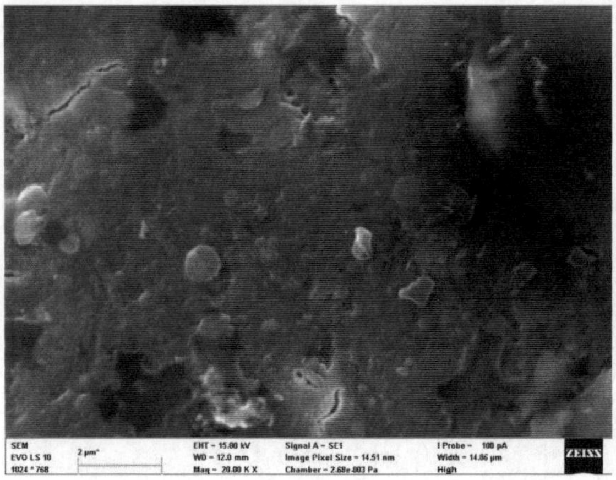

Figure 6. SEM picture of sample 3_Kb.

Figure 7. SEM picture of sample 3_KbUU.

Figure 6 and Figure 7 shows SEM pictures of the same mixtures with different time of sonification. Both pictures were taken at the magnification 20 000x. There is no significant proof of silica fume dispersion on figure 6. On the Figure 7 more homogenous distribution of silica fume particles can be observed.

4. Conclusion

This work focuses on the improvement of silica fume dispersion in cement pastes and mortars by using basic laboratory equipment such as ultrasonic disperser. Idea for using the ultrasonication was the low price of the equipment and its presence in every laboratory.

The main focus of this work was to show the effect of the sonification on the mechanical properties according to mixture composition. An assumption was that all mixtures will increase their mechanical properties (compressive strength) with increasing time of ultrasonication. From the measured data can be observed, that mixtures containing calcinated bauxite as an aggregate showed a different behavior than mixtures containing normalized sand and pastes.

The pastes and mortars with sand showed the lowest compressive strength after 5 minutes of ultrasonication. This results can be explained by the fact that during the first phase of ultrasonication agglomerates are binding with each other and form bigger agglomerates. In the second phase, the cavitation is overcoming the cohesive force of agglomerates and results in deagglomeration of silica fume particles.

In the mixtures with calcinated bauxite the different shape of calcined bauxite particles plays an important role. Normalized sand particles are spherical compared to plane particles of bauxite. These plane particles act as small blades which have a big impact on silica fume agglomeration during the mixing.

Continuous research will focus on the different phenomena of hydration curves between dispersed and undispersed silica fume particles and their puzzolan activity.

Acknowledgment

This outcome has been achieved with the financial support by the project: GA19-16646S "The elimination of the negative impact of zinc in Portland cement by accelerating concrete admixtures", with financial support from the Czech science foundation.

This work was supported by the project of specific research: FCH-S-18-5194 "Inovativní pokročilé materiály – stadium vlastností a možnosti aplikace" with the financial support of Brno University of Technology.

References

[1] Diamond S and Sahu S 2016 Densified silica fume: particle sizes and dispersion in concrete *Materials and Structures* **39** pp 849–859

[2] Aïtcin P C 2005 *Vysokohodnotný beton* (Praha: Pro Českou komoru autorizovaných inženýrů a techniků činných ve výstavbě).

[3] Collepardi M 2009 *Moderní beton* (Praha: Pro Českou komoru autorizovaných inženýrů a techniků činných ve výstavbě)

[4] Meng W, Lunkad P, Kumar A and Khayat K 2016 Influence of Silica Fume and Polycarboxylate Ether Dispersant on Hydration Mechanisms of Cement *The Journal of Physical Chemistry* **120** pp 26814–26823

[5] Kumar A, Oey T, Falzone G et al. 2017 The filler effect: The influence of filler content and type on the hydration rate of tricalcium silicate *Journal of the American Ceramic Society* **100** pp 3316–3328

[6] Oey T, Kumar A, Bullard J W, Neithalath N, Sant G and Scherer G 2013 The Filler Effect: The Influence of Filler Content and Surface Area on Cementitious Reaction Rates *Journal of the American Ceramic Society* **96** pp 1978–1990

[7] Qing Y, Zenan Z, Li S and Rongshen C 2006 A comparative study on the pozzolanic activity between nano-SiO2 and silica fume *Journal of Wuhan University of Technology-Mater. Sci. Ed* **21** pp 153–157

[8] Martínez-Velandia D, Payá J, Monzó J and Borrachero M V 2008 Granulometric activation of densified silica fume (CSF) by sonication *Advances in Cement Research* **20** pp 129–135

[9] Kusters K, Sotiris A, Pratsinis E, Thoma S G and Smith D M 1993 Ultrasonic fragmentation of agglomerate powders *Chemical Engineering Science* **48** pp 4119–4127

[10] Enomoto N, Maruyama S and Nakagawa Z 1997 Agglomeration of silica spheres under ultrasonication *Journal of Materials Research* **12** pp 1410–1415

[11] Durney T E and Meloy T P 1986 Particle shape effects due to crushing method and size *International Journal of Mineral Processing* **16** pp 109–123

[12] Molugaram K, Shanker J and Ramesh S A 2014 A Study on Influence of Shape of Aggregate on Strength and Quality of Concrete for Buildings and Pavements *Advanced Materials Research* pp 776–779

IOP Conf. Series: Materials Science and Engineering **583** (2019) 012023 doi:10.1088/1757-899X/583/1/012023

Effects of shrinkage-reducing admixtures on autogenous shrinkage in alkali-activated materials

J Kotrla[1], F Soukal[1], V Bilek[1] and M Alexa[2]

[1] Brno University of Technology, Faculty of Chemistry, Purkyňova 118, 612 00 Brno, Czech Republic

[2] Brno University of Technology, Faculty of Civil Engineering, Veveří 331/95, 602 00 Brno, Czech Republic

Email: xckotrla@fch.vut.cz

Abstract. This paper deals with possible solutions of chemical and autogenous shrinkage of alkali-activated blast furnace slag (AAS). Shrinkage is the main reason why AAS is not used in practice. One of the solutions how to reduce the shrinkage is the utilization of special chemical admixtures. In concrete technology, the use of chemical compounds based on hydrocarbon alcohols is well known, however the efficiency in AAS is still very poorly elucidated. Therefore, isopropanol was used in this study. The main aim of the work was to recognize and study the relations between chemical processes in the early stages of hydration. The effect of isopropanol on alkali activation was determined by calorimetric measurements and compared with the initial and final setting time of binder. It was found out that the presence of isopropanol additive had only insignificant effect on the setting time. On the other hand, the results indicated that an increasing amount of isopropanol deaccelerates the hydration process which was represented by smaller amount of primary CSH gel formation and delayed time of secondary binder phase formation.

1. Introduction

Current progress in development of new materials and technologies for reduction of waste forces cement industry to look for new building materials for example blast furnace slag (BFS). BFS is a secondary raw material that is produced in metallurgical industry and its bigger use in construction could reduce air pollution [1]. Although the cement industry uses a significant amount of BFS, still, enormous amount of BFS is not currently used. Therefore, its usage as alternative secondary binder is favorable action. Unfortunately, its utilization is limited by significant drying and autogenous shrinkage, which is substantially higher compared to OPC-based materials [1, 2]. Activated slag cement (ASC) which is a mixture of BFS and activator is one of the possibilities where to put BFS. Activator is a chemical compound that increases slag reactivity during hydration. Compared to OPC-based materials, ASC shows higher strength, lower porosity and heat evolution [3]. ASC exhibits higher chemical resistance, especially in acidic environment compared to OPC-based materials. This resistance seems to be based on the nature of the hydration products, rather than their porosity [4].

Concrete shrinkage is closely related to the increasing binder phase content. However, for some materials it is necessary to keep the content of binding phase, so it is favorable to use the shrinkage reducing admixture (SRA). The first mention about SRA reducing shrinkage of concrete during drying dates back in 1980's. However, the use of SRA in concrete may have a negative effect on the hydration of the binder phase. It is also important to think about long-term concrete properties as leaching of SRA can eliminate long-term reduction in shrinkage [5]. SRAs are surfactants that reduce the surface tension of the pore solution and water films that are formed on the solid surface during the drying of the cement material. In 1983, Sato et al. studied the effect of SRA on cement materials and concluded that the reduction in shrinkage due to SRA occurred due to the reduction of the surface tension of the pore solution, from which the capillary force theory of drying shrinkage was derived [6].

However, even a smaller addition of SRA strongly influences the ongoing hydration processes. Calorimetric measurement showed that the pre-induction period is changed and the delay of the main hydration peak is prolonged up to several days [7]. The study of Bílek and coauthors [7] also showed that with the increasing dose of SRA intensity of the initial peak is reduced and it is divided into two almost separated peaks during calorimetric measurement.

2. Materials and Methods

2.1. Materials and Sample Preparation

As a solid precursor for the reference binder for measurement blast furnace slag (BFS) with a volume diameter of 12 μm was used. The amount of amorphous phase, melilite, merwinite, β–C_2S and calcite was determined by X–ray diffraction. The chemical composition was determined by X–ray fluorescence using Ritvield method and it is presented in table 1.

Table 1. Chemical composition of the BFS determined by X-ray fluorescence analysis.

Chemical composition in wt. %									
CaO	SiO_2	MgO	Al_2O_3	SO_3	TiO_2	K_2O	MnO	Na_2O	Fe_2O_3
41.1	34.7	10.5	9.1	1.4	1.0	0.9	0.6	0.4	0.3

Water glass with silica modulus of 1.86 provided by Vodní sklo, a.s. was used as an activator. Isopropanol in amount of 0.10, 0.25, 0.50, 0.75 and 1.00 wt. % of slag was used to reduce AAS shrinkage and to modify the AAS hydration. Following procedure of preparing samples was used: Isopropanol was added to mixture of activator and water. After homogenization of the mixture the BFS was added. The measurements were made on ASS pastes with water-to-binder ratio (w/b) 0.35. After the mixing, material was cast into the molds and moist cured for 24 hours. After that, the prisms (40 mm × 40 mm × 160 mm for mechanical strength and 25 mm × 25 mm × 280 mm for shrinkage measurement) were removed from the molds and placed into autogenous storage. Samples were prepared and stored in mild conditions with temperatures 25°C.

2.2. Surface tension

Samples with weight of 20 g consisting of waterglass and water were placed to the six plastic vials. Isopropanol was added to each vial to give the resulting concentration of 0, 0.1, 0.25, 0.5, 0.75, 1.0 wt. %. The prepared solutions were homogenized for 3 minutes and the surface dynamic tension was measured on bubble-pressure tensiometer KSV Sigma 701.

2.3. Shrinkage measurement

Shrinkage tests were performed on 25 mm × 25 mm × 285 mm beams, which were stored at 25°C under autogenous conditions. Shrinkage was measured almost daily using to ASTM C490 apparatus, three samples from each mixture were prepared.

2.4. Mechanical strength testing

Samples for measurement of mechanical properties were stored under the same conditions as samples for shrinkage measurement. The bending strength was tested on mortar samples with dimensions 40 mm × 40 mm × 160 mm and compressive strength tests were performed on broken parts from flexural testing. Flexural and compressive strength were performed at the age of 1, 7 and 28 days.

2.5. Isothermal calorimetry
The effect of isopropyl alcohol on AAS hydration was studied by isothermal calorimetry (TAM Air, TA Instruments). BFS (4 grams) and mixture of liquids were tempered separately inside the calorimeter at 25°C, after that, components were mixed together and mechanically stirred for three minutes.

2.6. Thermogravimetric analysis
Using thermogravimetric analysis (DSC SDT Q600, TA Instruments), influence of SRA on hydration, rate and amount of hydration products was studied. The measurement conditions were as follows: temperature set to 30°C, temperature ramp 10°C per minute to 1 000°C. After compressive strength measurement part of the samples was ground and put into the isopropanol to stop hydration processes.

3. Results and discussion

3.1. Surface tension
Surface tension plays a very important role in many physical-chemical processes and therefore the influence of alcohols on surface tension has been extensively investigated [9, 10, 11]. The surface tension measurement in figure 1 illustrates that the surface tension is decreasing with the increasing addition of SRA.

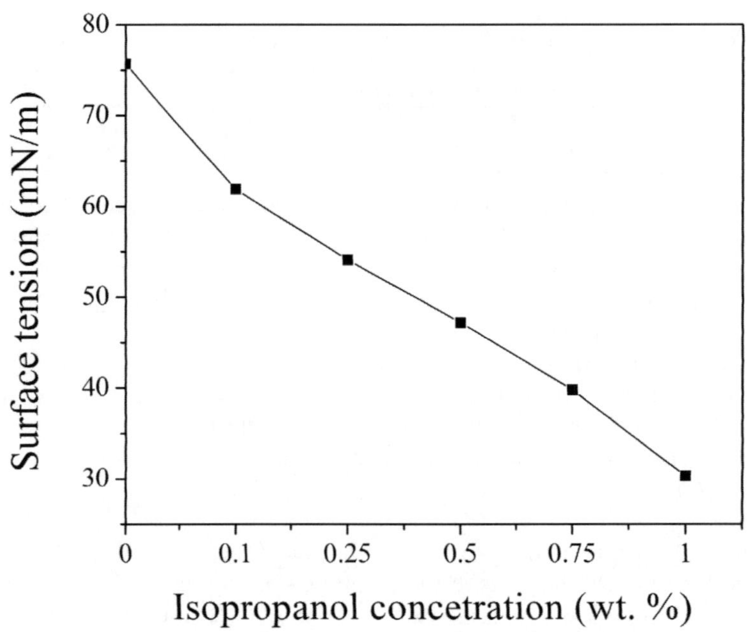

Figure 1. Effect of isopropanol addition on surface tension of porous solution.

3.2. Isothermal calorimetry
The heat flow curves for measured samples are shown in figure 2. The course of the hydration curves of the reference mixture corresponds to the description Shi and Day [8], where the first peak indicates the dissolution of the BFS. The second peak is assigned to the pre-induction period and the formation of the primary C-S-H gel, this pre-induction period lasts from 30 minutes to 9 hours of hydration. After that a secondary C-S-H gel is formed, which starts to form after 100 hours after hydration at the highest dose of isopropanol. The addition of isopropanol significantly influences the course of the hydration curves. While at the first peak, isopropanol influence is greatly varied due to a large number of possible acting factors. At the second peak, which is shown in more detail in figure 2

(right), it is evident that the heat flow decreases with the increasing SRA dose. Also, the distribution of this peak corresponding to primary C-S-H formation was observed here and the distribution was more apparent with the increasing dose of isopropanol. This may mean that there are 2 possible reactions during the pre-induction period, which overlap in the reference sample and therefore cannot be distinguished from each other. This phenomenon was also observed with the application of SRA based on hexylene glycol [7]. In both cases the first peak maximum is reached around 1 hour after the start of the mixing process and the second peak has a maximum about after 3 hours. However, while for hexylene glycol, the second peak was dominant, for the use of isopropyl alcohol the first peak was more dominant. The effect of increasing isopropanol additions on the change of position of the third peak can be observed. At low doses of isopropanol, only minimal retardation of hydration on formation of the secondary C-S-H gel occurs. With the increasing dose of isopropanol, the hydration retardation increases and at the maximum dose, hydration is delayed up to 160 hours. Also, with the increasing dose, not only the third peak is delayed, but also the main stage of overall secondary CSH gel formation takes longer time.

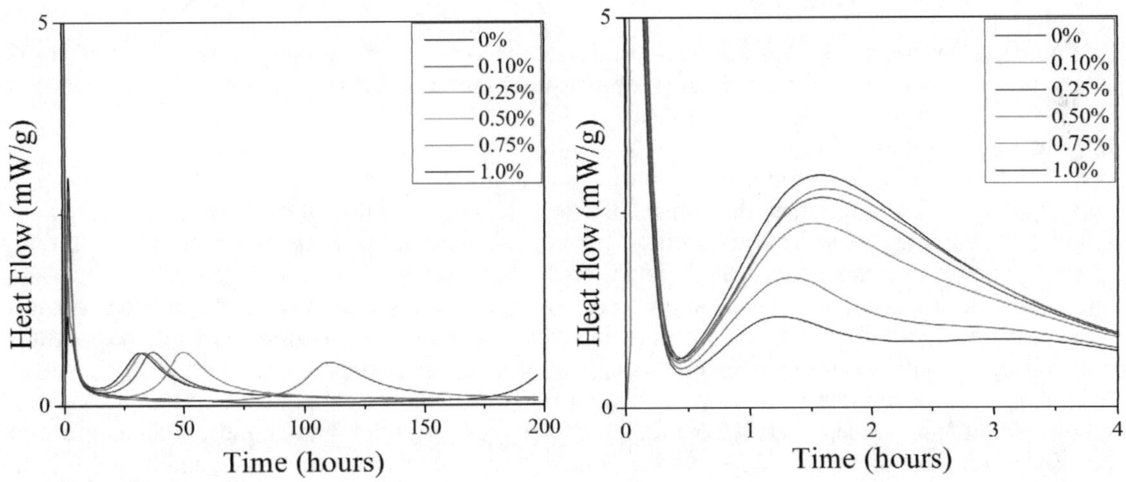

Figure 2. Influence of isopropanol addition on hydration of BFS and amount of heat flow during hydration.

3.3. Autogenous shrinkage

The shrinkage development depending on the age and the mortar composition during the first 50 hours is shown in figure 3 (left). It is clear that isopropanol at the dosages of 0.10% and 0.25% did not significantly affected shrinkage (there is a 20% reduction in shrinkage) of the samples and after 32 hours of hydration, the shrinkage of these samples was very similar to the reference sample. For samples with isopropanol addition of 0.50% and 0.75%, there were more than half shrinkage values compared to reference sample in the beginning. A similar course of shrinkage as in the reference was observed after 48 hours of hydration. The biggest differences in shrinkage values compared to the reference were recorded at the maximum dose of 1.0% SRA. At this concentration the shrinkage decreased by 87% after 48 hours. However, within a time interval of 48 to 120 hours, sample of isopropanol 1.0% performed high increase of shrinkage and after 12 days (from beginning of mixing) the shrinkage was same as for the reference sample, which could be observed in figure 3 right. It can also be concluded that the resulting shrinkage for mixtures of isopropanol 0.10%, 0.25% and 0.50% was the same (and had a similar course) as the reference.

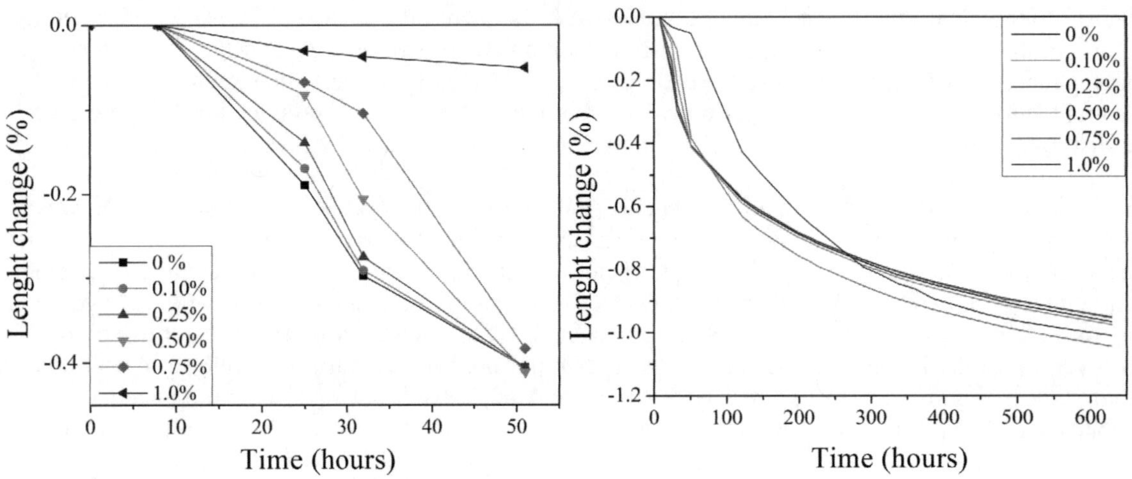

Figure 3. Development of shrinkage depending on isopropanol addition over 48 hours (left). Total shrinkage of measured samples (right).

3.4. Flexural and compressive strength

Figure 4 (left) shows the development of flexural strength and influence of SRA addition. As can be observed, there is only minimal difference between samples because they have a small flexural strength after 24 hour. At seven-day strength testing, it is already possible to observe a decreasing tendency with increasing isopropanol additions. The decline in strength is closely related to the hydration processes. The lowest strength is achieved in the sample with 1.0% of SRA, because secondary C-S-H gel formation has not occurred yet. After 28 days hydration of all the mixtures proceeded significantly resulting in a partial equalization of the strengths.

The compressive strength results given in figure 4 (right) showed a similar trend as the flexural strength. For strength values after 24 hours, there is a decreasing trend with an increasing isopropanol dose. These one-day strength values were affected by the delay of the pre-induction period and formation of the primary C-S-H gel. For seven-day strength measurements, the compressive strength values are similar for the samples apart from the sample with 1.0% isopropanol addition, where C-S-H gel is not formed yet. Measurement after 28 days shows, that samples have very similar compressive strength values and therefore the addition of isopropanol only affected the course of hydration, not the resulting mechanical properties.

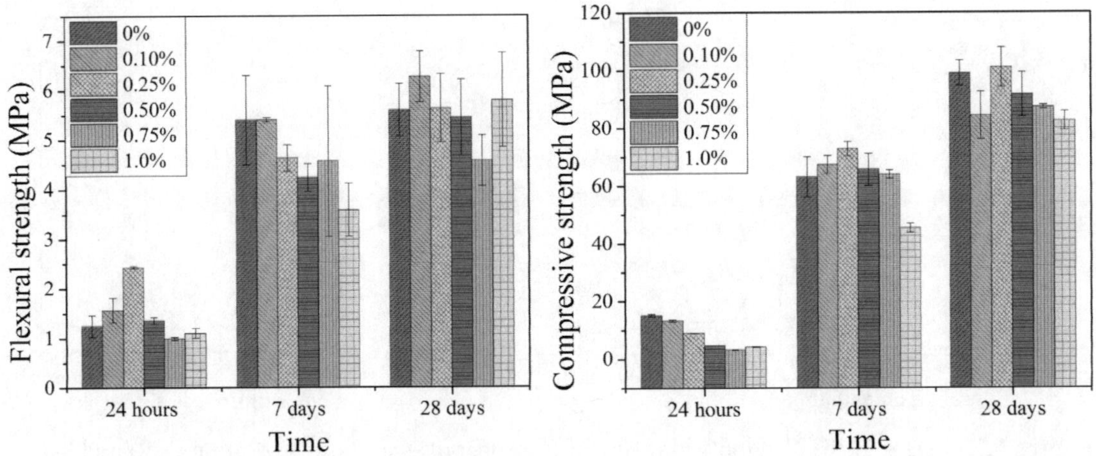

Figure 4. Development of mechanical properties of AAS in time depending on the addition of isopropanol.

3.5. Thermogravimetric analysis

TG-DTA analysis was used to determine amount of hydration products and ongoing chemical processes. From the TG curves, mass decay and decomposition temperatures were found. An example of a TG-DTA curve is shown in figure 5 (left). At an elevated temperature, the first weight loss occurs at 106°C , which is caused by the dehydration of water from the C-S-H gel [12, 13]. Another weight loss is caused by dehydroxylation of the C-S-H gel and decomposition of hydrotalcite [14]. This weight loss occurs very slowly and begins at 200°C and ends at about 600°C. Decreasing intensity of these peaks with increasing isopropanol dose is consistent with calorimetric outcomes. Mass looses taking place beyond 600°C are related to carbonates decomposition. In the case of present study, two separated peaks between 650–700°C and 750–800°C were observed. It was reported [15] that the region of carbonates decomposition in cementitous materials can be significantly modified when organic solvents are used for stopping the hydration resulting sometimes even in two separated peaks [16]. This would be also the case of the present study where isopropanol was used not only to stop the hydration but also as an additive during the mixing process of fresh pastes. It is worth mentioning that these weight losses increase with an increasing isopropanol dose, except for one-day samples, which would be related to different hydration degree of the samples at early ages. Nevertheless, further research is needed to deeply investigate this issue.

From figure 5, noticeable exothermic peak can be observed around 800°C which is likely related to recrystalization of decomposed C-S-H phase to Wollastonite structure [13, 17].

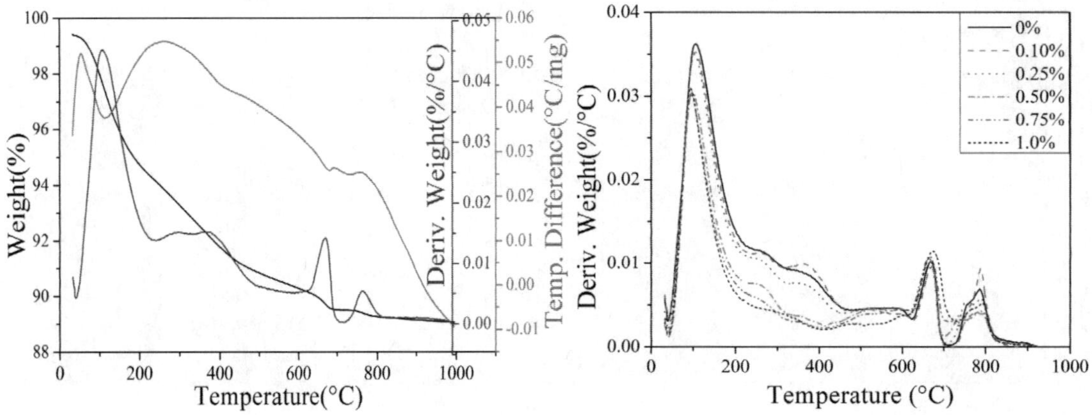

Figure 5. TG-DTA curve of sample with 1% of isopropanol after 7 day hydration (left) and DTG curves of samples after one day hydration.

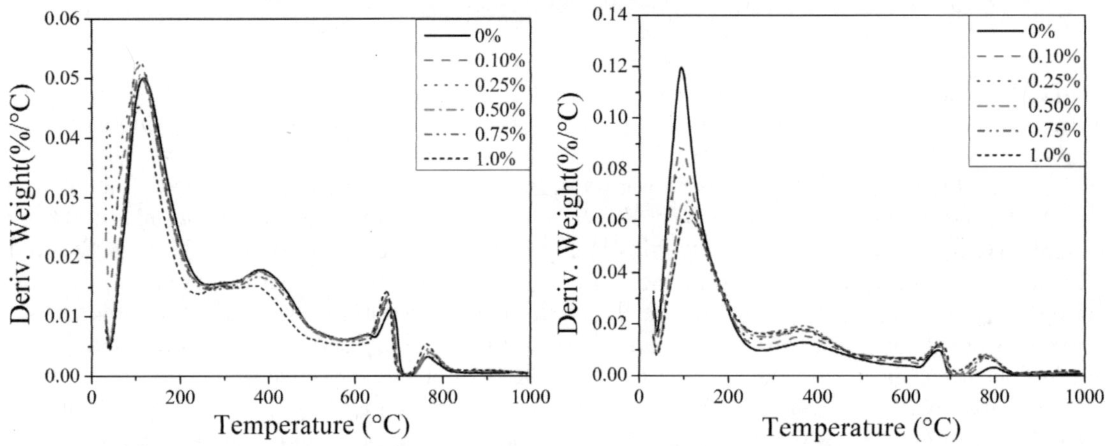

Figure 6. DTG curves of samples after 7 (left) and 28 (right) day hydration.

4. Conclusion

According to the interpretation of the obtained data, the addition of isopropanol has a significant effect on hydration of AAS. From the surface tension measurements, it is clear that isopropanol reduces the surface tension of the pore solution, but this addition does not have a significant effect on the autogenous shrinkage of AAS from the long-term point of view. Only in the initial stages, the autogenous shrinkage is reduced in presence of isopropanol, which is, however, not due to reduction of surface tension, but due to a considerable delay in hydration, that could be observed also by isothermal calorimetry and thermogravimetric analysis results. The difference is also evident in the development of strength values, with a difference during the initial days depending on the isopropanol addition, while the strengths after 28 days were very similar.

Acknowledgments

This study was done with the financial support by the project of specific research FAST/FCH-J-18-5595.

References

[1] Antonio A and Cincotto M A 2008 Drying and autogenous shrinkage of pastes and mortars with activated slag cement *Cement and Concrete Research* **38**(4) pp 565–574

[2] Bílek V Jr., Kalina L and Novotny R 2018 Polyethylene glycol molecular weight as an important parameter affecting drying shrinkage and hydration of alkali-activated slag mortars and pastes *Construction and Building Materials* 166(30) pp 564–571

[3] Douglas E, Bilodeau A and Malhotra V M 1992 Properties and Durability of Alkali-Activated Slag Concrete. *Material Journal* **89**(5) pp 509–516

[4] Shi C and Stegemann J A 2000 Acid corrosion resistance of different cementing materials *Cement and Concrete Research* **30**(5) pp 803–808

[5] Eberhardt A B 2011 *On the mechanisms of shrinkage reducing admixtures in self consolidation mortars and concretes* Ph.D. Thesis (Baubaus Universitat Weimar)

[6] Bílek V, Keršner Z and Šimonová H 2017 *Non-Traditional Cement and Concrete VI* (Brno: NOVPRESS)

[7] Bílek V Jr., Kalina L and Novotny R 2016 Some Issues of Shrinkage-Reducing Admixtures Application in Alkali-Activated Slag Systems *Materials* pp 1–12

[8] Shi C and Day R L 1995 A calorimetric study of early hydration of alkali-slag cements *Cement and Concrete Research* **25**(6) pp 1333–1346

[9] Mulero Á and Pierantozzi M 2017 An Artificial Neural Network for the surface tension of alcohols *Fluid Phase Equilibria* **449**(15) pp 28–40

[10] Concalves F, Trindade A R and Costa C S 2010 PVT, viscosity, and surface tension of ethanol: New measurements and literature data evaluation *The Journal of Chemical Thermodynamics* **42**(8) pp 1039–1049

[11] Lee J W and Park S B 2000 Densities, Surface Tensions, and Refractive Indices of the water + 1,3-Propanediol System *Journal of chemical and engineering data* **45**(2) pp 166–168

[12] Ashraf W 2018 Microstructure of chemically activated gamma-dicalcium silicate paste *Construction and Building Materials* **185** pp 617–627

[13] Mec P, Bohacova J and Turicova M 2016 Thermal Analysis Application in Study of Alkali-Activated Blast Furnace Slage *Applied Mechanics and Materials* **835** pp 547–553

[14] Neto A, Cincotto M and Repette W 2007 Drying and autogenous shrinkage of pastes and mortars with activated slag cement *Cement and Concrete Research* **38** pp 565–574

[15] Mitchell L D and Margeson J C 2006 The effects of solvents on C-S-H as determined by thermal analysis *Journal of Thermal Analysis and Calorimetry* **86** pp 591–594

[16] Zhang J and Scherer G W 2011 Comparison of methods for arresting hydration of cement *Cement and Concrete Research* **41** pp 1024–1036

[17] Rodriguez T, Garbev K and Merz D 2017 Thermal stability of C-S-H phases and applicability of Richardson and Groves´and Richardson C-(A)-S-H(I) models to synthetic C-S-H *Cement and Concrete research* **95** pp 45–56

IOP Conf. Series: Materials Science and Engineering **583** (2019) 012024 doi:10.1088/1757-899X/583/1/012024

Mechanical activation of two different types of fly ash and their effect on alkali activation.

M Marcin[1], M Sisol[1], I Brezani[1] and D Behunova[2]

[1] Technical University of Košice, Faculty of Mining, Ecology, Process Control and Geotechnologies, Institute of earth resources, Department of mineral processing, Park Komenského 19, 042 00 Košice, Slovak Republic
[2] Institute of Geotechnics SAS, Watsonova 45, 040 01 Košice, Slovak Republic

Email: michal.marcin@tuke.sk

Abstract. The article aim is to analyze the influence of mechanical activation of a fly ash, with different proportion of mechanically activated and not mechanically activated material on alkali activation. The paper describes fly ash based geopolymers and their mechanical properties. Fly ashes were mixed in four different ratios with mechanically activated fly ash. Activation solution was constant in all specimens. Compressive and flexural strengths of the hardened product were tested after 7 and 28 days of the storage in laboratory conditions and samples were cured after activation in hot air chamber for 6 hours in 80°C temperature. The results show how mechanical activation of materials affects their strengths with highest amount achieved was 44 MPa.

1. Introduction

Fly ash is a by-product derived from coal combustion in thermal power plants and heating plants. Fly ash is a fine powder containing predominantly spherical vitreous particles originated from the combustion of finely ground coal. It is considered as a material that predominantly contains SiO_2 and Al_2O_3. These properties make it an attractive raw material for secondary industrial processing [1]. Fly ash as an excellent source of aluminosilicate material is a raw material for production of alkali-activated binders known as geopolymers. Geopolymers exhibit many excellent properties, such as high compressive strength, good acid resistance, low shrinkage, low creep and fire resistance [2]. They are generally synthesized by mixing aluminosilicate-reactive material with strong alkali solutions. Alkali activating solution is important for the dissolution of Si and Al to form geopolymer precursors and finally aluminosilicate material [3]. Geopolymers are obtained from the chemical reaction of alumino-silicate oxides with sodium silicate solutions in a highly alkaline environment. As an alkali activating solution, a strongly alkaline aqueous solution of sodium or potassium hydroxide is most commonly used [3, 4]. Under a strong alkali solution, aluminosilicate-reactive materials dissolve and form free SiO_4 and AlO_4 tetrahedral units. With the progress of reaction, water is gradually removed, and the SiO_4 and AlO_4 tetrahedral clusters are linked to yield polymeric precursors through the sharing of all oxygen atoms between two tetrahedral units, thereby forming amorphous geopolymers [5].

The most commonly used fly ash for geopolymer production is Class F fly ash with low content of calcium. However, more and more researches are found, in which fluidized bed combustion fly ashes with high content of calcium are used for geopolymer production. Our research is focused on fly ashes without industrial utilization, deposited in landfills. The main reason why these fly ashes have no industrial applications is their high loss on ignition (LOI) that in some types of fly ash is almost 25%.

However, based on previous research [6], it was shown that high content of loss on ignition is no limitation for utilization of fly ashes in alkali activation. The geopolymers now represent a new group of inorganic substances, because they have a significant environmental and energy potential. They belong to a group of the inorganic polymer covalently bound macromolecules with the chain consisting of -Si-O-Al-O-.

The geopolymers find a broad range of applications in the field of transportation, emergency repairs, metallurgy, coating, membrane materials, and nuclear waste disposal. Despite a significant commercial and technological potential geopolymers' easy-brittle character limits their extensive applications where great efforts are made to overcome such a disadvantage. Numerous studies are dedicated to optimize the strength of geopolymer products and to understand the mechanism of the geopolymerization [7–10]. Recently, there have been many studies on the mechanical activation (MA) of solids for new materials or processes. The purpose of MA is to change the reactivity or physical-chemical properties of a solid. In particular, solid reactivity is induced by mechanical energy or a combination of mechanical and chemical effects through shearing, compression, extension, bending, and impact. Kumar et al. showed that mechanically activated slag could achieve complete hydration without any chemical addition, and MA had a positive effect on the early strength development of the cement [11, 12].

In the last years, different studies have investigated the possibility of using different types of wastes mixed with fly ash or slag as raw materials. The selection of the materials to create the geopolymers depends on several factors such as availability, disposal urgency, difficulty for recycling and final applications [13].

The aim of presented contribution is to point out to possibility of utilization of seemingly no utilisable by-products of coal combustion for production of alkali-activated materials – geopolymers.

2. Materials and methods

The materials used for the alkali activation were two types of fly ash (TO, TF). Fly ash TO is derived from combustion of black coal in pulverized coal boilers obtained from an upper layer of the coal-ash sludge bed. Fly ash TF is derived from the same combustion process as fly ash TO, except that it is obtained from a fly ash hopper.

Chemical composition of the fly ashes determined by X-ray fluorescence (XRF) spectrometer Spektro X-lab 2000 is shown in table 1. Phase composition of fly ashes performed using the X-ray diffractometer D8 Bruker 2 theta / 2 theta using CuKα radiation is shown in figure 1. Figure 2 shows morphology of fly ash particles performed by scanning electron microscopy (SEM) using the device MIRA 3 FE-SEM microscope (TESCAN, Czech Republic) equipped with a high-resolution cathode (Schottky field emitter) and with three-lens Wide Field Optics™ design.

Table 1. Chemical composition of fly ashes used in alkali activation.

Type of fly ash	Chemical composition [wt. %]					
	Al_2O_3	SiO_2	CaO	Fe_2O_3	Na_2O	LOI
TO	14.55	32.01	1.75	7.55	0.23	24.80
TF	24.01	44.27	2.36	9.67	0.90	19.81

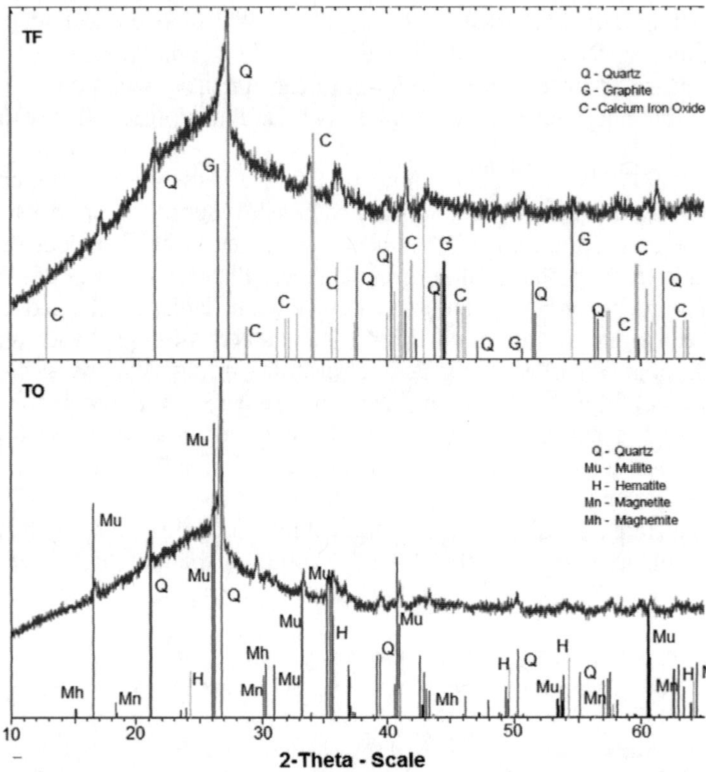

Figure 1. X-ray patterns of fly ashes used in alkali activation.

Based on results from performed analysis of fly ashes it can be concluded that morphology and phase composition of fly ashes are affected by combustion method, at which the fly ash is produced. The combustion of coal in pulverized coal boilers taking place at combustion temperature 1400–1700°C results in a melting of particles of combusted coal. Therefore, particles of fly ash have spherical shape and fly ash contains high content of vitreous phase (halo indicated between 2-Theta = 20°–35°).

Minority crystalline phase of used fly ashes is mainly composed of quartz, mullite, hematite, magnetite, maghemite and graphite.

Fly ashes TF and TO contain different phase composition, even though they are derived from the same process combustion. Fly ash TF contains despite of quartz also graphite and iron in the form of calcium iron oxide. Fly ash TO contains no graphite content and iron in three different forms as hematite, magnetite, and maghemite. Fluidized bed combustion fly ashes contain calcium compounds.

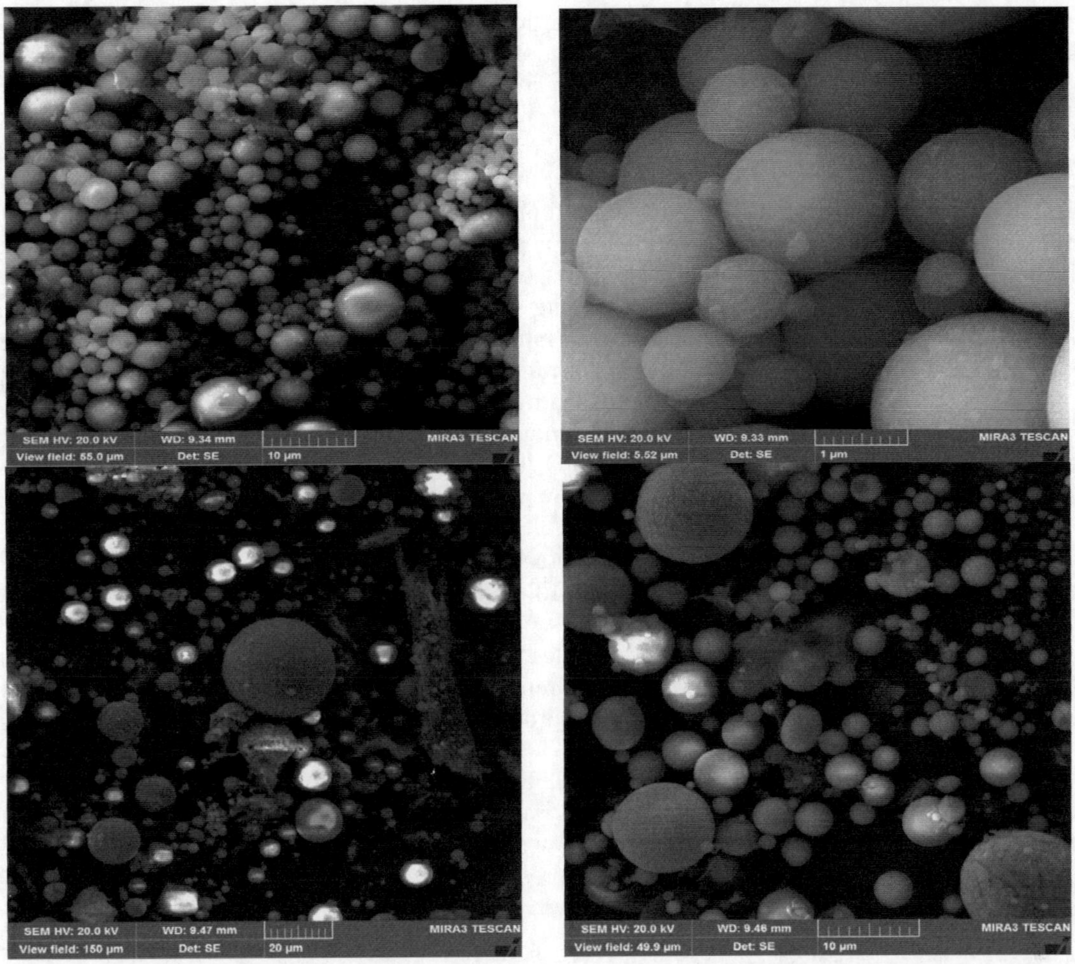

Figure 2. SEM patterns of fly ashes used in alkali activation (upper row TF, bottom row TO).

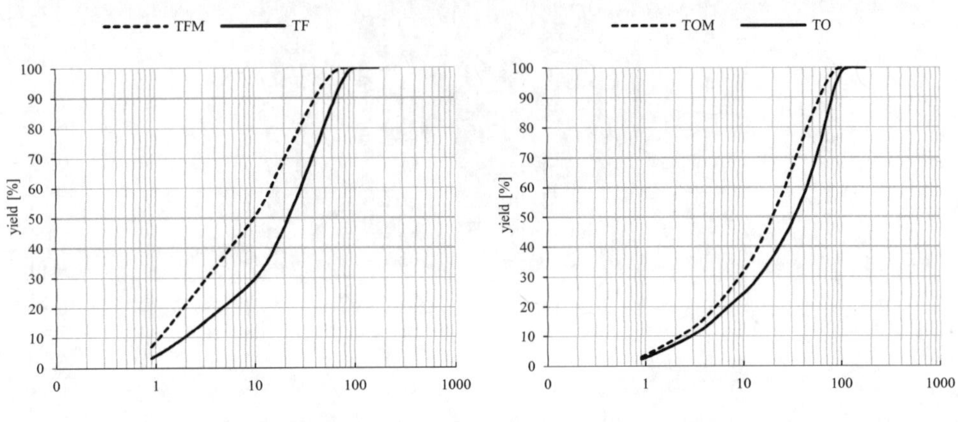

Figure 3. Particle size distribution of fly ashes before (TF, TO) and after milling (TFM, TOM).

Grain size analysis revealed that about 55% particles of fly ash TF, 70% of TO is passing through a sieve size of 45 microns. After mechanical activation (MA) it is about 75% particles of fly ash TF and 95% of TO, is passing through a sieve size of 45 microns. Particle size distribution before and after MA is shown in figure 3.

Milling was performed in laboratory ball mill with steel balls with different diameters. The laboratory mill used for mechanical activation had volume 30 dm^3 and was filled with steel ball with diameters ranging from 2 to 10 cm. Material for mechanical activation with steel balls formed 60% of mill volume. Revolution was adjusted to 90 per minute. The material was homogenised before alkali activation. No other treatment was applied to the material.

The activation solution was prepared by mixing solid NaOH pellets with sodium water glass and water. The sodium water glass from the Kittfort Praha Co. with the density of 1.328–1.378 g/cm^3 was used. It contains 36–38% Na_2SiO_3 and the molar ratio of SiO_2/Na_2O is 3.2–3.5. The solid NaOH with the density of 2.13 g/cm^3 was obtained from Kittfort Praha Co. containing at least 97–99.5% of NaOH.

The fly ash mixture samples with different amount of mechanically activated FA were prepared. Activation solution used has 8% of Na_2O, 1.25 SiO_2/Na_2O ratio and water-to-fly ash ratio was adjusted to 0.3. Material for alkali activation was created by mixing pure FA with 0, 50, 75 and 100% addition of mechanically activated FA.

FA and MAFA mixture were stirred with the activation solution for 10 minutes, until a creation of a homogenous mixture. The mixture was then filled into prismatic molds with the dimensions 40 x 40 x 160 mm and compacted on the vibration table VSB-40. The pastes were cured in a hot air drying chamber at 80°C for 6 hours. Thereafter, the samples were removed from the molds, marked, and stored in laboratory conditions till the moment of the strength test. The values of compressive and flexural strengths were determined after 7 and 28 days according to the Slovak Standard STN EN 12390-3.

3. Results and discussion

Properties and a composition of material play an important role in the alkali activation. The samples were synthetized from two different types of FA with an addition of MAFA by several proportions. The effect of mechanical activation on final compressive strength and flexural strengths as examined. The resulting strengths from TF fly ash are shown in figures 4 and 5.

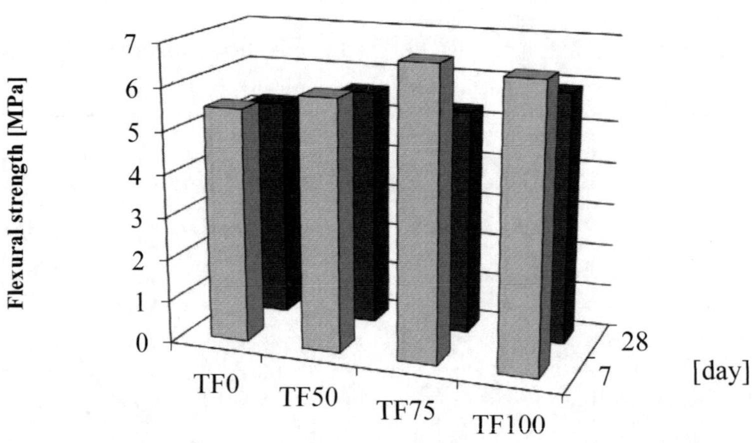

Figure 4. Flexural strengths of TF fly ash mixtures.

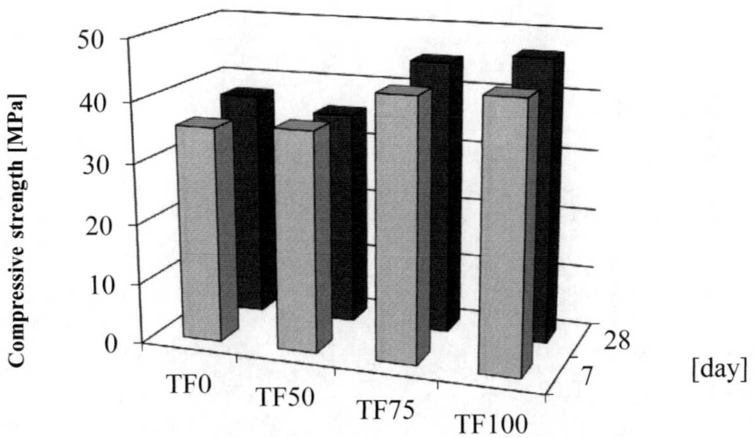

Figure 5. Compressive strengths of TF fly ash mixtures.

The flexural strength results show that maximum strength of all samples is not achieved in the mixture prepared with 100% mechanically activated fly ash, but with only 75% of mechanically activated FA. In all samples after 28 days the flexural strengths were lower from their first flexural strengths. In contrary compressive strengths results show that geopolymer mortar created only by mechanically activated fly ash has the highest strength 44 MPa after 28 days. All samples have the highest compressive strength values after 28 days.

TO based geopolymers mortar showed a little different result. These results are depicted in figures 6 and 7.

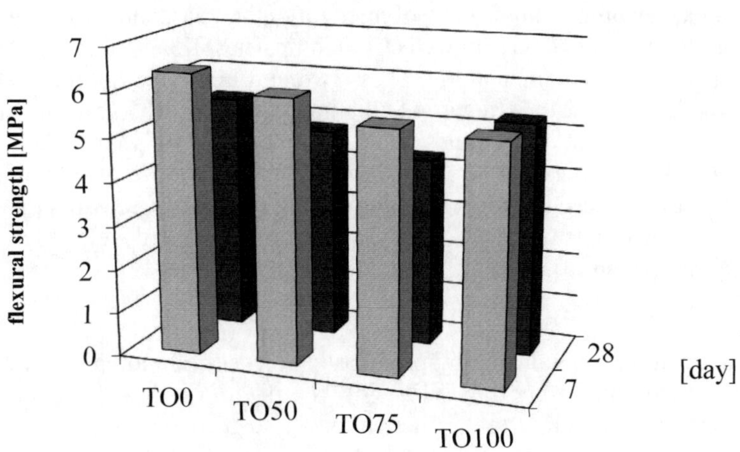

Figure 6. Flexural strengths of TO fly ash mixtures.

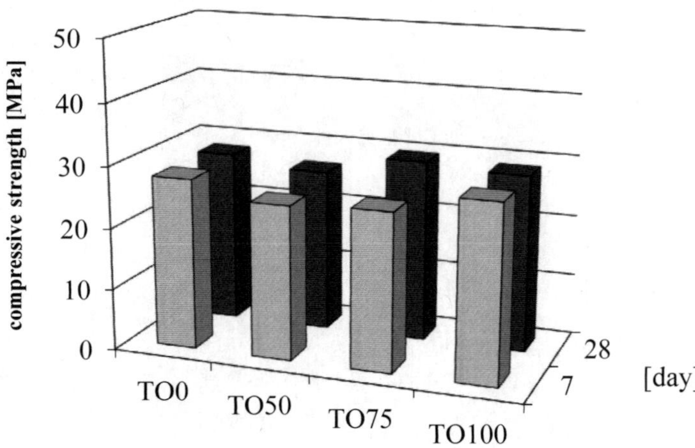

Figure 7. Compressive strengths of TO fly ash mixtures.

Flexural strengths were downgrading with higher amount of the mechanically activated fly ash used in geopolymer mortar. Common thing in these two fly ashes was that both have lower strengths after 28 days. Compressive strengths were also lower in comparison with TF mixtures. In this case the results were similar in all mixtures so we can conclude that the amount of the mechanically activated fly ash in this case is not that relevant and is not suitable or needed with TO fly ash.

Results of the previous work [6] found that increase in content of particles higher than 45 μm by grinding results in increase of mechanical strength alkali-activated fly ashes. Higher mechanical strength is result of improved dissolution of fly ash into alkaline liquid following a decrease particle size (an increase surface area); leading to improved polymerization and hardening of the geopolymer phase, therefore higher compressive strength is achieved [14]. Comparing strength of alkali-activated fly ashes TF and TO obtained by the same combustion process is found higher compressive strength of alkali-activated fly ash TF and higher flexural strength of alkali-activated fly ash TO. Compared to fly ash TO, fly ash TF contains higher content of aluminosilicate component and higher content of soluble phase that can explain its higher compressive strength. Content of aluminosilicate component in sources material is important factor because the silica and alumina are the main precursors for the geopolymeric reaction. Aluminosilicate materials with the higher extent of dissolution developed higher compressive strength after geopolymerization [3].

4. Conclusions

Fly ash from coal combustion is a valuable by-product that has still no complete industrial utilization. Therefore, it is necessary to find further possibilities of its utilization. Fly ash contains the high content of alumosilicate component; therefore, alkali activation is seen to be an applicable method for its utilization.

In this experimental programme, fly ash based geopolymer mortar samples were casted by varying the proportion of the mechanically activated fly ash such as 0%, 50%, 75% and 100% by weight of fly ash.

Mechanical activation in these types of fly ash had different effect on final strengths. In the first type strengths were little bit enhanced after the mechanical activation. The second type of geopolymers mortars showed that mechanical activation is not necessary and all samples had approximately same strengths.

Acknowledgments

This work was supported by the research grant project VEGA, No. 1/0472/18.

IOP Conf. Series: Materials Science and Engineering **583** (2019) 012024 doi:10.1088/1757-899X/583/1/012024

References

[1] Zhi-Tong Y 2014 Generation, characterization and extracting of silicon and aluminum from coal fly ash *Fly ash-Sources, applications and potential environmental impacts* chapter 1 pp 3–57

[2] Duxson P, Fernández-Jiménez A, Provis J L, Lukey G C, Palomo A and Van Deventer J S J 2007 Geopolymer technology: the current state of the art *J Mater Sci* **42** pp 2917–2933

[3] Xu H and van Deventer J S J 2000 The geopolymerisation of alumino-silicate minerals *Int J Miner Process* **59** pp 247–66

[4] Skvara F 2007 Alkali activated materials or geopolymers *Ceramics – Silikáty* pp 173–177

[5] Davidovits J 2008 *Geopolymer: Chemistry and applications* (France: Institut Géopolymère) 2nd ed. p 584

[6] Sisol M, Drabová M and Mosej J 2014 Alkali activation of fresh and deposited black coal fly ash with high loss on ignition *Mineral Resources Management* **30**(2) pp 103–116

[7] van Deventer J S J, Provis J L, Duxson P and Lukey G C 2007 Reaction mechanisms in the geopolymeric conversion of inorganic waste to useful products In *Journal of Hazardous Materials* vol 139 Issue 3 pp 506–513

[8] Zhang H Y, Kodur V, Qi S L and Wu B 2015 Characterizing the bond strength of geopolymers at ambient and elevated temperatures In *Cement and Concrete Composites* vol 58 pp 40–49

[9] He J, Jie Y, Zhang J, Yu Y and Zhang G 2013 Synthesis and characterization of red mud and rice husk ash-based geopolymer composites In *Cement and Concrete Composites* vol 37 pp 108–118

[10] Zhang J, He Y, Wang Y, Mao J and Cui X 2014 Synthesis of a self-supporting faujasite zeolite membrane using geopolymer gel for separation of alcohol/water mixture In *Materials Letters* vol 116 pp 167–170

[11] Bo W, Yimin Z and Shenxu B 2017 Preparation of geopolymers from vanadium tailings by mechanical activation In *Construction and Building Materials* vol 145 pp 236–242

[12] Kumar S, Kumar R, Bandopadhyay A, Alex T.C, Kumar B R, Das S K and Mehrotra S P 2008 Mechanical activation of granulated blast furnace slag and its effect on the properties and structure of portland slag cement In *Cement and Concrete Composites* vol 30 Issue 8 pp 679–685

[13] Toniolo N and Boccaccini A R 2017 Fly ash-based geopolymers containing added silicate waste A review *Ceramics International* **43** pp 14545–14551

[14] Temuujin J, Williams R P and van Riessen A 2009 Effect of mechanical activation of fly ash on the properties of geopolymer cured at ambient temperature *Journal of Materials Processing Technology* **209** pp 5276–5280

IOP Conf. Series: Materials Science and Engineering **583** (2019) 012025 doi:10.1088/1757-899X/583/1/012025

Properties of the composite with recycled concrete manufactured by modified technology

T Trtik[1], K Seps[1], R Chylik[1], J Fladr[1] and T Pavlu[1]

[1] Faculty of Civil Engineering, Czech Technical University in Prague, Thákurova 7, 166 29 Praha 6, Czech Republic

E-mail: tomas.trtik@fsv.cvut.cz

Abstract. The article describes basic mechanical, physical and durability characteristics of concrete with recycled aggregate. Recycled concrete aggregates from AZS s.r.o. are made using newly modified technology and properties for use in concrete have been tested. In this part of the research the compressive strength, tensile strength, modulus, freeze and thaw resistance are tested. The basic aim was to test two mixtures with fifty percent substitution of the coarse component of natural aggregate by recycled one (the standard EN 206 does not already allow this) and to prove that such a quantity of recycled aggregate can be used for concrete constructions. The properties were tested on 150 mm cubes and beams 100 x 100 x 400 mm after 28 days.

1. Introduction

Concrete, which is completely dependent on natural raw materials, is the most used building material due to its properties, whose production is about 10 billion tons a year [1]. These facts lead to over-exploitation of natural resources and high volumes of waste concrete, which constitutes around 40% of construction and demolition waste. The use of concrete waste as aggregate or admixtures for concrete is one of the most effective approaches to the recycling of construction and demolition waste. The aim of this work is to verify the properties of concrete, which can be classified as plain concrete with a minimum allowable amount of cement and the prescribed percentage substitution of recycled aggregate, which reaches acceptable material characteristics and verification of the two-cycle recycling process.

1.1. Percentage replacement of natural aggregates

The possibilities of substitution of the aggregate coarse fraction in concrete with recycled aggregate and requirements for the properties of recycled aggregate are set out in valid Czech standards [2, 3]. The standards reflected the years of research into the properties of recycled aggregate concrete and its impact on concrete properties. The use of a coarse fraction of recycled aggregate concrete usually leads to a deterioration of all properties of the concrete, so the maximum possible substitution is 30% [4] or 50% [4, 5] depending on the quality of recycled aggregate concrete and concrete class as it is stated in table 1.

The standard allows a maximum of 50% substitution of coarse aggregate, although there are studies where 100% coarse fraction of aggregate is replaced by recycled aggregate concrete [6, 7]. The coarse fraction of natural aggregate (4–8 mm and 8–16 mm) was partially replaced by recycled concrete aggregate. Replacement of aggregate (30% and 50%) in the concrete mix was performed for the most

IOP Conf. Series: Materials Science and Engineering **583** (2019) 012025 doi:10.1088/1757-899X/583/1/012025

frequently produced concrete classes C16 / 20 X0 and C25 / 30 XC2. Table 1 shows that for plain concrete, which is mostly specified with low exposure class, percentage substitution of recycled aggregate from an unknown source can be used. Structural concrete, which is used in load-bearing structures, must not contain a percentage of recycled aggregate.

Table 1. The maximum percentage of coarse aggregate replacement [2].

	X0	XC1,XC2	XC3, XC4, XF1, XA1, XD1
Type A	50%	30%	30%
Type B	50%	30%	0%

[a] Recycled aggregates of type A from a known source can be used for the degree of environmental impact for which the original concrete was designed.
[b] Recycled B-type aggregates are not used in concrete class > C30 / 37.

2. Material and Methods

The cooperation was established with the recycling company AZS 98 s.r.o. To achieve the highest quality of recycled aggregate with stable properties, the recycling process and technology were modified. Undesirable components such as reinforcement and clay components are first separated from recycled concrete. Pure 63–128 mm fragments of concrete enter the second cycle when pure concrete is crushed into fractions 0–4 mm, 4–8 mm, and 8–16 mm. The recycling device is designed to capture fine particles when crushing concrete fragments, which can be further used as an admixture or substitute for cement.

The properties of recycled aggregate (figure 1, figure 2) have been verified in the previous part of the authors' research and meet conditions for use in concrete [8].

Figure 1. Recycled aggregate concrete fraction (8–16 mm).

Figure 2. Recycled aggregate concrete fraction (4–8 mm).

Six series of concrete mixtures were designed (table 2). Three series of concrete mixtures (C2, C4 and C6) were designed for concrete class C16 / 20 with cement quantity of 300 kg/m^3, which is the minimum allowable quantity for the structural concrete [2]. Concrete mixtures (C1, C3 and C5) were designed for concrete class C 25 / 30. Two different types of cements with the same strength grade 42.5R were used in the experiment. Concrete mixtures (C1, C2, C3 and C4) had 30% substitute for coarse natural aggregate. Concrete mixtures (C5, C6) were designed with 50% replacement of coarse recycled

IOP Conf. Series: Materials Science and Engineering **583** (2019) 012025 doi:10.1088/1757-899X/583/1/012025

aggregate (table 3). Two types of cement were used in the experiment This type of cement was supplied by recycling company. Cement Mokrá has been used in our previous research and series with this cement were added for comparison. Both cements have the same strength class.

Table 2. Components of concrete for each mixture.

	C1 (kg/m^3)	C2 (kg/m^3)	C3 (kg/m^3)	C4 (kg/m^3)	C5 (kg/m^3)	C6 (kg/m^3)
CEMENT	380 [c]	300 [c]	380 [d]	300 [d]	380 [d]	300 [d]
WATER	149	149	149	149	149	149
NA [a] (0-4mm)	800	850	800	850	800	850
NA [a] (4-8mm)	175	175	175	175	125	125
NA [a] (8-16mm)	525	525	525	525	375	375
PLASTICIZER [e]	2	2	2	2	2	2
RAC [b] (4-8mm)	75	75	75	75	125	125
RAC [b] (8-16mm)	225	225	225	225	375	375

[a] Natural aggregate
[b] Recycled concrete aggregate
[c] Cement Mokrá (42.5 R)
[d] Cement Radotín (42.5 R)
[e] Polycarboxylate base plasticizer (Stachement 2180)

Table 3. Percentage replacement of coarse fraction with recycled aggregate for each mixture.

	C1	C2	C3	C4	C5	C6
Percentage replacement	30%	30%	30%	30%	50%	50%

3. Results and discussion

3.1. Compressive strength

The compressive strength tests were performed with cube specimens 150/150/150 mm by method specified in ČSN 73 EN 12390-3 [12]. Five samples were made for each series of concrete mixtures. Average compressive strengths are given in figure 3.

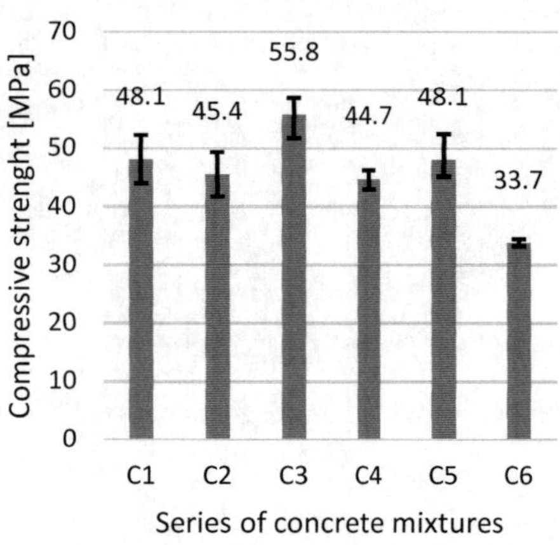

Figure 3. Average compressive strengths.

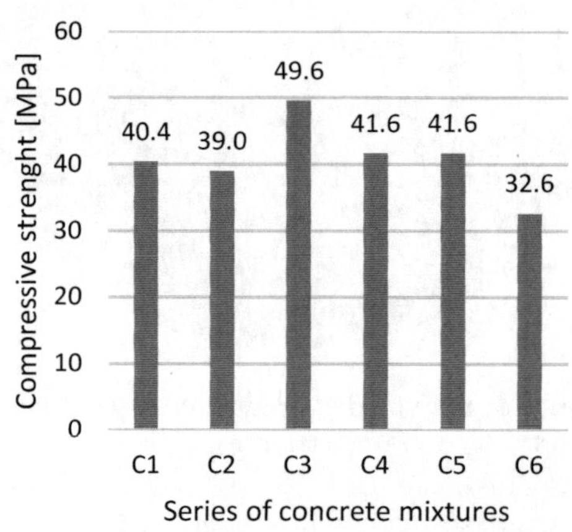

Figure 4. Characteristic compressive strengths.

Comparison series (C1 vs C3, C2 vs C4) of concrete mixtures showed that the mixtures with Radotín cement have higher compressive strengths and that is the reason why were used for further research.

Characteristic compressive strength of specimens are given in figure 4. All series of concrete mixtures can be classified to the C25 / 30 concrete class, including the C6 series, which was designed to meet the C16 / 20 concrete class. Comparison of results from our research with other researchers is in table 4.

Table 4. Comparison of compressive strength with other research groups.

	Cement content (kg/m^3)	Compressive Strength (MPa)
Our research	380; 300	48.1; 33.6
[14]	394	42; 41.5
[15]	575	33
[16]	410	47.4

Compressive strength is similar or higher to samples made [14, 15, 16] with similar or higher dosages of cement as you can see in.

3.2. Tensile strength

After evaluation of compressive strength, further research was concentrated on concrete mixtures C5 and C6, where the maximum allowable substitution by recycled aggregate (50%) is allowed. Results of flexural strength are similar to obtained in other study [16] and it is showed in table 5.

Table 5. Comparison of flexural strength with other research.

	Cement content (kg/m^3)	Flexural Strength (MPa)
Our research	380; 300	5.32; 4.3
[16]	410	4.78[a]

[a] after 56 days

3.3. Testing the freeze – thaw resistance of the RAC (recycled aggregate concrete)

The freeze thaw resistance tests of RAC were performed with specimens 100/100/400 mm by method specified in ČSN 73 1322 [9].

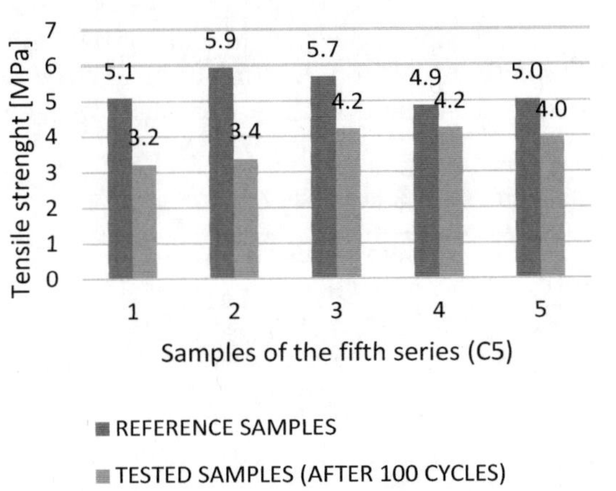

Figure 5. Comparing tensile strengths of concrete after 100 cycles for series C5 of concrete mixtures.

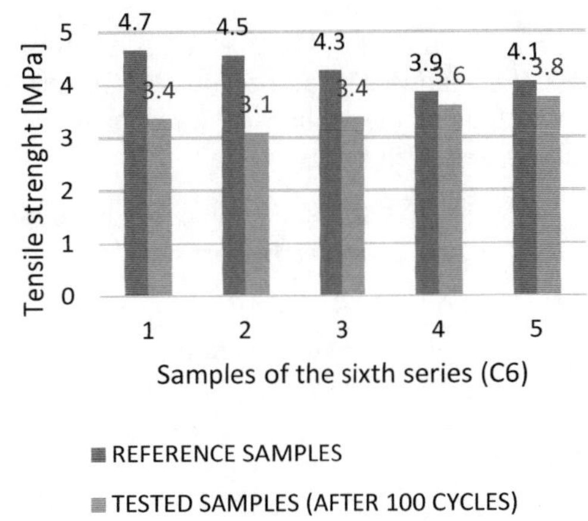

Figure 6. Comparing tensile strengths of concrete after 100 cycles for series C6 of concrete mixtures.

Results were compared with reference samples that were not subjected to freeze thaw cycling. Comparing tensile strengths of concrete after 100 cycles for series C5 of concrete mixtures are given in figure 5 and for series C6 of concrete mixtures are given in figure 6. Freeze thaw resistance test is evaluated according to the standard [9]. The ratio between reference samples that were not subjected to freeze cycling and tested samples is calculated. If this ratio is greater than 0.75, frost resistance is proven. According to results, the ratio is 0.79 for concrete mixture C5 and 0.72 for concrete mixture C6. Average tensile strength values of 3.8 MPa for C5 and 3.4 MPa for C6 meet the tensile strength for the C 35 / 45 concrete class and exceed their grades.

3.4. Modulus of elasticity

Materials with a higher modulus of elasticity require higher stresses to achieve the same deformation. The static and dynamic modulus of elasticity was measured on 100/100/400 mm beams (figure 8) according to the standard [10, 13].

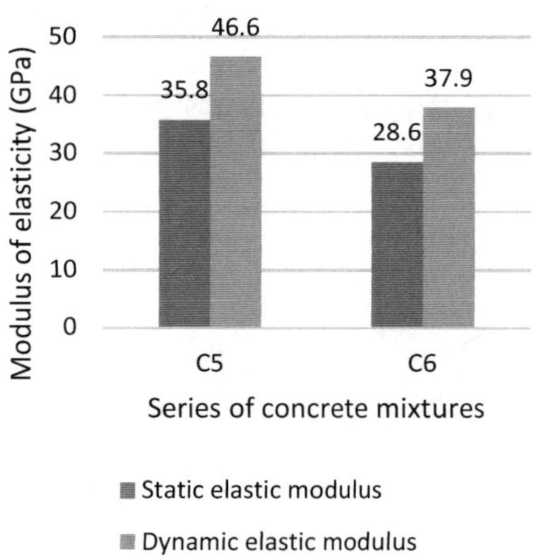

Figure 7. The average values of the static and dynamic modulus.

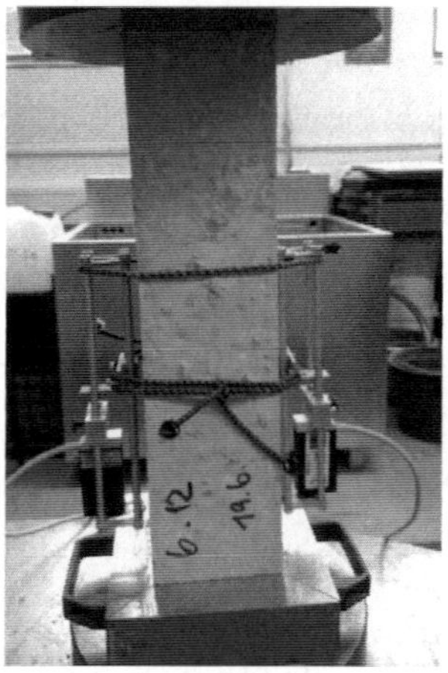

Figure 8. Measurement of the modulus of elasticity.

The static modulus of elasticity has similar results to the compressive strength. For concrete mixture C6, the static elastic modulus decreases by 20% compared to concrete mixture C5 (figure 7). Nevertheless, the value of 28.6 GPa is quite sufficient and measured values were also observed by another researchers (see table 6).

Table 6. Comparison of modulus of elasticity with other research groups.

	Cement content (kg/m^3)	Modulus of elasticity (GPa)
Our research	380; 300	35.8; 28.6
[15]	575	29
[16]	410	24.07

4. Conclusions

Two concrete mixtures which were tested in detail with fifty percent substitution of recycled aggregate concrete. Concrete mixtures were designed as C25 / 30 and C16 / 20 concrete class, both were tested for compressive strength, tensile strength, dynamic and static modulus of elasticity and frost resistance.

The results of compressive strength of concrete shows that the type of cement has an influence on properties. Significant decline in compressive strength was at 50% replacement of recycled aggregate and use the minimum amount of cement (300 kg/m^3). All series of concrete mixtures can be classified to the C25 / 30 concrete class, including the series, which was designed to meet the C16 / 20 concrete class. Compressive strength, flexural strength and modulus of elasticity are similar or higher to results get by other research programs (made with equal or higher dosages of cement). The average tensile strength after 100 cycles of freeze thaw resistance test had values (3.8 MPa and 3.4 MPa), which is a sufficient value for structural concrete. Achieved results can be considered very good.

Recycled aggregate concrete is better suited for lower class of concrete. The results further show that the recycling of recycled aggregate concrete by two-cycle recycling can produce significantly better results than the aggregate produced in a conventional one-cycle recycling process. It is necessary to continue in this area in order to ensure even more recycled concrete utilization, thus saving natural resources.

Acknowledgments
This paper was prepared thanks to the support of the Technology Agency of the Czech Republic (TAČR), project "Environmentally Efficient Construction and Demolition Waste for Structures" (No. TH02030649), project (No. SGS18/115/OHK1/2T/11) and project (No. SGS19/149/OHK1/3T/11)

References

[1] Evangelista L, Guedes M, Brito J, Ferro A C and Pereira M F 2015 Physical, chemical and mineralogical properties of fine recycled aggregates made from concrete waste *Construction and Building Materials* **86** pp 178–188
[2] Czech Standardization Agency 2018 *ČSN EN 206 + A1 Concrete – Specification, performance, production and conformity* (Prague)
[3] Czech Standardization Agency 2008 *ČSN EN 12620 + A1 Aggregates for concrete* (Prague)
[4] Verian K P, Ashraf W and Cao Y 2018 Properties of recycled concrete aggregate and their influence in new concrete production *Resources, Conservation and Recycling* **133** pp 30–49
[5] Kou S, Poon Ch and Agrela F 2011 Comparisons of natural and recycled aggregate concretes prepared with the addition of different mineral admixtures *Cement and Concrete Composites* **33** pp 788–795
[6] Zaharieva R, Buyle-Bodin F, Skoczylas F and Wirquin E 2003 Assessment of the surface permeation properties of recycled aggregate concrete *Cement and Concrete Composites* **25** pp 223–232
[7] Yang J, Du Q and Bao Y 2011 Concrete with recycled concrete and crushed clay bricks *Construction and Building Materials* **25** pp 1935-1945
[8] Trtík T, Bílý P, Fládr J and Vašková J 2018 Comparison of properties of recycled concrete aggregate with natural aggregate *Proceedings of the 12th Fib International PhD Symposium in Civil Engineering* pp 1201–1206
[9] ČSN 73 1322:2018 *Determination of frost resistance of concrete* (Prague: Czech Standardization Agency)
[10] ISO 1920-10:2016 *Determination of static modulus of elasticity in compression*
[11] Šeps K, Fládr J and Broukalová I 2016 Resistance of recycled aggregate concrete to freeze-thaw and deicing salts *Procedia Engineering* **151** pp 329–336
[12] EN 12390-3:2009 *Testing hardened concrete – Part 3: Compressive strength of test specimens*
[13] ČSN 73 1372:2012 *Non-destructive testing of concrete – Testing of concrete by resonance method* (Prague: Czech Standardization Agency)
[14] Puthussery J V, Kumar R and Garg A 2017 Evaluation of recycled concrete aggregates for their suitability in construction activities: An experimental study *Waste Management* vol 60 pp 270–276
[15] Corinaldesi V and Moriconi G 2009 Influence of mineral additions on the performance of 100% recycled aggregate concrete *Construction and Building Materials* **23** pp 2869–2876
[16] Surya M, Kanta Rao V V L and Lakshmy P 2013 Recycled aggregate concrete for Transportation Infrastructure *Procedia - Social and Behavioral Sciences* **104** pp 1158–1167

IOP Conf. Series: Materials Science and Engineering **583** (2019) 012026 doi:10.1088/1757-899X/583/1/012026

The products of a high temperature reaction of fly ash with lime and soda

D Beranova[1], L Galvankova[1], J Masilko[1] and P Ptacek[1]

[1] Brno university of Technology, Faculty of Chemistry, Institute of Material Chemistry, Purkyňova 464/118, Brno 612 00, Czech Republic

E-mail: xcberanovad@fch.vut.cz

Abstract. A composition of coal combustion fly ash depends on a coal composition and conditions of a combustion process. The main components are SiO_2 and Al_2O_3. An amount of Al_2O_3 in coal combustion fly ash is relatively high, about 30–40%, and for this reason, the fly ash could be potentially used as a raw material source for alumina production. There are several ways to get alumina. The fly ash could be acid leached or sintered with a suitable additive. One of the possible sintering processes is lime-soda sinter process. The article deals with products of lime-soda sinter process of the fly ash produced from brown coal in Czech thermal power plant. The phase composition of sintered material was determined by X-ray diffraction analysis (XRD). The presence of sodium aluminate and dicalcium silicate was confirmed.

1. Introduction

Global production of alumina (Al_2O_3) was 132 million tonnes in 2017 [1]. Over 95% of the alumina is made from bauxite by Bayer process. Bauxite ores contain about 30–60% of aluminium oxide [2]. Bayer process is a complex process which involves five major steps: grinding and pre-treatment of bauxite, pressure digestion in caustic liquor, separation of red mud (insoluble residues) from a caustic solution containing alumina, precipitation of alumina hydrate and its separation from impurities, calcination at high temperature [3].

Fly ash is a by-product of a coal combustion process in thermal power plants. Its composition depends on the type and composition of coal and combustion process conditions. Majority of fly ashes produces in the Czech Republic are made from brown coal. The major components of high- temperature coal combustion fly ashes are SiO_2, Al_2O_3, Fe_2O_3, TiO_2, CaO. An amount of aluminium oxide in Czech fly ashes reaches about 30% [4, 5]. Therefore, fly ash could be used as a raw material for alumina production.

There are several types of processes of alumina production from coal combustion fly ash: leaching processing, whereas leaching agent is usually used acid [6, 7], sintering processing used several types of sintering agents [8, 9] and combinations of leaching and sintering processes or other special processes [10–12].

Sintering processes are high-temperature reactions of the alumina ores with sintering additives. The sinter is leached to extract soluble Al_2O_3 compounds. The Al_2O_3 precipitate from leach liquors in a purified form [13]. Lime-soda sinter process used as sintering additives $CaCO_3$ (lime) and Na_2CO_3 (soda). The reaction is carried out at temperatures above 1100°C. The main products of the reaction are water-soluble sodium aluminate ($NaAlO_2$) and water-insoluble ($2CaO{\cdot}SiO_2$) [14]. Sintered material is solved in water or diluted alkali solution [15]. The reaction of sintering additives with fly ash is quite similar to the reaction of kaolinite (1) [16]:

$$\begin{aligned} &Al_2O_3 \cdot 2SiO_2 \cdot 2H_2O + 4CaCO_3 + Na_2CO_3 \rightarrow \\ &\rightarrow Na_2O \cdot Al_2O_3 + 2(2CaO \cdot SiO_2) + 5CO_2 + 2H_2O \end{aligned} \quad (1)$$

In addition to the mentioned main phases, others such as calcium aluminates, CaO-Al_2O_3-SiO_2 and Na_2O-Al_2O_3-SiO_2 compounds may arise [16]. The low C/S (CaO/SiO_2) molar ratio can lead to insoluble $Na_2O \cdot Al_2O_3 \cdot 2SiO_2$. On the contrary, the high C/S molar ratio, in turn, promotes the formation of insoluble $4CaO \cdot Al_2O_3 \cdot Fe_2O_3$. Both cases lead to losses of Al_2O_3 [9].

The paper studied the phase composition of the sintered material and the influence of the ratio of sintering additives to it.

2. Experimental

2.1. Materials

High-temperature brown coal combustion fly ash from Czech power plant, soda (Na_2CO_3) and lime ($CaCO_3$) were used for samples preparation. Phase composition of fly ash was determined by XRD on the device Empyrean (Panalytical) (see table 1). The chemical composition of used fly ash was examined by X-Ray fluorescence on the device Xenemetric EX-6600 SSD. The content of major oxides is given in table 2, other oxides were present in an amount less than 0.2%.

Table 1. The phase composition of fly ash.
(+++ majority phase, ++ significant phase, + minority phase)

Mullite	Quartz	Hematite	Magnetite	Amorphous
+++	++	+	+	+++

Table 2. The chemical composition of fly ash (major oxides) [wt. %].

SiO_2	Al_2O_3	CaO	Na_2O	K_2O	MgO	SO_3	Fe_2O_3	TiO_2	P_2O_5
47.3	29.8	4.31	0.88	1.51	1.29	1.06	19.2	1.43	0.27

2.2. Samples preparation

Fly ash was mixed with sinter additives, sodium and calcium carbonates. The amount of additive was such that N/A molar ratio was between 1.3–1.8, where N means Na_2O from Na_2CO_3 and A means Al_2O_3 from the fly ash. The value of C/S molar ratio was between 2.0–2.3, where the C means CaO from $CaCO_3$ and S means SiO_2 from the fly ash. In the sample name, the first number means N/A, the second number means C/S. The list of prepared samples is summarized in the table 3. The table 4 shows weight of sinter additives for individual molar ratios.

Table 3. The list of prepared samples (N/A= Na_2O/Al_2O_3 molar ratio, C/S= CaO/SiO_2 molar ratio).

Name	N/A	C/S	Name	N/A	C/S	Name	N/A	C/S	Name	N/A	C/S
P_1.3_2.0	1.3	2.0	P_1.3_2.1	1.3	2.1	P_1.5_2.2	1.5	2.2	P_1.5_2.3	1.5	2.3
P_1.4_2.0	1.4	2.0	P_1.4_2.1	1.4	2.1	P_1.6_2.2	1.6	2.2	P_1.6_2.3	1.6	2.3
P_1.5_2.0	1.5	2.0	P_1.5_2.1	1.5	2.1	P_1.7_2.2	1.7	2.2	P_1.7_2.3	1.7	2.3
P_1.6_2.0	1.6	2.0	P_1.6_2.1	1.6	2.1	P_1.8_2.2	1.8	2.2	P_1.8_2.3	1.8	2.3
P_1.7_2.0	1.7	2.0	P_1.7_2.1	1.7	2.1						
P_1.8_2.0	1.8	2.0	P_1.8_2.1	1.8	2.1						

Table 4. Weight of sinter additives for individual molar ratios.

N/A	1.3	1.4	1.5	1.6	1.7	1.8
m (Na_2CO_3) [g/1 g fly ash]	1.41	1.52	1.63	1.73	1.84	1.95
C/S	2.0	2.1	2.2	2.3		
m ($CaCO_3$) [g/1 g fly ash]	5.52	5.79	6.07	6.35		

The mixture was homogenized in a vibratory mill for 60 s. Tablets with diameter 13 mm and high 5 mm were pressed from the homogenized mixture. The compressing pressure was 4 MPa and the compressing time was 3 min. The tablets were sintered on a platinum pod in a laboratory furnace at 1250°C for 30 min (heating rate 5°C/min; cooling rate 3°C/min to 500°C and then natural cooling).

The morphology of sintered tablets was determined by scanning electron microscope (SEM) on the device ZEISS EVO LS 10. The composition of sintered material was analyzed by X-Ray diffraction (XRD).

3. Results and discussion

The mixture of fly ash, lime and soda was pressed to tablets for studying the products of lime-soda sinter process. The figure 1 shows that tablets shrunk and changed the colour after the sintering process. The shrinking of tablets is a result of sodium and calcium carbonate decomposition. Because of the high-temperature of reaction carbonate oxide evaporate and the resulting structure is sintered. The morphology of sintered tablets was observed by SEM and the example can be seen in figure 2.

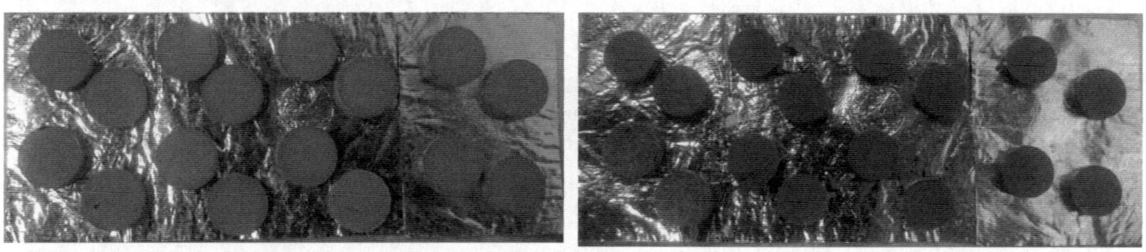

Figure 1. Tablets before (left) and after (right) sintering.

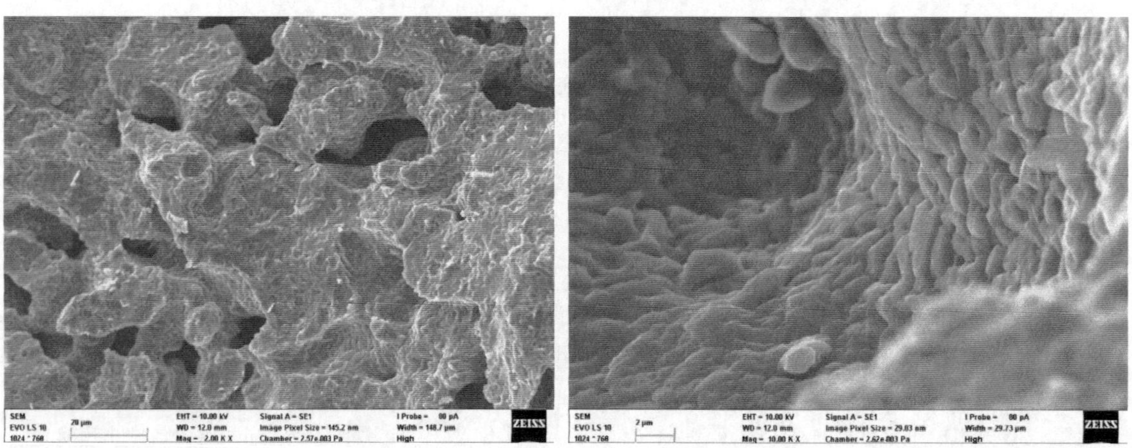

Figure 2. The microstructure of sample P_1.6_2.0, 10 000 × (left) and 2 000 × (right) (SEM).

The phase composition of sintered material was studied on four series of samples. The difference was in the N/A and C/S molar ratios. The results of XRD analyses are shown in figures 3–6.

Figure 3 shows the phase composition of sintered samples with the C/S molar ratio of 2.0. The N/A molar ratio differ from 1.3 (bottom) to 1.8 (up). The major phase in all samples is Ca_2SiO_4 ($2CaO{\cdot}SiO_2$), which is a reaction product of silicon phases from fly ash and CaO [17]. It is very slowly soluble in water, what makes it possible to separate it from alumina phases [18]. The next silicate phase which is present in all samples is $Na_2Ca_3Si_2O_3$. Its content decreases with increasing N/A molar ratio. In samples with N/A molar ratio from 1.3 to 1.7 is present SiO_2 (quartz). Quartz is an unreacted fly ash component. Its content increase with increasing N/A molar ratio to N/A = 1.6, its amount is minimal in sample P_1.7_2.0. In the sample P_1.8._2.0, there is not any unreacted phase from fly ash or sintering additives. In samples with N/A = 1.3–1.7 are present aluminium-silicate phases as $NaAlSiO_4$, $Na_8Al_4Si_4O_{18}$, $Na_{1.95}(Al_{1.95}Si_{0.05}O_4)$. The occurrence of such phases is undesirable because it decreases alumina recovery [16]. The phase which is desired because it is well soluble in water and is thus suitable for the separation of alumina is $NaAlO_2$ [18]. Sodium aluminate is present in sample P_1.7_2.0 and P_1.8_2.0, whereas in sample P_1.8_2.0 $NaAlO_2$ is the only aluminium phase. So for the C/S = 2.0 is the ideal N/A molar ratio of 1.8.

In the series with C/S = 2.1 which is shown in figure 4 the situation is quite similar. The main phase in all samples is Ca_2SiO_4. Quartz is present samples with N/A = 1.3–1.7, whereas the highest amount of it is in P_1.5_2.1. Sodium-calcium silicates ($Na_2Ca_3Si_2O_8$, Na_2CaSiO_4) are present in all samples. Their amount decrease with increasing N/A molar ratio. Aluminium-silicate phases $NaAlSiO_4$, $Na_8Al_4Si_4O_{18}$ were detected just in samples with low N/A molar ratio, N/A = 1.3 and 1.4, but $Na_{1.95}(Al_{1.95}Si_{0.05}O_4)$ is in all samples except P-1.8_2.1. Sodium aluminate, as desired aluminium phase, is present from N/A = 1.5. Its amount increases with increasing N/A molar ratio. In samples with N/A = 1.5–1.7 $NaAlO_2$ exists next to $Na_{1.95}(Al_{1.95}Si_{0.05}O_4)$, in P_1.8_2.1 the sodium aluminate is the only aluminium phase. So the ideal N/A ratio for C/S = 2.1, similar to C/S = 2.0, is 1.8.

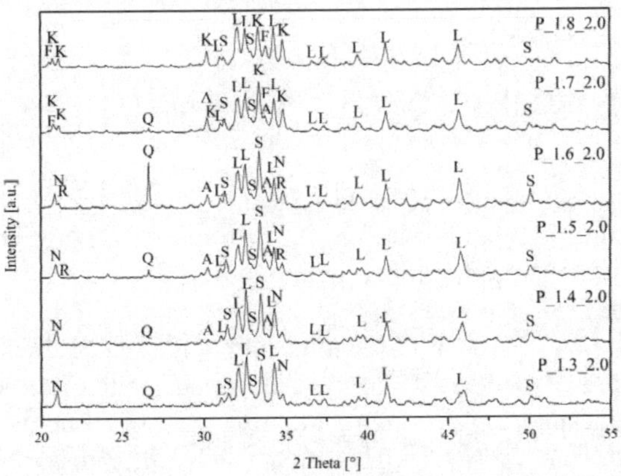

Figure 3. The phase composition of sintered samples with N/A = 1.3–1.8 and C/S = 2.0 (XRD).

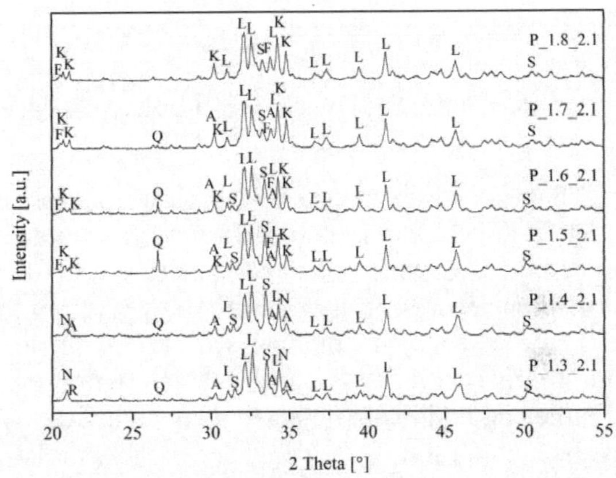

Figure 4. The phase composition of sintered samples with N/A = 1.3–1.8 and C/S = 2.1 (XRD).

K = $NaAlO_2$, F = Na_2CaSiO_4, L = Ca_2SiO_4, S = $Na_2Ca_3Si_2O_8$, Q = SiO_2, N = $NaAlSiO_4$, A = $Na_{1.95}(Al_{1.95}Si_{0.05}O_4)$, R = $Na_8Al_4Si_4O_{18}$

The results described above show, that N/A = 1.3 and 1.4 are too low. The samples contain a large number of phases. Aluminium is located in various aluminium-silicate phases and $NaAlO_2$ is not present.

The series with C/S = 2.2 contain four samples with N/A from 1.5 to 1.8. All samples of the series contain Ca_2SiO_4, Na_2CaSiO_4 and $NaAlO_2$. P_1.5_2.2 contains quartz, as unreacted phase, $Na_2Ca_3Si_2O_8$, $Na_2Ca_3Si_3O_{10}$ and aluminium-silicate phases $NaAlSiO_4$ and $Na_{1.95}(Al_{1.95}Si_{0.05}O_4)$. Although the sample contains $NaAlO_2$ which allows the separation of aluminium, the chosen ratio is not suitable since the presence of aluminium-silicates will reduce the efficiency of the process. Combination of N/A = 1.6 and C/S = 2.2 is not suitable, too. The sample still contains small amount of aluminium-silicate $NaAlSiO_4$ and quartz, which amount is higher than in sample P_1.5_2.2. The amount of $NaAlO_2$ is higher and it increases with increasing N/A molar ratio. The phase composition of samples P_1.7_2.2 and P_1.8_2.2 seems to be very similar. There are present only three phases, Ca_2SiO_4, $NaAlO_2$ and

Na_2CaSiO_4. $NaAlO_2$ is the only aluminium phase, thus means both samples are suitable for alumina recovery. The last series, with the C/S = 2.3, contains four samples. Two samples with higher N/A molar ratio (N/A = 1.7 and 1.8) contain just desired phases, Ca_2SiO_4, $NaAlSiO_4$ and Na_2CaSiO_4. In the samples with lower N/A molar ratio (N/A = 1.5 and 1.6), there are present $NaAlSiO_4$ and quartz. Their amount decreases with increasing N/A molar ratio.

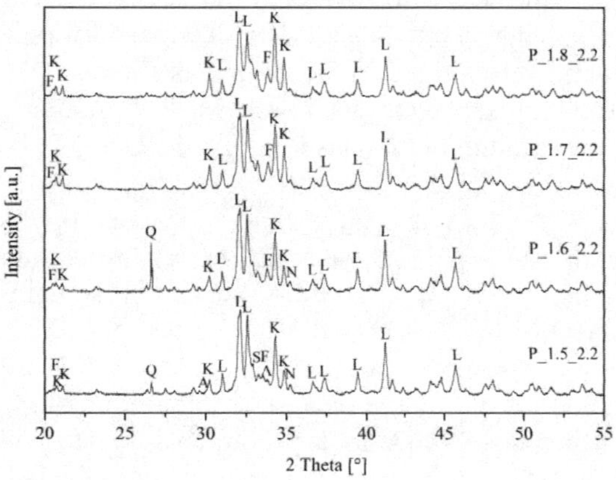

Figure 5. The phase composition of sintered samples with N/A = 1.5–1.8 and C/S = 2.2 (XRD).

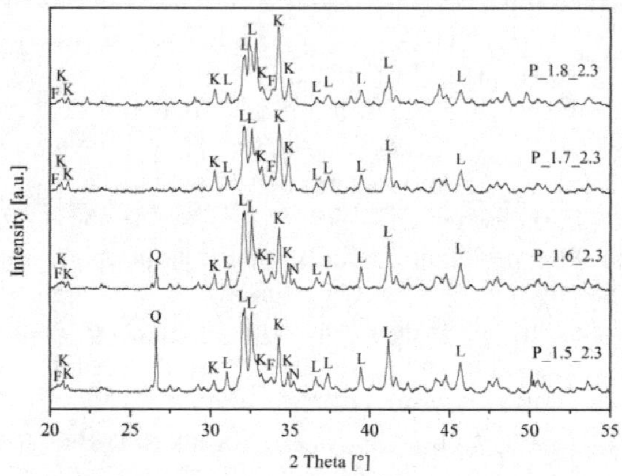

Figure 6. The phase composition of sintered samples with N/A = 1.5–1.8 and C/S = 2.3 (XRD).

K = $NaAlO_2$, F = Na_2CaSiO_4, L = Ca_2SiO_4, S = $Na_2Ca_3Si_2O_8$, Q = SiO_2, N = $NaAlSiO_4$, A = $Na_{1.95}(Al_{1.95}Si_{0.05}O_4)$

The results of our experiments show, that for sintering temperature 1250°C and sintering time 30 min is for the lime-soda sinter process the most suitable sample P_1.7_2.2 or P_1.8_2.0. Samples with higher ratio have the same phase composition, but they are not economically advantageous. An overview of the phases present in each sample is given in the table 5.

Our results do not coincide with the conclusions of the work [16]. Padilla and Sohn achieved the best aluminium recovery at comparable temperature and time (1200–1250°C and 20–30 min), but the ratios of sintering additives to fly ash were significantly lower. According to the results of their experiments, the best conditions were described as N/A molar ratio of 1.3 and C/S molar ratio of 1.7–1.8. Bai et. al. in their article [9] write, that the best conditions of the process are 1200°C, 60 min, C/S = 2.0 and N/A = 0.98. Even if the sintering time is twice time longer, N/A molar ratio is half in according to our experiments. It is likely that the lime-soda sinter process will be heavily influenced by the specific composition of fly ash, so it will not be possible to generalize the ideal conditions.

Table 5. The list of phases present in sintered samples.

Phase	P_1.3_2.0	P_1.4_2.0	P_1.5_2.0	P_1.6_2.0	P_1.7_2.0	P_1.8_2.0	P_1.3_2.1	P_1.4_2.1	P_1.5_2.1	P_1.6_2.1	P_1.7_2.1	P_1.8_2.1
$NaAlO_2$					×	×			×	×	×	×
$Na_{1.95}Al_{1.95}Si_{0.05}O_4$		×	×	×	×		×	×	×	×	×	
$NaAlSiO_4$	×	×	×	×			×	×				
$Na_8Al_4Si_4O_{18}$			×	×			×	×				
SiO_2	×	×	×	×	×		×	×	×	×	×	
Ca_2SiO_4	×	×	×	×	×	×	×	×	×	×	×	×
$Na_2Ca_3Si_2O_8$	×	×	×	×	×	×	×	×	×	×	×	×
Na_2CaSiO_4					×	×			×	×	×	×

Phase	P_1.5_2.2	P_1.6_2.2	P_1.7_2.2	P_1.8_2.2	P_1.5_2.3	P_1.6_2.3	P_1.7_2.3	P_1.8_2.3
$NaAlO_2$	×	×	×	×	×	×	×	×
$Na_{1.95}Al_{1.95}Si_{0.05}O_4$	×							
$NaAlSiO_4$	×	×			×	×		
$Na_8Al_4Si_4O_{18}$								
SiO_2	×	×			×	×		
Ca_2SiO_4	×	×	×	×	×	×	×	×
$Na_2Ca_3Si_2O_8$	×							
Na_2CaSiO_4	×	×	×	×	×	×	×	×
$Na_2Ca_3Si_3O_{10}$	×	×			×	×		

4. Conclusion

The presented work focused on the lime-soda sinter process as a possible method for aluminium recovery from the brown coal combustion fly ash. Lime and soda were used as sintering additives. The essence of the process is to convert the aluminium components of the fly ash to water-soluble sodium aluminate and the silicate phases of the fly ash to insoluble dicalcium silicate. Sintering temperature 1250°C and sintering time 30 min were used in experiments. Sodium aluminate as the only aluminium phase in the sintered material was in samples with N/A = 1.8 and C/S = 2.0, 2.1, 2.2, 2.3 and in samples with N/A = 1.7 and C/S = 2.2, 2.3. The next phases present in the said samples were dicalcium silicate and sodium-calcium silicates. With regard to the economic aspect, the best samples for alumina recovery are sample with N/A = 1.7 and C/S = 2.2 and sample with N/A = 1.8 and C/S = 2.0.

Acknowledgement

This study was supported by the project of specific research FCH-S-18-5194 with the financial support of Brno University of Technology.

References

[1] *World aluminium ...the website of the International Aluminium Institute* 2018 (*Online* http://www.world-aluminium.org/statistics/alumina-production/#data)
[2] *Bauxite Residue Management: Best Practice* 2015 (International Aluminium Institute)
[3] Ouellet V, Bergeron S and Verville D 2007 Bayer Process Control at Alcan Vaudreuil Works *IFAC Proceedings Volumes* **40** pp 25–8
[4] Skvara F, Svoboda P, Doležal J, Kopecký L, Pawlasová S, Myšková L, Lucuk M, Dvořáček K, Beksa M and Šulc R 2006 Concrete based on fly ash geopolymer *The Tenth East Asia-Pacific*

Conference on Structural Engineering and Construction pp 407–12

[5] Beranova D, Opravil T, Ptacek P and Snop R 2018 Release of ammonia from conventional power plant fly ash after the introduction of SNCR process *IOP Conference Series: Mater. Sci. Eng.* **379**

[6] Nayak N and Panda C 2010 Aluminium extraction and leaching characteristics of Talcher Thermal Power Station fly ash with sulphuric acid *Fuel* **89** pp 53–8

[7] Li L, Wu Y, Liu Y and Zhai Y 2011 Extraction of Alumina from Coal Fly Ash with Sulfuric Acid Leaching Method *The Chinese Journal of Process Engineering* **11** pp 254–8

[8] Wang X, Wang B, Xiao Y, Liu X, Zhaou L, Yu G and Sun Q 2014 The Optimization of Sintering Process for Alumina Extraction from Fly Ash *Advanced Materials Research* **878** pp 264–70

[9] Bai G, Teng W, Wang X, Qin J, Xu P and Li P 2010 Alkali desilicated coal fly ash as substitute of bauxite in lime-soda sintering process for aluminum production *Trans. Nonferrous Met. Soc. China* **20** pp 169–75

[10] Sun Y, Liang Z, Sun F, He S, Wang B and Wang Y 2017 Recovery of Alumina from Coal Fly Ash by $CaCl_2$ Calcination Followed by H_2SO_4 Leaching *J. Environ. Anal. Toxicol.* **07**

[11] Shemi A, Ndlovu S, Sibanda V and van Dyk L D 2015 Extraction of alumina from coal fly ash using an acid leach-sinter-acid leach technique *Hydrometallurgy* **157** pp 348–55

[12] Shemi A, Mpana R N, Ndlovu S, van Dyk L D, Sibanda V and Seepe L 2012 Alternative techniques for extracting alumina from coal fly ash *Miner. Eng.* **34** pp 30–7

[13] Goodboy K 1976 Investigation of a Sinter Process for Extraction of Al_2O_3 From Coal Wastes *Metall. Trans. B* **07** pp 716–8

[14] Cservenyak F 1947 Recovery of alumina from Kaolin by the Lime-Soda Sinter Process: Report of Investigations

[15] Padilla R and Sohn H 1985 Sodium Aluminate Leaching and Desilication in Lime-Soda Sinter Process for Alumina from Coal Wastes *Metall. Trans. B* **16** pp 707–13

[16] Padilla R and Sohn H 1985 Sintering Kinetics and Alumina Yield in Lime-Soda Sinter Process for Alumina from Coal Wastes *Metall. Trans. B* **16** pp 385–95

[17] Goodboy K 1976 Investigation of a Sinter Process for Extraction of Al_2O_3 From Coal Wastes. *Metall. Trans. B*. 1976 **7** pp 716–8

[18] Grim R, Machin J and Bradley W 1945 *Alumina Extraction: Amenability of Various Types of Clay Minerals to Alumina Extraction by the Lime Sinter and Lime-Soda Sinter processes* (Urbana: Illinois State Geological Survey) pp 47

IOP Conf. Series: Materials Science and Engineering **583** (2019) 012027 doi:10.1088/1757-899X/583/1/012027

Rheological properties of belite-rich cement doped with sulfur

M Bohac[1], D Kubatova[1], A Rybova[1], A Zezulova[1] and T Stanek[1]

[1] Research Institute for Building Materials, Hněvkovského 30/65, Brno 617 00, Czech Republic

Email: bohac@vustah.cz

Abstract. Reactive belite cement based on clinker doped with sulfur is a new type of low-energy cement that is still in the laboratory testing phase. In terms of application, one of the desired properties of new cement is its well-defined rheological behavior in fresh state. The chemical composition of the belite cement was determined by wet chemistry analysis, the phase composition of the belite cement clinker by quantitative X-ray powder diffraction analysis and microscopic point counting and specific surface by Blaine method. Rheological properties were determined by rotational rheometer in flow and oscillation regime. Shear and time dependent rheological parameters were calculated from flow curves. Stability of belite cement paste was judged based on the critical strain determination in amplitude sweep test. Setting was evaluated by Tussenbrock needle penetration test and oscillation time test on rheometer. The properties of the laboratory belite cement were correlated to industrial cement CEM I, CEM II/A-LL and CEM III/A. Reactive belite cement pastes show good flow properties both due to small amount of alite and larger amount of rounded belite grains and clusters. Belite rich cement show decent stability comparable to CEM III and very good dispersing efficiency of PCE based HRWR.

1. Introduction

The cement industry is one of the primary producers of greenhouse gas CO_2. Around 7% of global anthropogenic CO_2 emissions are related to the production of Portland cement, mainly due to the limestone and clay burning at around 1450°C [1].

There are several alternative clinker systems that deserve serious attention with respect to global reductions in concrete-related CO_2 emissions. Reactive belite-rich Portland cement clinkers (RBPC) is one of them. RBPC belong to the same family as ordinary Portland cement (OPC) in terms of clinker phase composition, which comprises C_2S-C_3S-C_3A-C_4AF phases. Unlike sulfobelite cements the RBPC doesn't contain Klein's compound. The main difference in phase composition between RBPC and OPC lies mainly in the belite/alite ratio. For RBPC the belite is the most abundant phase, as opposed to alite in OPC.

The RBPC manufacture leads to lower specific energy consumption and CO_2 emissions. Furthermore, there is an advantage of requiring less high-grade limestone as a raw material. The burning temperature for RBPC is about 100°C lower than for OPC, which can lead to somewhat lower kiln heat consumption and permit more use of low-grade kiln fuels [2].

Staněk and Sulovský [3] discussed the mechanism of activation during preparation of belite-rich clinkers with an increased Ca:Si ratio in the structure of C_2S and partial substitution of SiO_4^{4-} by SO_4^{2-}. Activation was done by sulfate ions, which substitute SiO_4 in the structure of belite causing an increase of Al_2O_3 incorporation in the belite and increased CaO:SiO_2 ratio in belite. The sulfur doping stabilizes alite M1 modification in the clinker.

There are several papers on rheology of sulfobelite cement but only few about belite cement without Klein's compound. The behavior of belite-cement in presence of polynaphthalene sulfonate and polycarboxylate types of high-range water-reducers (HRWR) was already discussed in [4]. Authors reported that lower reactivity and rounded particles of belite cement compared to ordinary cement causes lower shear thickening response regardless of the w/c and type of HRWR in use.

2. Experimental

2.1. Materials and methods

Belite clinker doped with sulfur was prepared in laboratory scale. Common materials for cement production were used for raw meal preparation: limestone, clay shale, quartz sand, iron ore tailings and FGD (flue gas desulfurization) gypsum as the source of SO_3. Raw materials were ground in laboratory ball mill in amount of 10 kg on fineness characterized by sieve residue 12% on 0.09 mm sieve. Tablets were prepared from homogenized raw meal and burnt in Kanthal furnace at 1350°C for 60 min. Clinker was quenched rapidly in the air on piece of metal outside the furnace. Quantitative phase composition of clinker determined by XRD with Rietveld refinement and by point counting in light microscope is given in table 1. In case of point counting method, the quantitative composition (wt. %) was calculated from area and specific gravity of clinker phases.

Specific gravity of the clinker determined by pycnometer method is 3299 kg/m^3. Cement was prepared by grinding the clinker with 2% of gypsum in laboratory ball mill. Specific gravity of the cement was 3259 kg/m^3 and specific surface 400 m^2/kg. SO_3 content in the cement was 3.93 wt. %.

Particle size distribution (PSD) of the belite cement was determined by a laser diffraction method using CILAS 920L laser particle size analyzer. The range of the analyzer is 0.3–400 μm. Sample was treated with ultrasound (60 s) before measurement and dispersing medium was isopropyl alcohol. PSD of RBPC is given in figure 1.

Setting of the belite-rich cement was monitored by oscillation test monitoring evolution of complex modulus (G^*/t) on rotational rheometer, Tussenbrock needle penetration test (τ/t), semiadiabatic (T/t) and isothermal (heat flow/t) calorimetry. Calorimetric methods were used for correlation of rheological properties with heat evolution during early hydration.

Flow parameters calculated using Bingham [5] and Herschel-Bulkley model were tested by rotational rheometer DHR-1 (TA Instruments) using geometry of coaxial cylinders. Samples were pre-sheared with 150 s^{-1} for 60 seconds and the logarithmic sweep was used with 5 points/decade, 5 sec of steady state sensing and 5 sec of averaging time. Yield stress - τ_y (Pa), plastic viscosity ν_{pl} (Pa.s) and thixotropy/rheopexy (Pa/s) were determined from hysteresis loops of ascending and descending curves in 1 s^{-1} – 150 s^{-1} – 1 s^{-1} shear rate region.

Table 1. Quantitative phase composition of the clinker.

	XRD	Point counting
	vol. %	wt. %
β-C_2S	86.5	84.7
C_3S	7.4	8.2
C_3A cub.	1.4	0.5
C_3A orth.	0.9	-
C_4AF	2.4	5.1
CaO	1.4	1.5

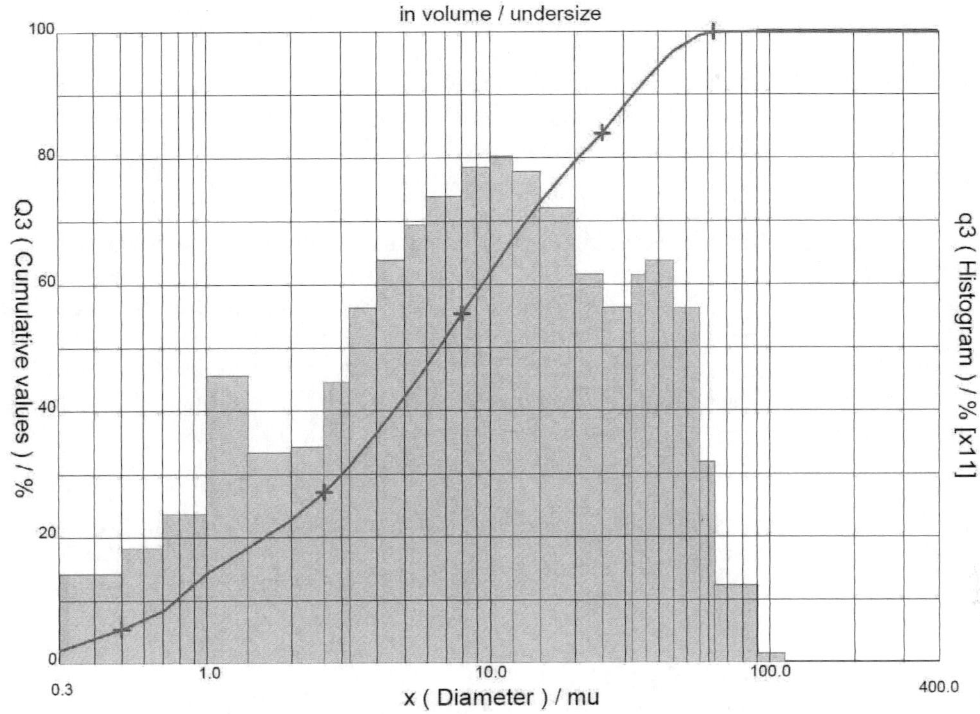

Figure 1. PSD of belite cement doped with S.

Linear viscoelastic region and critical strain (%) was determined in oscillation regime by monitoring storage and loss moduli (G' and G") in chosen region of oscillation strain γˆ(%) and frequency 1 Hz. Geometry of parallel plates with cross-hatched surface was used. All rheological and calorimetric testing was done on cement pastes with w/c = 0.32 at 25°C. From two to four measurements were performed for each sample to keep standard deviation below 5%. For each sample, the measurement started after 5 minutes after adding of water.

3. Results and Discussion

3.1. Setting of RBPC

Setting of the RBPC paste was followed by combination of needle penetration Tussenbrock method, oscillation test on rotational rheometer monitoring evolution of complex modulus (*G*/t*), semiadiabatic (*T/t*) and isothermal (*heat flow/t*) calorimetry (figure 2 and 3).

In case of Tussenbrock test, threshold shear stress is related to penetration resistance calculated from the depth of indentation of needle. It shows slow increase of shear stress up to 190 min followed by steeper trend till the end of setting at 270 min. Different trend comparing to needle penetration test was found for setting test done on rotational rheometer (figure 2). Evolution of *G** related to setting f cement paste is strongly affected by structural break down, since the measurement is continuous and the bonds formed during setting is continuously broken by oscillatory movement of the upper plate. The values are more related to dynamic than to static yield stress; as can be seen in the case of needle penetration test. Both static and dynamic yield stresses are critical for many field applications [6]. Despite the fact that geometry of parallel plates with cross-hatched surface was used, the test is limited by the adherence of the measured material to the plate. The course of the curve is gradual without significant acceleration of the increase in the values associated with the onset of setting.

Two different calorimetric methods were used to describe the beginning of the hydration of RBPC (figure 3). The belite cement was compared to three industrially produced cements CEM I 42.5 R, CEM II/A-LL 42.5 R and CEM III/A 42.5 N (Hranice plant). Due to preparation of the sample, the semiadiabatic method measurement starts up to 5 minutes after the addition of water. In case of isothermal calorimetry, the measurement starts within few seconds after addition of water and the large exothermic signal related mainly to initial reactions of aluminate phase and tempering heat is monitored. The period of slow reactions characterized both by slow development of T and heat flow resp. in time is followed by main exothermic peak related to hydration of C_3S and formation of portlandite (CH) and C-S-H. The onset of the C-S-H formation is related to the start of the setting of cement pastes. In terms of length of dormant period and start of the setting, the RBPC is very similar to CEM I, it means there is no dilution effect of supplementary cementitious materials as it is for CEM II and III. The important fact is that the RBPC contains small amount of C_3S which is the main contributor of the heat at this acceleration period of hydration. Belite hydration takes place at later times and thus not affecting setting process. Isothermal calorimetry is able to monitor very subtle differences of the rate in which the heat is dissipated due to its sensitivity, which is in the range of μm, and the ability to separate individual exotherms compared to semiadiabatic calorimetry. On the other hand, semiadiabatic calorimetry reflects more the autocatalytic nature of the reactions.

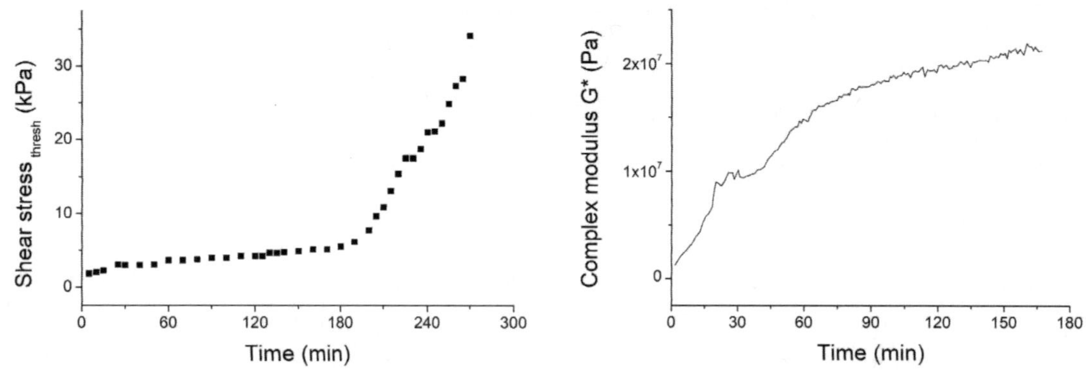

Figure 2. Setting of the RBPC, Tussenbrock needle test and oscillation time test.

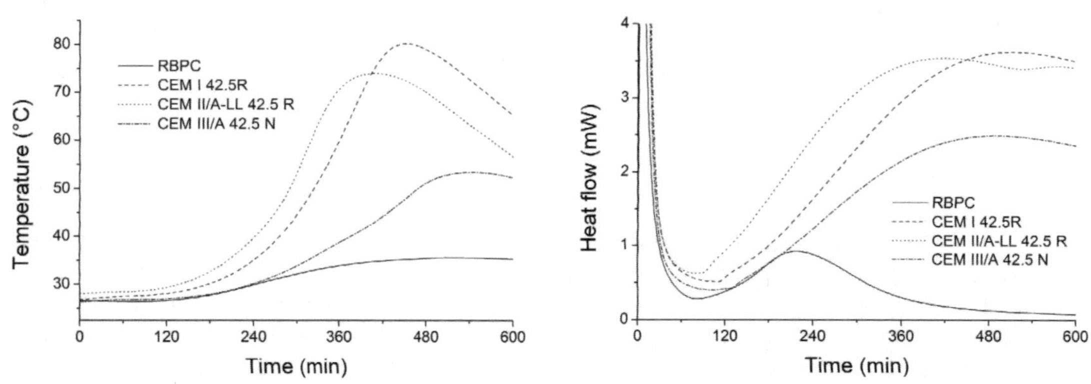

Figure 3. Period of slow reactions, semiadiabatic and isothermal calorimetry.

3.2. Rheological parameters and stability of RBPC compared to industrial cements

Rheological parameters yield stress and plastic viscosity were calculated from flow curves using the Bingham model (figure 4). Stability of the cement paste was tested by amplitude sweep test to determine critical strain (%) [7], which is defined as the cut-off point between linear and nonlinear regions of strain sweep (figure 4). The mechanisms for the strain of the yield stress and the critical strain of linear viscoelastic region of cement pastes were investigated by Roussel [8]. RBPC parameters were correlated to industrial type of cements. It has been previously published [9–13] that in case of OPC strain was 10^{-5} to 10^{-1}, while the frequency was kept constant at 1 Hz until the structure broke down. In view of previous studies, the RBPC show decent values of critical strain revealing promising stability of the material. Results of amplitude sweep show that stability of the RBPC paste is closer to CEM II and CEM III than to CEM I.

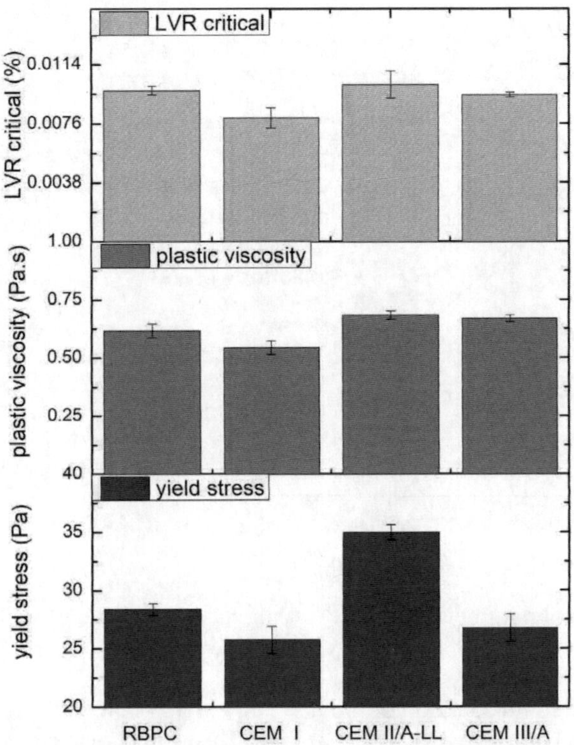

Figure 4. Yield stress, plastic viscosity and critical strain of RBPC and industrial cements.

Yield stress, as an important parameter of cementitious material qualifying flowability [14], of RBPC is close to CEM I and CEM III and plastic viscosity is similar to CEM III. RBPC, same as other belite cements [4], is expected to improve rheology of cement-based materials because of the rounded grains of C_2S and their relatively lower reactivity at early age compared to C_3S.

3.3. Rheology of RBPC with PCE superplasticizer

The role of various dosage of PCE superplasticizer on RBPC rheology was determined by rotational rheometer calculating rheological parameters yield stress, plastic viscosity and rate index with Herschel-Bulkley model from flow curves. SP positively decreases values of yield stress, plastic viscosity and thixotropy with increasing dosage (figure 5). Several instrumental approaches how to determine thixotropy have been proposed through last decades [8, 15–18]. Shear-thinning behavior of the paste [4, 19, 20] (rate index values >1) is more pronounced with increasing dosage of SP. In terms of plasticization, PCE SP works similarly to OPC. RBPC show very good dispersing efficiency of PCE based HRWR, which compatibility with OPC based materials has been discussed intensively [21–23].

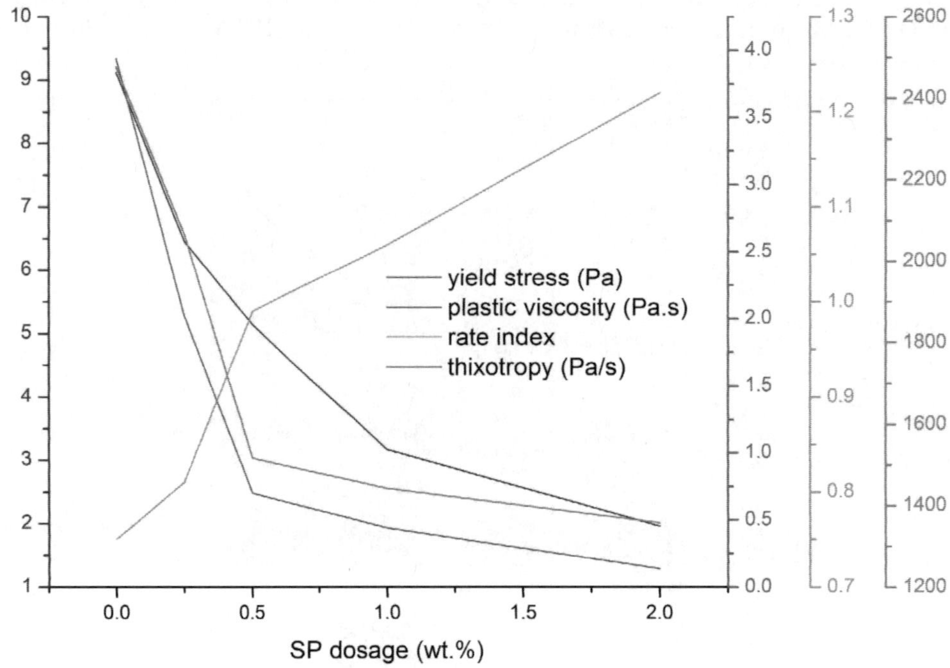

Figure 5. Rheological parameters of RBPC with various dosage of PCE superplasticizer.

4. Conclusion

Prior pilot testing and market introduction of the new low-energy cement, the good workability in fresh state is one of the crucial parameters of the material. This study proved that RBPC has good flow properties comparable to industrial cement due to its high belite content. Setting and heat evolution during period of slow reactions of RBPC is ruled by the presence of alite and its relatively small content. Setting reactions were followed by four different methods showing pros and cons of each. Amplitude sweep tests revealed improved stability of RBPC compared to CEM I. Testing PCE based HRWR compatibility, the RBPC showed decent dispersing efficiency and predictable shear-dependent behavior.

Acknowledgements

Thanks are due to project No. 19-05762S financed by the Czech Science Foundation.

References

[1] Worrell E, Price L, Martin N, Hendriks C and Meida L O 2001 Carbon dioxide emissions from the global cement industry *Annu. Rev. Energy Environ.* **26** pp 303–329
[2] Gartner E and Sui T 2018 Alternative cement clinkers *Cem Concr Res* **114** pp 27–39
[3] Staněk T and Sulovský P 2015 Active low-energy belite cement *Cem Concr Res* **68** pp 203–210
[4] Yahia A and Tanimura M 2015 Rheology of belite-cement – Effect of w/c and high-range water-reducer type *Constr Build Mater* **88** pp 169–174
[5] Bingham E C 1922 Fluidity and plasticity (McGraw-Hill Book Company, Incorporated)
[6] Roussel N 2006 A thixotropy model for fresh fluid concretes: theory, validation and applications, *Cem Concr Res* **36** pp 1797–1806
[7] Yuan Q, Zhou D, Khayat K H, Feys D and Shic C 2017 On the measurement of evolution of structural build-up of cement paste with time by static yield stress test vs. small amplitude oscillatory shear test *Cem Concr Res* **99** pp 183–189
[8] Roussel N, Ovarlez G, Garrault S and Brumaud C 2012 The origins of thixotropy of fresh cement pastes *Cem Concr Res* **42** pp 148–157

[9] Yuan Q, Lu X, Khayat K H, Feys D and Shi C 2017 Small amplitude oscillatory shear technique to evaluate structural build-up of cement paste *Mater Struct* **50** pp 1–12
[10] Papo A and Caufin B 1991 A study of the hydration process of cement pastes by means of oscillatory rheological techniques *Cem Concr Res* **21** pp 1111–1117
[11] Schultz M A and Struble L J 1993 Use of oscillatory shear to study flow behavior of fresh cement paste *Cem Concr Res* **23** pp 273–282
[12] Nachbaura L, Mutina J C, Nonata A and Choplin L 2001 Dynamic mode rheology of cement and tricalcium silicate pastes from mixing to setting *Cem Concr Res* **31** pp 183–192
[13] Betioli A M, Gleize P J P, Silva D A, John V M and Pileggi R G 2009 Effect of HMEC on the consolidation of cement pastes: isothermal calorimetry versus oscillatory rheometry *Cement and Concrete Research* **39** pp 440–445
[14] Ferraris C F 1999 Measurement of the rheological properties of high performance concrete: state of the art report *J. Research-National Inst. Stand. Technol.* **104** pp 461–478
[15] Qian Y and Kawashima S 2018 Distinguishing dynamic and static yield stress of fresh cement mortars through thixotropy *Cem Concr Comp* **86** pp 288–296
[16] Lapasin R, Longo V and Rajgelj S 1979 Thixotropic behaviour of cement pastes *Cem Concr Res* **9** pp 309–318
[17] Wallevik J E 2005 Thixotropic investigation on cement paste: experimental and numerical approach *J Newt fluid Mech* **132** pp 86–99
[18] Mahaut F, Mokeddem S, Chateau X, Roussel N and Ovarlez G 2008 Effect of coarse particle volume fraction on the yield stress and thixotropy of cementitious materials *Cem Concr Res* **38** pp 1276–1285
[19] Struble L J and Schultz M A 1993 Using creep and recovery to study flow behavior of fresh cement paste *Cem Concr Res* **23** pp 1369–1379
[20] Larrard F De, Ferraris C and Sedran T 1998 Fresh concrete: a Herschel-Bulkley material *Mater Struct* **31** pp 494–498
[21] Anderson P J and Roy D 1988 The effect of superplasticizer molecular weight on its adsorption on, and dispersion of cement *Cem Concr Res* **18** pp 980–986
[22] Shaughnessy R and Clark P E 1988 The rheological behavior of fresh cement paste *Cem Concr Res* 18(3) pp 327–341
[23] Jolicoeur C and Simard M A 1998 Chemical admixture–cement interactions: phenomenology and physico-chemical concepts *Cement Concr Compos* **20** pp 87–101

The influence of specimen size on the dynamic properties of lightweight concrete determined by the resonance method

D Kocab[1], P Zitt[1], P Danek[1], M Canek[1], M Alexa[1] and P Hanus[1]

[1] Brno University of Technology, Faculty of Civil Engineering, Veveří 331/95, Brno, Czech Republic

Email: dalibor.kocab@vutbr.cz

Abstract. The paper examines the way specimen size influences measurements by the resonance method. The experiment was performed with a lightweight concrete with a porous aggregate (Liapor). Several sets of specimens were made; they differed both in size and type (drilled cylindrical cores, cast prisms). All the specimens were tested for the natural frequency of longitudinal, transverse, and torsional vibration, which were then used to calculate values of the dynamic modulus of elasticity. The outcome of this experiment is a statistical analysis of the test results.

1. Introduction

Concrete, being the most widely used building material, is subject to strict requirements set by the modern society. For some time now, its quality has not been assessed on the basis of compressive strength only. Today, there are also environmental factors or matters of sustainability to consider [1, 2], which is naturally reflected in the requirements. Both the industry and researchers are still more frequently concerning themselves with properties that have to do with concrete durability (and by extension the service life of concrete structures), such as surface permeability or shrinkage [3, 4].

The modulus of elasticity (Young's modulus) plays an important role in this regard. A current trend in civil engineering is to build very slender and elegant structures, something which cannot be achieved without prestressing. Whole structures or their parts are thus susceptible to various strains and may experience vibrations, deflections, cracking, or creep. Young's modulus is closely linked with all these phenomena [5, 6]. Young's modulus also functions in the most advanced model of concrete shrinkage, which is a numerical model designed by a team led by professor Z. P. Bažant [7].

Thanks to a wide choice of available lightweight aggregate (with a broad range of bulk density, particle size, and strength), a high-quality lightweight concrete can be designed with a great spectrum of compressive strength and bulk density. Lightweight concrete can be used as an open-structure concrete for thermal insulation as well as dense lightweight structural concrete [8, 9]. Lightweight concretes have been used in Scandinavia and Canada for decades; however, in recent years lightweight concrete structures have been spreading all over the world [8, 10, 11]. One of the benefits of lightweight concrete is that it reduces the vertical load of the structure, precast elements are less costly to transport, requiring less heavy machinery for handling. Its other strong points are good thermal insulation properties (proportional to bulk density), better durability (determined by less severe shrinkage and lower permeability, stronger interfacial transition zone, and excellent freeze-thaw resistance), and, when the lightweight aggregate is made from industrial by-products, it is more eco-friendly. On the other hand, some downsides are that the concrete can be brittle if designed as a structural concrete with higher

IOP Conf. Series: Materials Science and Engineering **583** (2019) 012028 doi:10.1088/1757-899X/583/1/012028

strength (due to the high strength of the cement paste), higher temperature during ageing due to more hydration heat, or difficulties with pouring or pumping because of the high water absorption of the lightweight aggregate [8, 12].

2. Experiment

The experiment used the resonance method to determine selected dynamic parameters of lightweight concrete using specimens of varying size. The goal was to ascertain to what extent the specimen size can affect the value of the modulus of elasticity.

2.1. Material

A lightweight concrete with a lightweight aggregate (Liapor) was designed. Both fractions of this aggregate were pre-soaked in water for 24 hours prior to mixing. The composition of the lightweight concrete is detailed in table 1. The water/cement ratio was 0.46 and the concrete was defined using fresh-state and 28-day properties; see table 2.

Table 1. Fresh concrete composition.

Component	Content per 1 m^3 of fresh concrete (kg)
Cement CEM II/B-S 32.5 R	450
Sand 0/4 mm (Bratčice)	776
Aggregate Liapor 1/4 mm	52
Aggregate Liapor 4/8 mm	198
Admixture Sika VZ 10	0.68
Admixture Sika Stabilizer-4R	0.45
Water	208

Table 2. Concrete properties in fresh state (FS) and hardened state (HS) at 28 days of age; the number in brackets is the sample standard deviation.

Property	Value
Flow value (FS)	450 mm
Bulk density (FS)	1,620 kg/m^3
Cubic compressive strength (HS)	23.4 (1.0) N/mm^2
Cubic tensile splitting strength (HS)	2.40 (0.08) N/mm^2
Bulk density (HS)	1,640 (28) kg/m^3
Water content (HS)	13.4 (0.1)%

2.2. Test specimens

The experiment used 6 cubes of 150 mm, 8 prisms with the nominal dimensions of 100 × 100 × 400 mm (henceforth P100), 3 prisms of 150 × 150 × 600 mm (P150) and one prism of 300 × 300 × 900 mm (P300). All the specimens except for the largest prism were compacted on a vibration table; prism P300 was compacted with an immersion vibrator (figure 1). When the moulds were filled with the fresh concrete, they were covered with a PE sheet and stored on a level surface under standard laboratory conditions. The concrete aged in the moulds for 3 days, after which it was removed and placed under water. The exception was once again prism P300, which was left in the mould - it was thoroughly sprinkled with water, covered with several layers of wet fabric and wrapped in a PE sheet. At the age of 28 days all the specimens were removed from the water and prism P300 was demoulded.

Figure 1. Manufacturing of a prism of 300 × 300 × 900 mm (P300).

Figure 2. Drilling cores from prism P300.

Next, 3 cube specimens were tested for compressive strength and the other 3 for shear splitting strength; the strength data was supplemented with a determination of bulk density. Moisture content was also measured on all 6 cubes after the the strength tests. Table 2 lists all this data. The other specimens were left in standard laboratory conditions without any curing for 15 more months; the dynamic properties were therefore measured at the age of approximately 16 months. After the measurements by the resonance method, core samples were drilled from prism P300 - 6 had the nominal dimensions of 150 × 300 mm (figure 2), 6 of 100×300 mm, and 6 of 75 × 300 mm. The cores were left in laboratory conditions for a week so that all the water that was used during the drilling could evaporate. Afterwards the cores (henceforth CS) were tested by the resonance method.

2.3. Resonance method

Any solid body begins to vibrate when excited by an external impulse. This vibration can take several forms; however, the assessment of the dynamic material properties of regular bodies is performed only with the natural frequencies of longitudinal (f_L), torsional (f_t), and flexural (f_f) vibration [13, 14]. During resonance the amplitude of the vibration in the specimen rapidly increases at the moment when the excitation force frequency is identical to the natural (i.e. resonance) frequency of the specimen – this holds as long as the external impulse is continuous (it lasts for a longer time). However, the basic resonance frequently can also be determined using interrupted (impulse) vibration – in this case the impulse is typically a mechanical strike of an impact hammer [15]. Using an accelerometer and a Fourier analyzer the natural frequency of the specimen's vibration can be easily determined. The position of the accelerometer and the point of impact is shown in figure 2.

The actual measurement of natural frequencies was performed using a Handyscope HS4 oscilloscope with an acoustic emission sensor. The frequencies were evaluated using a software (an accessory to the oscilloscope), which operates on the basis of fast Fourier transformation. Figure 4 shows the determination of f_L on prism P300, figure 5 shows the determination of f_t of prism P100. Following the standard ČSN 73 1372 [16], Young's modulus E_{crL} the dynamic shear modulus G_{cr} were calculated.

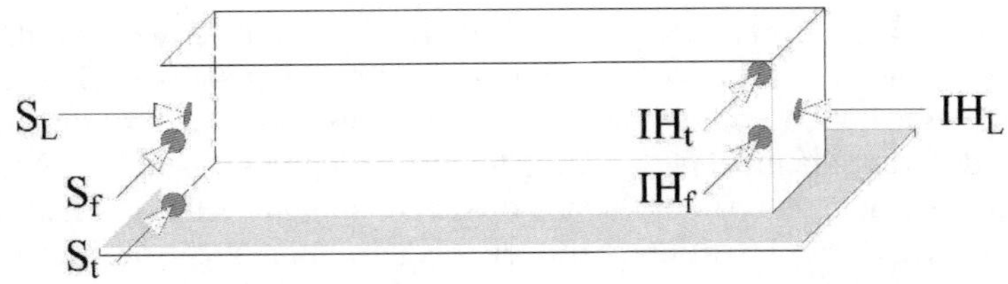

Figure 3. Position of the sensors (S) and the impact hammer strike point (IH) during the measurement of the natural frequencies of longitudinal (L), flexural (f), and torsional (t) vibration.

Figure 4. Determination of the natural frequency of longitudinal vibration in prism P300.

Figure 5. Determination of the natural frequency of torsional vibration in prism P100.

3. Results

The values of the dynamic modulus of elasticity E_{crL} and G_{cr} for the lightweight concrete with Liapor are plotted in figure 6 and 7. Results for the prism of 300 × 300 × 900 mm are only represented by one value, because there was only one specimen of these dimensions. Results for the prism of 150 × 150 × 600 mm are represented by three values, results for prism P100 by eight values, and results for all the cores are represented by six values. The graphs show the average values for each specimen set and the error bar represents sample standard deviation (except for prism P300, of course).

All data sets (i.e. the values of the dynamic parameters measured on all specimens – again with the exception of set P300, which consisted of a single prism) were first checked for data normality on a significance level of 0.05. The results for each set were statistically compared using a two-sample *t*-test, again on a level of significance of 0.05. The statistical test was set so as to compare the equality of the mean values of the test results. If the hypothesis is rejected, the result sets can be considered different with a statistical significance.

The statistical analysis shows that Young's modulus E_{crL} as well as the shear modulus G_{cr} measured on the core specimens are differ with a statistical significance from the values measured on the prism specimens. Results for prism P300 could not be included in the statistical comparison – this is why it is only possible to say that the dynamic parameters reached the lowest values; however, when compared to prisms P150 this difference is only very small. Although the values of dynamic parameters measured for prisms P100 are on average higher than prisms P150, they do not differ with a statistical significance. Similarly, the results for each test set of core specimens do not bear a statistically significant difference either. The above applies to both properties being examined. Basically the same conclusions were drawn by [17, 18], the only difference being that they determined the Young's modulus either by the ultrasonic pulse velocity test or the static test. The influence of specimen size was investigated e.g. by [19], and with similar results, only they used a different property as the evaluation criterion.

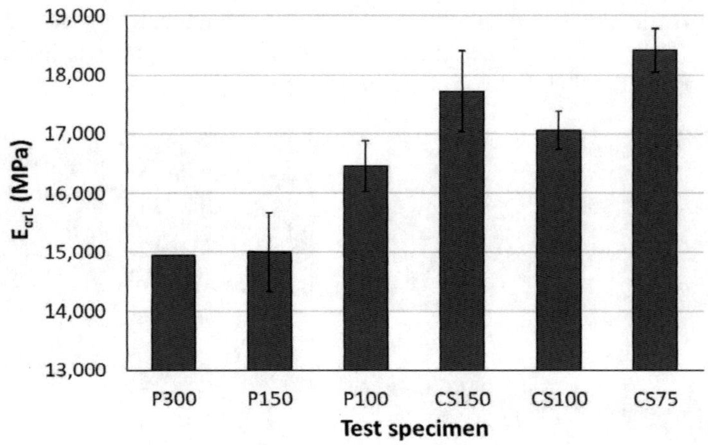

Figure 6. A bar chart of the average values of dynamic Young's modulus; the error bar represents the sample standard deviation.

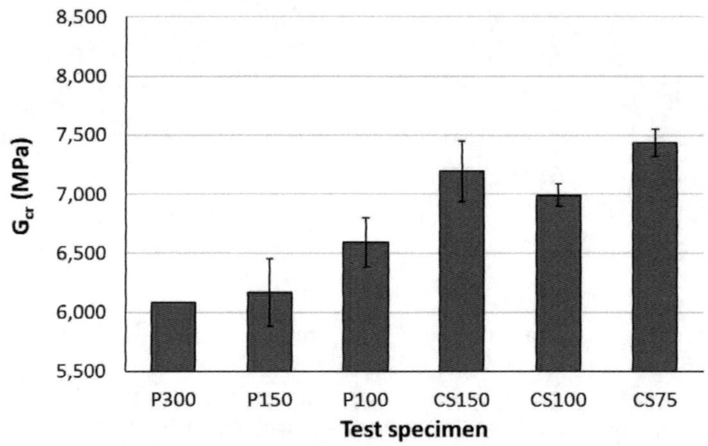

Figure 7. A bar chart of the average values of dynamic shear modulus; the error bar represents the sample standard deviation.

4. Conclusion

The experiment results indicate that the larger the prism specimen, the smaller (but only slightly) the value of E_{crL} and G_{cr} when measured by the resonance method. In cast specimens, this conclusion applies only to the mean values of the data sets, because this decrease is not statistically significant. It must also be emphasised that the difference in the specimen size was not particularly large – it would therefore be interesting to expand the experiment by specimens that have a volume in orders of cubic metres.

In the case where core specimens were drilled from a larger specimen (a prism with the dimensions of 300 × 300 × 900 mm), the influence of specimen size on the results is statistically significant on a significance level of 0.05. The moduli of elasticity E_{crL} and G_{cr}, determined by the resonance method on the core specimens, reached 114 to 123% the value of the same parameter determined by the same method on prism P300, from which they were drilled.

Acknowledgements

This work was supported by the GA0 - Czech Science Foundation [grant No. GA17-14302S].

References

[1] Teplý B, Vymazal T and Rovnaníková P 2017 Introduction to an approach to performing sustainability quantification of concrete structures *Solid State Phenomena* **272** pp 273–279
[2] fib Bulletin No. 28 2004 *Environmental Design* (Lausanne, Switzerland: fib) 80
[3] Han S, Park W and Yang E 2013 Evaluation of concrete durability due to carbonation in harbour concrete structures *Construction and Building Materials* **48** pp 1045–1049
[4] Yang J, Wang Q and Zhou Y 2017 Influence of Curing Time on the Drying Shrinkage of Concretes with Different Binders and Water-to-Binder Ratios *Advances in Materials Science and Engineering* **2017** pp 1–10
[5] Neville A 2012 Properties of Concrete (New Jersey: Prentice Hall)
[6] Kockal N U 2015 Validity and reliability of estimated modulus of elasticity of cementitious materials *IOP Conf. Series: Materials Science and Engineering* **103** pp 1–6
[7] RILEM draft recommendation: TC-242-MDC multi-decade creep and shrinkage of concrete 2015 *Materials and Structures* **48**(4) pp 753–770
[8] Hubertová M 2012 Lehký beton | lightweight concrete *Beton TKS* **7** pp 107–119
[9] Ke Y, Beaucour A L, Ortola S, Dumontet H and Cabrillac R 2009 Influence of volume fraction and characteristics of lightweight aggregates on the mechanical properties of concrete *Construction and Building Materials* **23** pp 2821–2828
[10] *Wellington Stadium* 2019-02 (*Online:* https://www.escsi.org/wp-content/uploads/2017/10/4800-Wellington-Stadium-NEW-VERSION.pdf)
[11] Chandra S and Berntsson L 2002 *Lightweight Aggregate Concrete: Science, Technology and Applications* (USA: William Andrew Publishing)
[12] Brooks A L, Zhou H and Hanna D 2018 Comparative study of the mechanical and thermal properties of lightweight cementitious composites *Construction and Building Materials* **159** pp 316–328
[13] Park S-J, Yim H J and Kwak H-G 2014 Nonlinear resonance vibration method to estimate the damage level on heat-exposed concrete *Fire Safety Journal* **69** pp 36–42
[14] ASTM C215-14 2014 *Standard Test Method for Fundamental Transverse, Longitudinal, and Torsional Resonant Frequencies of Concrete Specimens* (West Conshohocken, PA: ASTM International)
[15] Park S-J and Yim H J 2016 Evaluation of residual mechanical properties of concrete after exposure to high temperatures using impact resonance method *Construction and Building Materials* **129** pp 89–97
[16] ČSN 73 1372: 2012 *Non-destructive testing of concrete – Testing of concrete by resonance method* (Prague: ÚNMZ)
[17] Lee B J, Kee S-H, Oh T and Kim Y-Y 2015 Effect of Cylinder Size on the Modulus of Elasticity and Compressive Strength of Concrete from Static and Dynamic Tests *Advances in Materials Science and Engineering* **2015** ID 580638 p 12
[18] Matulić P, Juradin S, Marušić E and Domazet A 2016 Effect of test specimen size on mechanical properties of shotcrete *GRAĐEVINAR* **68**(4) pp 301–309
[19] Wang S, Zhang M and Quek S T 2011 Effect of Specimen Size on Static Strength and Dynamic Increase Factor of High-Strength Concrete from SHPB Test *Journal of Testing and Evaluation* **39**(5) pp 898–907

Proposal of methodology for testing resistance of building materials against mold infestation

I Kukletova[1] and I Chromkova[1]

[1] Research Institute for Building Materials, Hněvkovského 30/65, Brno 617 00, Czech Republic

Email: kukletova@vustah.cz

Abstract. Molds attack buildings on a broad scale, cause significant damage and carry a number of negative effects on human health. The design of the methodology is based on EN 60068-2-10 "Environmental Impact Testing – Parts 2-10: Tests – Test J and Instructions: Mold Growth" and ČSN 72 4310 "Testing the Resistance of Building Products and Materials against Mold". However, the first of these standards is intended for electrical industry and the other was published in 1975 with no later revisions. The design of the new methodology aims at simplifying and updating and, last but not least, imitating the success of already implemented methodology for testing resistance to green algae growth. The proposed methodology was successfully verified on both standards and real samples.

1. Introduction

Due to biological infestation, properties of building materials can change. The process of reducing the qualitative properties of building materials as a result of life processes and activities of living organisms is called bio-corrosion and the organisms are called biodeteriogens. In building materials such as building stone, concrete, plaster, or plastics, a high degree of bio-corrosion occurs. Biodeteriogens need appropriate conditions for their life, such as substrate humidity, heat, nutrients, favorable pH of substrate (acidity or alkalinity), and oxygen (except anaerobic bacteria). Some species need light [1, 2].

The entire complex of biodeteriogens that interact with each other usually enters the bio-corrosion process of materials. It is a combination of physical and chemical actions, sometimes even other influences. Growth processes of microorganisms cause pressures that disturb the integrity of matter. Products of metabolic transformation affect stability of mineral components of the building material and dissolve them. This results in an ion exchange between the organism and the material. Practical consequences of these phenomena are destructive changes of the material that manifest themselves as functional (physicochemical, chemical, etc.) and morphological (color spots, pitting, etc.) [3].

A noteworthy group of organisms involved in bio-corrosion are microscopic fibrous fungi. They are multiplied by mycelium growth and by spores. Spores are mold reproductive units that germinate in wet places with nutrients, then grow and produce reproductive organs. From them, ripe spores are released into the environment and being very small and light, carried by the air over long distances [2, 4]. In suitable conditions for sporulation, they grow on walls or other surfaces in the indoor and outdoor environments of buildings. The limiting factor for mold growth on walls and various substrates in houses and apartments is moisture. However, high humidity is unnecessary as long as the water is bioavailable. Based on the species representation we can deduce the substrate moisture level and vice versa. In places where moisture is evident at first glance, hydrophilic genera is found most often, such as *Alternaria, Epicoccum, Fusarium, Phoma, Stachybotrys, Trichoderma, Ulocladium*. In humid areas, genera *Cladosporium, Mucor, Rhizopus* often occur. *Aspergillus* and *Penicillium* are xerophilic and usually found in drier areas [5, 6].

Generally, microscopic fungi in the interior spaces, in addition to disrupting the substrate on which they grow, bring further undesirable consequences. They can have a very negative impact on human health, as they are significantly involved in a wide range of allergic and fungal diseases. They create specific proteins, glycoproteins, polysaccharides and many other substances in their hyphae and spores, which can cause allergic reactions in sensitive individuals after inhalation. The most common fungus in the air and on the walls, such as *Cladosporium* and *Alternaria*, are also the most important allergens [5]. Other products of these microscopic fungi may be mycotoxins, which are less likely to enter the atmosphere but may have carcinogenic and toxic effects. Most of these fungi can also have pathogenic effects on immune-impaired individuals. Some species also release volatile metabolites into the environment, such as secondary alcohol 1-octen-3-ol (characteristic fungal odor), various aldehydes, ketones, mono- and sesquiterpens, often foetid. These substances often cause headaches and eye burning. Microscopic fungi may be one of the factors contributing to the syndrome of diseased buildings [4, 6].

Testing the resistance of building materials against mold growth is dealt with by ČSN 72 4310 Testing the resistance of building products and materials against mold [7], but this standard was issued in 1975 and the design of the tests is time consuming and technically demanding. The design of the new methodology is based on this standard and newer standards EN 60068-2-10 Environmental Impact Testing - Parts 2-10: Testing - Test J and Instruction: Mold Growth [8], which is intended for the electrical industry and EN 15458 Paints - Laboratory method for testing the efficacy of preservatives in the coating against water algae [9].

The design of the new methodology aims at simplifying and updating the test and, last but not least, imitating the success of already implemented methodology for testing resistance to green algae growth. The method mainly uses macroscopic observations of mold growth on building materials under defined conditions. Several variants were examined to find the least technical demanding and yet functional method.

2. Methodology proposal

Test specimens or filter paper with tested substance are placed on the surface of Sabouraud agar in individual Petri dishes. The test is performed in 2–3 parallel determinations. Two types of culture controls are prepared - a nutrient medium dish and a dish with nutrient medium and sample-sized filter paper on the surface of the agar. It is also possible to use a negative control - a filter paper impregnated with a biocidal substance. The samples are sterilized by a UV-C lamp prior to the testing.

Both the samples and the surface of the agar are covered with a thin layer of the mixed mold culture. A maximum of 0.5 ml of suspension is applied per 100 cm^2 of the surface of the specimen. The inoculation mold culture is prepared by washing the individual mature mold cultures prepared from the collection strains. Strains can be selected according to source standards, taking into account the strains typical of the material and environment. At least 5 species of strains are used. The suspension is prepared by washing the spores away from the culture with mineral solution (3 g of $NaNO_3$, 1 g of KH_2PO_4, 0.5 g of $MgSO_4 \cdot 7H_2O$, 0.5 g of KCl, 0.01 g of $FeSO_4 \cdot 7H_2O$ in 1000 ml of distilled water, sterilization for 30 minutes at 0.1 MPa). Inoculation takes place under sterile conditions. The resulting spore concentration in the solution should be $1–2 \cdot 10^6$ spores / ml.

The incubation was carried out at 24 ±2°C and relative humidity 96 ±2%. Petri dishes were placed in a sealed cultivation vessel, where high relative humidity was maintained by a saturated Na_2SO_4solution. A desiccator was used as a suitable vessel, as its shape eliminates dripping of the condensing water onto the samples. For the same reason, Petri dishes were tested open. he vessels were placed in a cultivation box. The test lasted 28 days. During the incubation, the culture vessel was opened once a week in the biohazard box, exchanging air and documenting the condition.

During the test and at the end of the test, the growth of fungi on the test bodies and the nutrient medium in the vicinity was visually evaluated. Mold spores in both positive controls must be germinated within 6 days of incubation.

Resistance is then assessed as follows:

- fungicidal material - mold does not develop, inhibitory zone is visible on substrate agar,
- fungistatic material - mold does not develop on specimen, colonies are not visible in microscope, the mold grows on agar.
- Resistant material - Quantification 1–5 by scale:

1 - slight growth, growth visible only in microscope, scattered colonies
2 - gradual growth - numerous small colonies, up to 25% of the surface
3 - intense growth, up to 50% of the surface
4 - very intense growth, up to 75% of the surface
5 - full growth - 100% of the surface

3. Testing procedure

A number of samples were used to verify the design of the methodology - pulp, filter paper impregnated with Ajatin, fiber-cement board, standard concrete and a set of concretes with various additives. The testing bodies were sized 65 × 30 × 10 mm or 50 × 50 × 10 mm. The spore suspension was prepared from a collection of strains kept in the laboratory, using the following strains: *Aspergillus versicolor, Penicilium purpurogeum, Phoma sp., Alternaria alternata, Cladosporium cladosporoides.*

The tests were carried out according to the described test design and were incubated and monitored for 6 months.

Figure 1. Petri dishes in the cultivation vessel.

4. Results and evaluation

Tests of all materials were conducted according to the described methodology and their results corresponded to the composition of the sample. Parallel determinations showed insignificant differences and were evaluated consistently. Spore germination was observed on day 3 after inoculation. The prolonged cultivation time confirmed no changes in molding and change in overall material evaluation(see figures 2 and 3). Therefore, it is unnecessary to maintain the standard testing period of 3–6 months.

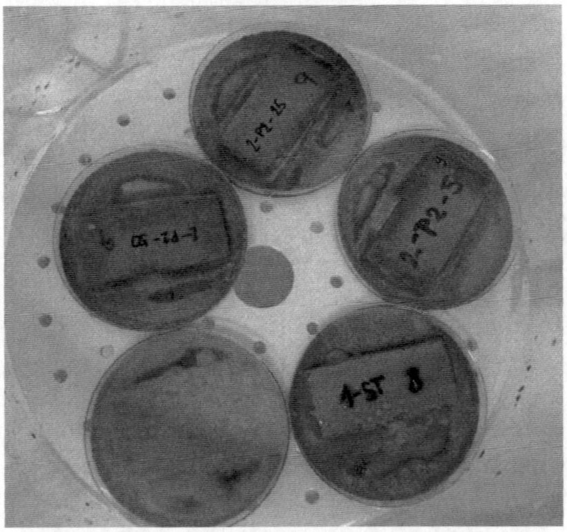

Figure 2. Samples after 28 days of incubation.

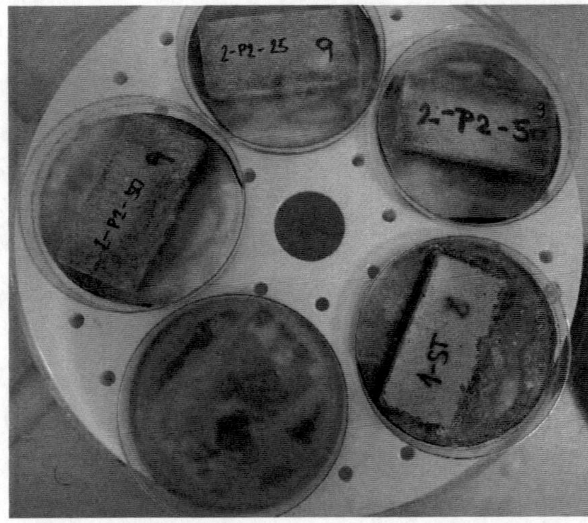

Figure 3. Samples after 6 months of incubation.

The results of individual sample testing are given in table 1. Figures 4 and 5 are photographs of samples graded 1 and 5, respectively.

Table 1. Evaluation of the resistance against mold testing.

Sample	Evaluation
Ajatin 6%	fungicidal
Pulp	5
Standardconcrete	1
glass-fiber-cement board	1
Concrete with admixture 1	2
Concrete with admixture 2	4
Concrete with admixture 3	3
Concrete with admixture 4	1
Concrete with admixture 5	2
Concrete with admixture 6	4
Concrete with admixture 7	5
Concrete with admixture 8	3
Concrete with admixture 9	4

Figure 4. Sample of glass-fiber-cement board grade 1.

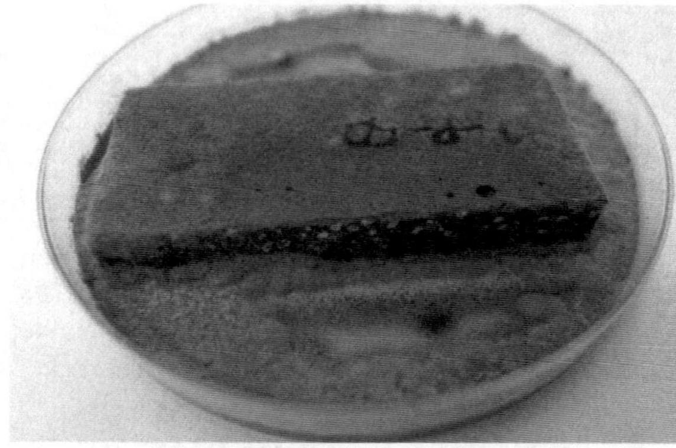

Figure 5. Sample of concrete with admixture 7 grade 5.

All grades of non-resistant materials were observed in the evaluation, only the biocidal product was fungicidal. The materials evaluated as grade 1 showed fungistatic effect on smooth surfaces, but small colonies grew on the areas damaged by cutting during sample preparation (see figure 6). This should be taken into account during preparation of testing specimens and their evaluation.

Figure 6. Growth of small colonies on cut surfaces.

5. Discussion

For testing of mold resistance in laboratory conditions, a number of standardized methods are available, apart from those mentioned above [7, 8, 9], mainly the ASTM C1338 [10] and ASTM G21-96 [11] methods and their derivatives. Some are intended for a particular type of material, others for a whole

IOP Conf. Series: Materials Science and Engineering **583** (2019) 012029 doi:10.1088/1757-899X/583/1/012029

group of materials. The principles of these methods are generally the same: mold spores are introduced onto surface of test material samples, incubated in a suitable environment and after several weeks, typically four, the surfaces of the test specimens are analyzed for mold growth. The methods differ particularly in the mold species used, the density of spores and the method of inoculation, the conditions of cultivation (relative humidity and temperature), the use of nutrient media and the final evaluation of the test. Some factors and their influence on the final results are discussed by professional public. Typically, five species of microscopic fungi are used to inoculate, and the methods combine different species. However, each mold has different ability to grow on different building materials, even under the same conditions, and also has different moisture requirements [6]. The use of nutrient media is also discussed as the real building materials are less nutritive, so the representativeness of the test decreases in presence of the medium [6]. Standard test protocols rely on the high relative humidity of the environment as a source of moisture, but it can be assumed that building materials will occasionally be exposed to direct humidity, however briefly. Real moisture and temperature fluctuations also occur in real constructions [12]. Furthermore, the microscope evaluation method is questionable, as the observer can overlook the mold growth without pigmentation [12]. Given all these controversies described, literary sources recommend caution when interpreting the results because in real applications, conditions are uncontrollable [12].

6. Conclusions

While mold resistance tests provide an overview of short-term performance of the samples under controlled conditions, they offer little information on the durability of long-term product performance in real-world applications. This must be taken into account when extrapolating the test results into real conditions. Also, the test results represent no metric value of absolute mold resistance for comparison between tests or between products.

This work aimed to propose a simple, technically low-demanding methodology for testing the resistance of building materials against mold infestation. The proposed methodology was validated on real samples and is suitable for screening material resistance, especially during research and development of new materials.

Acknowledgements

This paper was elaborated with the institutional support for long-term development of research organizations by the Ministry of Industry and Trade of the Czech Republic.

References

[1] Wasserbauer R 2001 *Biologické znehodnocení staveb* (Praha: ABF) p 257
[2] Büchli R and Raschle P 2004 *Řasy a houby na fasádách* (Ostrava: MISE, s.r.o.) p 108
[3] Ledererová J 2009 *Biokorozní vlivy na stavební díla* (Praha: Silikátový svaz) p 273
[4] Klánová K 2013 *Plísně v domě a bytě: odstraňování a prevence* (Praha: Grada) p 104
[5] Hanuláková J 2014 Plísně našich domácností a jejich odstraňování. Alergie a alergeny *13thSymp. lékařů ČR a SR Litomyšl* pp 33–38
[6] Flannigan B, Samson R A a Miller J D 2001 *Microorganisms in home and indoor work environments: diversity, health impacts, investigation and control* (Boca Raton, CRC Press) p 529
[7] ČSN 72 4310 (724310):1975 *Zkoušení odolnosti stavebních výrobků a materiálů proti plísním* (Praha: Úřad pro normalizaci a měření) p 11
[8] EN 60068-2-10 (345791):2008 *Environmental Impact Testing – Part 2-10: Tests – Test J and Instructions: Mold Growth* p 36
[9] EN 15458 (672032):2015 *Paints and varnishes – Laboratory method for testing the efficacy of film preservatives in coating against algae* p 16
[10] ASTM C1338 2014 *Standard test method for determining fungi resistance of insulation materials and facings* (ASTM International) p 3

IOP Conf. Series: Materials Science and Engineering **583** (2019) 012029 doi:10.1088/1757-899X/583/1/012029

[11] ASTM G21-96 2015 *Standard practice for determining resistance of synthetic polymeric materials to fungi* (ASTM International) p 6
[12] Johansson P 2014 *Determination of the Critical Moisture Level for Mould Growth on Building Materials* (Byggnadsfysik LTH, Lunds Tekniska Högskola, Report TVBH) vol 1020 p 208

IOP Conf. Series: Materials Science and Engineering **583** (2019) 012030 doi:10.1088/1757-899X/583/1/012030

The determination of measurement errors in impedance spectroscopy

M Lunak[1], I Kusak[1] and Z Chobola[1]

[1] Institute of Physics, Faculty of Civil Engineering, Brno University of Technology, Veveří 95, 602 00 Brno, Czech Republic

Email: kusak.i@fce.vutbr.cz

Abstract. The paper deals with the methodology, quantification and comparison of measurement deviations for two types of the ceramic shard. The ceramic material has better polarization properties than concrete material. For impedance measurement the ceramic shard is more appropriate. The IS method diagnostics based on processing of measured data which aims evaluating the absolute error and the relative error (deviation) of the loss factor values. Results show the spectrum parts which are useful in the lower frequency range only. When Agilent 33220A and Agilent 54645A are used to carry out the IS, the most important factor from the viewpoint of the tg$\delta(f)$ determination accuracy is the accuracy of the phase shift between the excitation signal and that at the measured system output. Other factors are the measurement accuracy of the electric capacity and resistance of the oscilloscope input channel.

1. Introduction

Impedance measurements are feasible with specialized devices (such as Agilent 4294A, 4284A), which are in most cases acquisition price exacting. Fortunately, there are cases where the impedance measurements can be carried out with alternative device blocks consisting of an exciting signal generator, a recording device and a data processing device. An impedance measurement apparatus consisting of a sine-wave voltage, a double-channel oscilloscope and a PC has been implemented at the Faculty of Civil Engineering of the University of Technology of Brno. Until now, experimental data in the form of the loss factor spectra versus frequency plots have been acquired and other frequency and impedance plots have been derived [1, 2].

The measurement reproducibility has been verified on various materials. The absolute errors of thus acquired physical quantity of the frequency spectrum have been assessed on the basis of mutual deviations of the spectrum individual points. The reliability of frequency interval readings has been established based on the observed deviations. So far, the measurement absolute error has not been determined from the input quantity accuracy.

The loss factor values [3–5] have been calculated from the formula:

$$\mathrm{tg}\delta = \frac{1}{\mathrm{tg}\varphi_1} \tag{1}$$

where:

$$\varphi_1 = \varphi_2 + \mathrm{arctg}\frac{\sin\varphi}{\frac{U_2}{U_1} - \cos\varphi} \tag{2}$$

φ is the phase difference between the voltages U_1 and U_2 of the respective exciting and the attenuated signal at the specimen output, φ_1 is the phase shift between the voltage and current vectors in the impedance consisting of the electric resistance R and the capacitance C of the oscilloscope input channel at the exciting signal frequency *f*.

φ_2 is determined from:

$$\varphi_2 = \mathrm{arctg}\frac{1}{2\cdot\pi\cdot f\cdot R\cdot C} \tag{3}$$

The values of *f*, φ, U_1 and U_2 are calculated as arithmetic means of the quantity measured. The measurement repetition number can be set in the automated measurement software, which controls Agilent 33220A generator, Agilent 54645A oscilloscope and the PC. In compliance with prof. Horák's Practical Physics textbook (basic principle of physical measuring in Czech language), the absolute error of tgδ of each point of the frequency spectrum is determined as follows [6]:

$$\delta_A(\mathrm{tg}\delta) = \sqrt{\begin{array}{l}\left(\frac{\partial \mathrm{tg}\delta}{\partial f}\cdot\delta_A(f)\right)^2 + \left(\frac{\partial \mathrm{tg}\delta}{\partial R}\cdot\delta_A(R)\right)^2 + \left(\frac{\partial \mathrm{tg}\delta}{\partial C}\cdot\delta_A(C)\right)^2 + \\ + \left(\frac{\partial \mathrm{tg}\delta}{\partial \varphi}\cdot\delta_A(\varphi)\right)^2 + \left(\frac{\partial \mathrm{tg}\delta}{\partial U_2}\cdot\delta_A(U_2)\right)^2 + \left(\frac{\partial \mathrm{tg}\delta}{\partial U_1}\cdot\delta_A(U_1)\right)^2\end{array}} \tag{4}$$

where $\delta_A(f)$ is the absolute error of the frequency which is used in the calculation. Absolute errors of the quantities *f*, φ, U_1 and U_2, *R*, *C are defined as follows*:

$\delta_A(f)$=*arithmetic mean of the measured frequency* / 10 000............Hz
$\delta_A(R)=R/100$ (corresponds to 1% accuracy)...................................Ω
$\delta_A(C)=C/100$ ((corresponds to 1% accuracy)...................................F
$\delta_A(\varphi)=1$..degree
$\delta_A(U_2)=0.0035/\sqrt{5}$ (corresponds to the amplitude measurement accuracy for 5 repeated measurements)......V
$\delta_A(U_1)=0.0035/\sqrt{5}$ (corresponds to the amplitude measurement accuracy for 5 repeated measurements)......V

Let the loss factor absolute error be denoted ERR and let the second powers of the terms in the parentheses under the second root sign be related to the variables under question and be denoted FREQ, CAPAC, REZIST, PHAS, U2, U1, respectively. The values of the terms will be expressed in numbers and their squared roots will be shown in diagrams for the sake of discussion. Furthermore, we will compare the incidence of the deviation with the loss factor values for two different specimens of the ceramic shard. The partial derivatives in the rem 4 have been calculated by Mathcad system.

2. Materials and methods

The ceramic specimens have been manufactured using Chvaletice fly ash, their firing temperature being 1 050°C. The firing losses amounted to 11.2%. The bulk density was about 2,005 kg/m^3. The water absorption capacity: 7.5%. The bending strength: 24 MPa. The granulometry: about 5% on a 63 mm sieve. The ceramic shard specimens differ from each other by the admixture:

Bentonite ($Al_2O_3\cdot 4SiO_2\cdot 9H_2O$) - clay. It is a white to light-yellow powder material featuring a specific chemical composition. It is used as ceramics plasticizing agent in the ceramic industry. In this research, a non-activated bentonite, in which the calcium and magnesium ions have not been replaced by activation by sodium ions, has been used. The apparent volume density was 936 kg/m^3, pH = 8.7. Bentonite admixture: 5%.

Water glass – sodium water glass Na_2SiO_3 of a silicate module of 1.6 and a density of 1560 kg/m^3 (KOMA a.s.) has been used to prepare the specimens.

The specimens were inserted between brass plate electrodes to be IS-tested. Agilent 33220A and Agilent 54645A were used to carry out the IS [7–10].

IOP Conf. Series: Materials Science and Engineering **583** (2019) 012030 doi:10.1088/1757-899X/583/1/012030

3. Results and discussion

The diagrams in figure 1 and figure 3 show the square roots of FREQ, CAPAC, etc. For the ceramic shard with 5% of bentonite (figure 1), the highest values are acquired by the terms representing the effect of the determination accuracy of the phase difference between U_1 and U_2 on the total absolute error of the loss factor. In the frequency range from 10^3 to 10^5 Hz an increased accuracy contribution is manifested by the determination of the electric capacitance and resistance, the oscilloscope input channel capacitance plus the electrode and coaxial cable capacitance. The remaining terms acquire negligible values.

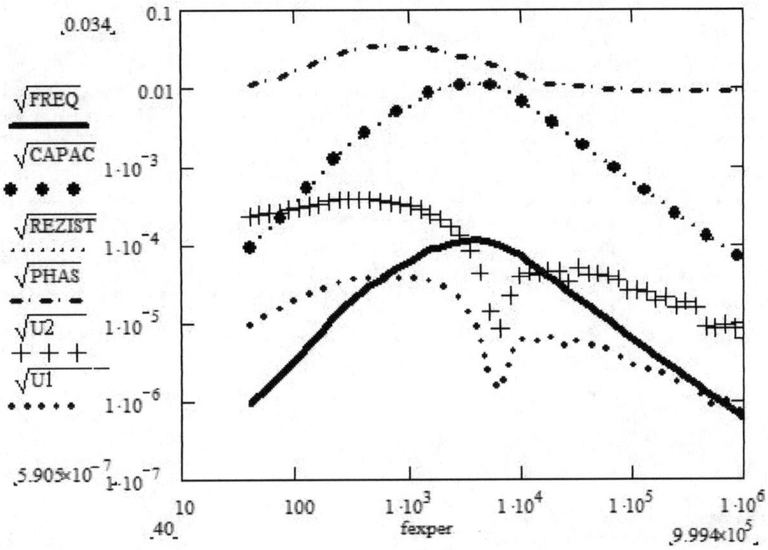

Figure 1. Square roots of FREQ, CAPAC, REZIST, PHAS, U2, U1 for 5% - bentonite ceramic specimens. They are expressed in relative frequency units, related to the experiment frequency of electric field in Hz.

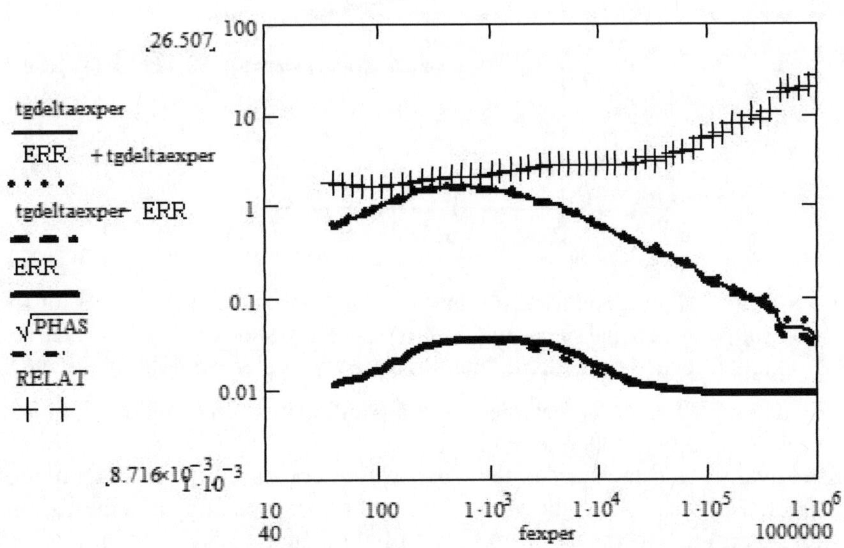

Figure 2. Loss factor and its absolute and relative errors versus frequency plots for the frequency spectrum and 5% - bentonite ceramics. They are expressed in relative frequency units, related to the experiment frequency of electric field in Hz.

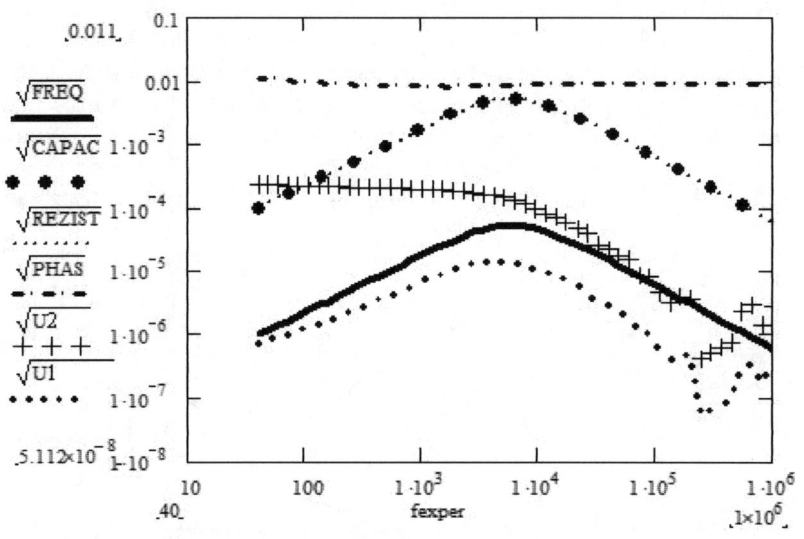

Figure 3. Square roots of FREQ, CAPAC, REZIST, PHAS, U2, U1 for sodium-water-glass ceramic specimens. They are expressed in relative frequency units, related to the experiment frequency of electric field in Hz.

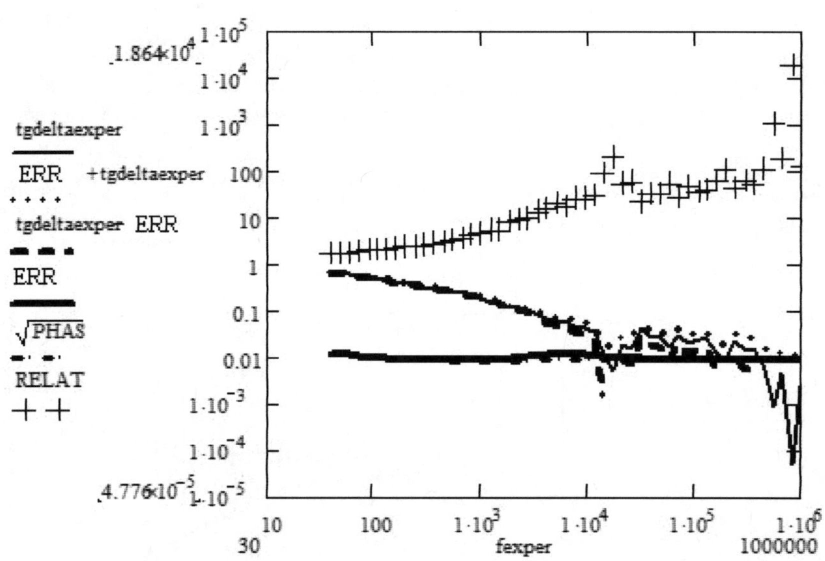

Figure 4. Loss factor and its absolute and relative errors versus frequency plots for the frequency spectrum and sodium-water-glass ceramic specimens. They are expressed in relative frequency units, related to the experiment frequency of electric field in Hz.

In the case of sodium-water-glass ceramic shard, a similar situation takes place, however, the values of all terms are several orders of magnitude lower.

The loss factor values and their absolute errors can be compared in the diagrams of figure 2 and figure 4. The bentonite ceramic shard shows the loss factor to exceed unity and the absolute error (ERR) of the loss factor determination ranges from 0.01 to 0.05. The loss factor values do not fall into this interval throughout the entire frequency spectrum. The absolute error (ERR) is growing slowly, exceeding 10% for tgδ below 0.1. The envelope of the tgδ(f) curve, which arose by adding and subtracting the absolute error to/from the loss factor values, forms a narrow band which broadens in the highest frequency range only.

For the sodium-water-glass ceramic specimens, the tg$\delta(f)$ spectrum shows a decreasing tendency, the respective values being below 0.7. For this material, the values of ERR do not fall below 0.01, however, they are significantly lower than those for bentonite ceramics. In figure 4, the loss factor gets very low values close to the absolute error value. In consequence, the relative error is growing up. Based on the relative deviation values, the loss factor value reliability and the experimental data from various parts of the frequency range can be assessed.

The values of accuracy are better for low frequency than for right side of frequency spectra from 10^4 Hz to 10^6 Hz. This is only for our measured data specific. Other scientific equipment as impedance analysers are, have typical accuracy 0.02% for precision bridge, but only up to 10^5 kHz or $2 \cdot 10^5$ kHz.

4. Conclusion

The IS method diagnostics based processing of intermediate data which aims at evaluating the absolute error and the relative error (deviation) of the loss factor values shows the spectrum parts to be applicable in the lower frequency range only. When the above mentioned instruments are used to carry out the IS, the most important factor from the viewpoint of the tg$\delta(f)$ determination accuracy is the accuracy of the phase shift between the excitation signal and that at the measured system output. Other factors are the measurement accuracy of the electric capacity and resistance of the oscilloscope input channel.

The paper compares two distinct spectra of tg$\delta(f)$ and its determination accuracy for two ceramic shard specimens. Different values of tg$\delta(f)$ have also been accompanied by different values of the standard deviation, which in this case have been directly proportional to the loss factor values.

In conclusion, it can be stated that the IS method measurement by means of Agilent 33220A generator, Agilent 54645A oscilloscope and plate electrodes can be considered satisfactory unless the loss factor value falls below 0.1.

Acknowledgments

This paper has been worked out under the project GAČR No. 19-11516S supported by Czech Science Foundation and project No. LO1408 "AdMaS UP - Advanced Materials, Structures and Technologies", supported by Ministry of Education, Youth and Sports under the „National Sustainability Program I".

References

[1] Tupy M, Sotiriadis K, Kusak I, Lunak M, Stefkova D and Petranek V 2015 *J. of Mater in Civil. Eng.* **28** 04015171
[2] Barylo G, Holyaka R, Hotra Z, Virt V, Vezyr F and Maria V 2017 Method of analysis errors of measurement conversion of impedance spectroscopy with activation nonharmonic signals *7th International Conference on Nanomaterials: Applications and Properties*
[3] Mentlik V 2006 *Dielektrické prvky a systémy* (BEN)
[4] Macdonald J R 1987 *Impedance spectroscopy, emphasizing solid materials and systems, bibliography*
[5] Lunak M, Kusak I and Chobola Z 2013 *Appl. Mech. Mater.* **446-447** pp 1389–1394
[6] Horak Z 1958 *Praktická fysika* (SNTL)
[7] Matsutani S, Shimosako Y and Wang Y 2012 *Physica A* **391** p 5802
[8] Fernandez-Jimenez A 1999 *Cement Concrete Res.* **29** p 1313
[9] Saafi M, Andrew K, Mcghon D, Taylor S, Rahman M, Yang S, Zhou X and Tang P L 2013 *Construction and Building Materials* **49** pp 46–55
[10] Monica J H and Harris A T 2013 *Progress in Materials Science* **58** pp 1056–1102

Chemical elimination of asbestos materials

K Necasova[1] and P Buchta[1]

[1] Research Institute for Building Materials, Hněvkovského 65, 617 00 Brno, Czech Republic

Email: necasova@vustah.cz

Abstract. Asbestos was a frequently used material in the past, mainly in the construction industry, on account of its excellent thermal and chemical resistance. Nowadays, all uses, handling and disposal of asbestos materials are subject to strict legislation due to health hazard. This work is a research of an efficient chemical way to decompose these materials with the lowest energy costs possible so that the method could compete with patented thermal decomposition and other currently common methods. Decompositions were performed at room temperature with acids, namely hydrochloric, sulfuric, phosphoric, acetic and oxalic acid and the results were evaluated based on mass analysis. Asbestos rope and asbestos-cement roof tile were used as samples. From the experiments carried out, the best decomposing agent was hydrochloric acid for the rope sample and acid mixture (hydrochloric and acetic acid in a 1:1 ratio). for the roof tile sample

1. Introduction

Asbestos is a common title for many natural silicate materials. It includes two large groups of minerals, amphibols and serpentines, all with fibrous structure. According to the composition, these large groups contain different types of asbestos materials. From the amphibols group, it is for example cummingtonite, amosite and crocidolite. The serpentines group is more voluminous and includes for example chrysotile, which is contained in 95% of all the asbestos waste [1]. Waste containing asbestos fibers represents a major environmental burden. This waste has a strong negative effect on living organisms, so its handling is subject to strict legislation [2]. There is a wide range of asbestos containing wastes, like coatings, facades, seals, ropes, tubes, slabs or roofing. One of the most widespread products is asbestos-cement roofing, which was widely used during the 1970s and 1980s. It is known under one of the trade names as Eternit [3]. Its former abundant use provided large amounts of this waste for the present. Therefore, it is important to find an effective way to get rid of harmful asbestos fibers in waste materials or to transform them into a safe form, eventually further applicable, for example in the building industry or in ceramics.

The most used possibility of the fiber destruction is thermal decomposition, which removes bound water and therefore transforms hazardous asbestos fibers into non-fibrous structures. The temperature changes during transformation were observed and analyzed using TG and DTA [4]. According to XRD analysis, the fibrous structure of asbestos is completely removed at around 1000°C [5]. Low-melt glass fragments are added to the mixture to lower the melting point of the asbestos-containing material and save energy [6].

Another possibility of asbestos fiber decomposition studied by other authors is microwave decomposition. According to Leonilli et al, microwave decomposition is a quick and efficient method of generating high temperature inside the material. The fiber degradation was proved by X-Ray Powder

Diffraction Analysis (XRD) and Scanning Electron Microscope (SEM). The resulting harmless material can be further used in ceramics, since it contains predominantly magnesium silicate [7]. High-energy milling can also eliminate the fibrous structure of asbestos. According to the study of Iwaszko J. et al., the original asbestos structure is decomposed within 2 hours at 650 rpm. Structure change was observed using Fourier-Transform Infrared Spectroscopy (FTIR) [8].

Finally, chemical ways of transformation were studied. It was found that different substances such as bases or organic and inorganic acids can be used for chemical decomposition, eventually their mixtures and waste solutions from chemical treatments. For example, in the work of Tabata M. et al., the waste solution after fly ash purification from heavy metals was used for decomposition. This waste solution was a mixture of sulfur and calcium hydroxide. Using this solution together with the ball mill, 99% of asbestos decomposition was achieved [9]. One of the works depicts decomposition with 30% phosphoric acid at high temperatures. The products were mainly phosphates (calcium phosphate, magnesium phosphate), further usable in production of fertilizers [10]. Also, organic acids can be used for asbestos fiber elimination, such as acetic acid and oxalic acid. As oxalic acid is naturally present in the environment, for example in plants and fruit, its use is preferable. According to the study of Valouma et al., decomposition by 0.05 M solution of oxalic acid at 26°C for 8 days is effective and economical. The main product is magnesium oxalate, which can be thermally transformed into magnesium oxide, an important substance not only for building industry, but also for healthcare and paint industry [1].

An innovative way is asbestos fiber decomposition by acid gases, as published in Kazumichi Y. et al. The acid gases HF and HCl emerge during thermal decomposition of freons, namely chlorodifluoromethane ($CHClF_2$), and cause a complete destruction of asbestos fibers. Therefore, two groups of dangerous waste substances are disintegrated [12].

2. Experimental

2.1. Materials and methods

Two types of asbestos waste materials were used for testing decomposition, a rope and a roof tile (see figures 1 and 2). The rope was formerly used frequently as sealing and insulation of boiler and fireplace doors. The roof tile sample was obtained during reconstruction of a holiday cottage.

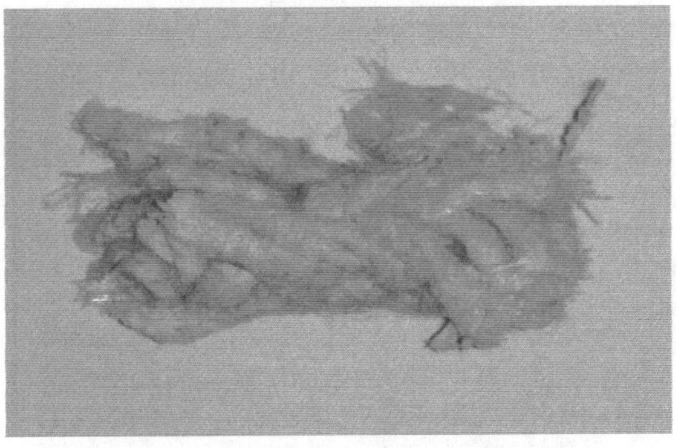

Figure 1. Original sample of rope.

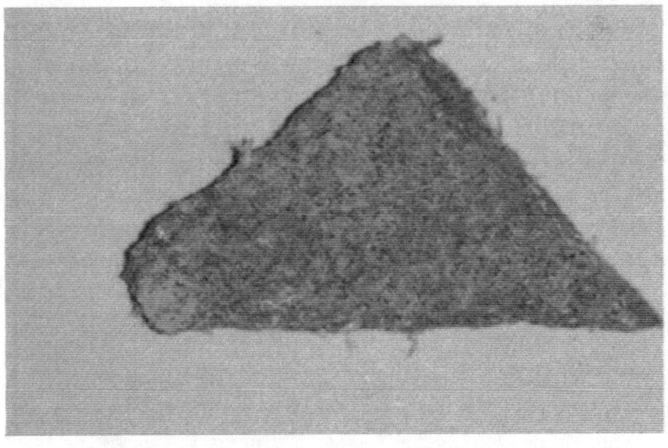

Figure 2. Original sample of roof tile.

Both materials were characterized by X-ray powder diffraction. The analysis was carried out on Bruker D8 Advance apparatus with Cu anode (λ Kα = 1.54184 Å) and variable divergence slits at Θ-Θ Bragg-Brentano reflective geometry. Sample of roofing was dissolved in 35% hydrochloric acid, while the rope was measured in its original state. All samples were analyzed on zero background sample holder, in an alcohol medium. The measurement records of rope and roof tile are depicted in figures 3 and 4, respectively.

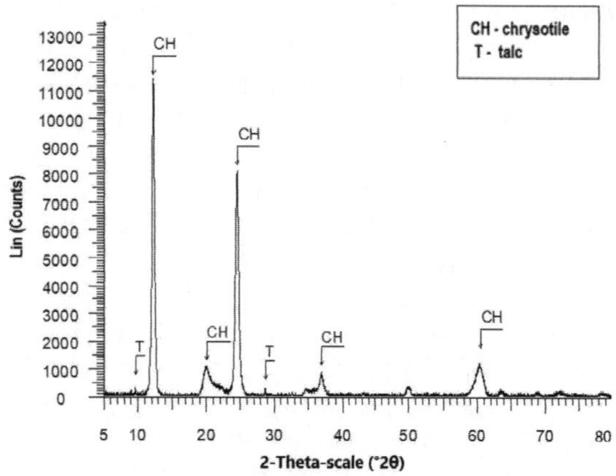

Figure 3. X-ray powder diffraction analysis of rope.

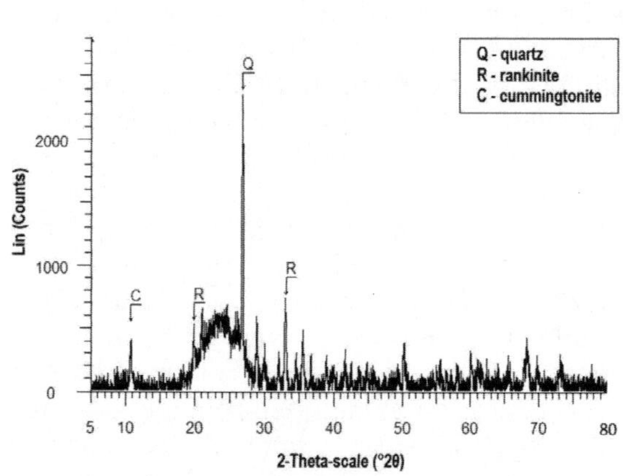

Figure 4. X-ray powder diffraction analysis of roof tile.

Based on the rope analysis it can be said that it is mainly composed from chrysotile and talc. Chrysotile is one of the asbestos materials ($Mg_3Si_2O_5(OH)_4$) and talc is hydrated magnesium silicate ($Mg_3Si_4O_{10}(OH)_2$). Compared to that, the roof tile is composed mainly from cummingtonite, quartz and rankinite. Cummingtonite is an asbestos material containing prevalently magnesium silicate hydroxide ($Mg_7Si_8O_{22}(OH)_2$). Quartz is silicon dioxide, the most abundant mineral on Earth, and rankinite is calcium silicate.

The microstructure of decomposed samples was studied by petrographic polarizing microscopy using Eclipse LV100ND by Nikon.

All experiments were conducted at room temperature (approximately 22°C) to save energy. Each sample was placed into a glass container with 150 ml of decomposition agent (sulfuric acid – H_2SO_4, phosphoric acid – H_3PO_4, oxalic acid –$(COOH)_2$, hydrochloric acid - HCl, acetic acid – CH_3COOH and combination of the latter two). Every decomposition test lasted 24 hours and was performed with constant agitation

3. Results and discussion

Generally, the acids caused significant surface damage and release of the fibers of the rope samples, while the roofing surface was damaged only partly and its color changed. The results of weight loss at the end of the tests are listed in table 1.

Table 1. Weight loss after acid decomposition.

Sample	Laboratory temperature Reagent		Decrease [%]
Rope	*Sulfuric acid*	H_2SO_4	-56.1
	Phosphoric acid	H_3PO_4	-132.7
	Hydrochloric acid	*HCl*	42.4
	Oxalic acid	*30% solution* $(COOH)_2$	14.2
	Acetic acid	*CH3COOH*	-30.1
	Mixture acid	*CH3COOH + HCl*	32.5
Roof tile	*Sulfuric acid*	H_2SO_4	-4.3
	Phosphoric acid	H_3PO_4	12.9
	Hydrochloric acid	*HCl*	62.4
	Oxalic acid	*30% solution* $(COOH)_2$	-0.8
	Acetic acid	*CH3COOH*	1.4
	Mixture acid	*CH3COOH + HCl*	67.1

3.1. Decomposition in sulfuric acid

The rope sample turned black and hardened after acid treatment, but the fibers remained unchanged. The material would be suitable for wet milling.

The roof tile sample changed color as well and the surface was mildly damaged. As seen from negative values in the table 1, weight of the samples increased after the test, which is probably due to formation of solid reaction products, for example magnesium sulfate. No observable decomposition occurred, therefore, sulfuric acid seems to be unsuitable for acid decompositions of asbestos.

3.2. Decomposition in phosphoric acid

The rope sample showed surface damage and release of a limited number of fibers. The weight increase of the rope sample listed in table 1 is caused probably by formation of unspecified solid substances.

The roof tile sample exhibit color change and mild surface damage. As opposed to the rope sample, weight loss occurred in the roof tile sample, indicating partial decomposition. As a conclusion, phosphoric acid decomposition could be applicable for asbestos roof tile. Emerging phosphates, mainly calcium phosphate, can be further used for fertilizing, as published in Pawełczyk A. et al. [10].

3.3. Decomposition in hydrochloric acid

A complete disintegration of the rope sample occurred and the fibers were released.

The roof tile sample decomposed to individual layers (see Figure 5) and part of the sample transformed into mash with fibers, which was most probably a dissolved carbonate structure. As depicted in table 1, weight loss occurred in both samples and approximately 50% decrease in weight indicates a well-chosen agent.

Figure 5. Roof tile sample treated by HCl.

3.4. Decomposition in 30% solution oxalic acid

The rope samples exhibited surface damage and fiber release.

The roof tile sample changed color and the surface was affected only slightly. As seen in table 1, weight loss was observed in the rope sample, while the weight of the roof tile sample increased about 0.8%. This could be due to reaction of oxalic acid with some cement components, as the cement represents around 90% of the total roof tile mass [3]. The most probable reaction is with calcium oxide which produces non-soluble calcium oxalate. Compared to the work of Valouma A. et al. [11], the results were less satisfactory, probably due to higher concentration of the acid and shorter reaction time.

3.5. Decomposition in acetic acid

The surface was disintegrated and the fibers were released in the rope sample (see figure 6). To illustrate the efficiency, the structure of the original sample is in figure 7.

The roof tile exhibited color change and surface damage which exposed the fibers and the sample volume considerably increased, separating some of the structure layers as seen in figure 8. While the rope sample showed a significant increase in weight, probably caused by formation of a precipitate. The roof tile sample exhibited little loss of weight indicating decomposition. Given that the layers of the roof tile were partly separated after the test, acetic acid could represent a suitable pre-treatment before decomposition by other, more convenient agent.

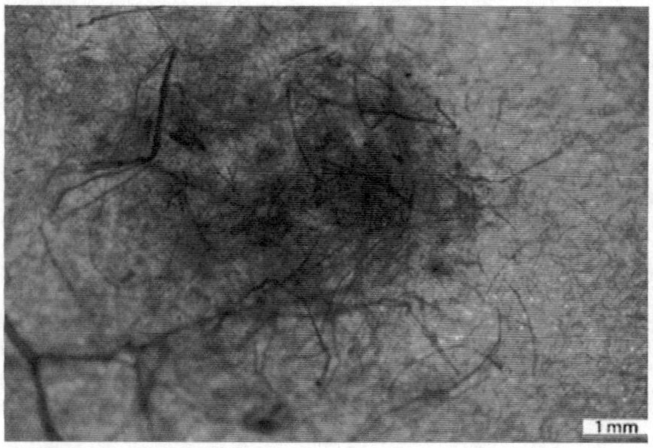

Figure 6. Rope decomposed in CH_3COOH (image from the optical microscope).

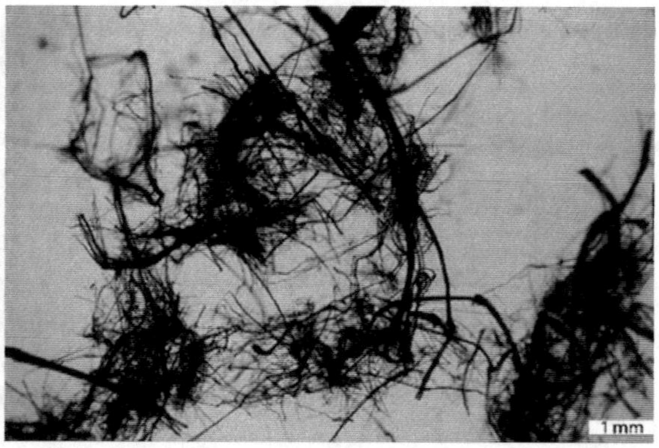

Figure 7. Original sample of the asbestos rope (image from the optical microscope).

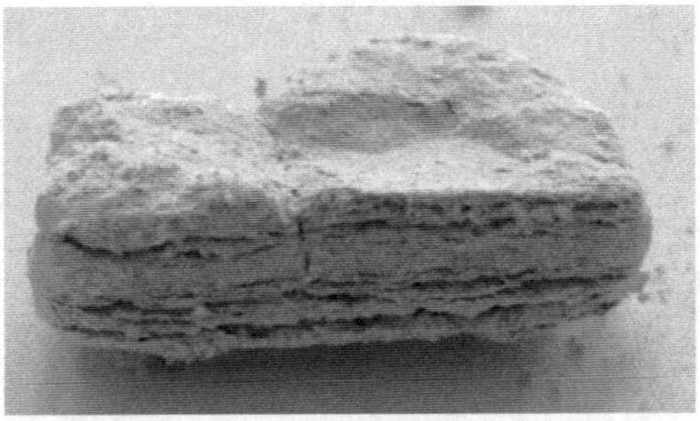

Figure 8. Roof tile decomposed in CH_3COOH.

3.6. Decomposition in acetic acid and hydrochloric acid combined

Based on the previous results, acetic acid was combined with hydrochloric acid in 1:1 ratio and the samples were exposed to this mixture. In the rope sample, short fibers were released and the whole structure hardened with no visible loss of material. However, the mass analysis showed 33% weight loss, so the hard and fragile structure of the treated sample could be suitable for wet milling and repeated decomposition.

The roof tile exhibited significant surface damage, change in color and 67% weight loss which indicates a great decomposing potential of the mixture. Presumed reaction products are calcium chloride and calcium acetate. Calcium chloride could be further used as admixture in road salt or desiccant. Calcium acetate could be used to produce flammable gel for portable cookers [13].

4. Conclusion

The aim was to find a decomposition agent suitable for asbestos treatment in laboratory conditions. Sulfuric acid, phosphoric acid, hydrochloric acid, oxalic acid and acetic acid were used and after evaluation of the efficiency, the samples were exposed to a mixture of acetic acid and hydrochloric acid. The best results were provided by oxalic acid, hydrochloric acid and the acid mixture. For the asbestos rope, the highest weight loss (42%) was achieved using hydrochloric acid. Decomposition in oxalic acid caused only 14% decrease in mass, but changes in material structure were obvious. After treatment by the acid mixture, 32% weight loss occurred.

For the roof tile samples, the most suitable was the acid mixture, which etched the individual layers from each other and decomposed 67% of the material. Using hydrochloric acid, 62% weight loss was achieved. After treatment in oxalic acid, precipitate emerged which increased the mass of the sample by 0.8%.

In further research, decomposition of asbestos in mixtures of acids will be examined, using preferably waste mixtures from chemical productions and pre-treatments, for example acid degreasing bath.

References

[1] Petránek J et al. *Encyklopedie geologie* (Česká geologická služba)
[2] Adámková M 2001 *Příručka pro nakládání s chemickými látkami a přípravky včetně nebezpečných* (Verlag Dashofer)
[3] *Problematika azbestu ve vnitřním prostředí škol– přednášky* 2012 (SZÚ)
[4] Kusiorowski R et al. 2015 Study on the thermal decomposition of crocidolite asbestos *Journal of thermal analysis and calorimetry* [1388-6150] 120 iss 3 pp 1585–1595
[5] Kusiorowski R et al. 2013 Thermal decomposition of asbestos-containing materials *Journal of Thermal Analysis & Calorimetry* vol 113 issue 1 pp 179–188
[6] Alessandro F Gualtieri et al. 2008 In situ ESEM study of the thermal decomposition of chrysotile asbestos in view of safe recycling of the transformation product *Journal of Hazardous Materials* vol 156 issues 1–3 pp 260–266
[7] Leonelli C et al. 2006 Microwave thermal inertisation of asbestos containing waste and its

recycling in traditional ceramics *Journal of Hazardous Materials* vol 135 issues 1–3 pp 149–155

[8] Iwaszko J et al. 2018 Influence of high-energy milling on structure and microstructure of asbestos-cement materials *Journal of Molecular Structure* vol 1155 pp 51–57

[9] Tabata M et al. 2016 Decomposition of asbestos by a supernatant used for immobilization of heavy metals in fly ash *The Journal of Material Cycles and Waste Management* (Dordrecht) issue 18 no. 3 pp 483–492 (doi:10.1007/s10163-016-0491-2)

[10] Pawełczyk A 2017 Chemical elimination of the harmful properties of asbestos from military facilities *Waste Management* vol 61 pp 377–385

[11] Valouma A et al 2017 Magnesium oxide production from chrysotile asbestos detoxification with oxalic acid treatment *Journal of Hazardous Materials* vol 336 pp 93–100

[12] Kazumichi Y et al. 2009 A novel decomposition technique of friable asbestos by $CHClF_2$-decomposed acidic gas *Journal of Hazardous Materials* vol 163 issues 2–3 pp 593–599

[13] Klikorka J, Hájek B and Votinský J 1982 *Obecná a anorganická chemie I a II* (Praha: Státní nakladatelství technické literatury)

IOP Conf. Series: Materials Science and Engineering **583** (2019) 012032 doi:10.1088/1757-899X/583/1/012032

Study of the efficiency and durability of hydrophobization modifications of building elements

V Novak[1] and J Zach[1]

[1] Brno University of Technology, Faculty of Civil Engineering, Centre AdMaS, Purkyňova 139, 612 00 Brno, Czech Republic

Email: novak.v@fce.vutbr.cz

Abstract. One way how to limit the effect of increased moisture on building constructions is the hydrophobization of these constructions or parts thereof. However, hydrophobization protection may be subjected of degradation effects that may reduce its effectiveness. The contribution deals with monitoring the degradation degree of hydrophobization protection of ceramic masonry elements caused by climatic effects. There is a comparison available between accelerated/simulated degradation and non-accelerated real-time degradation. The results are compared with the real capillary absorption of the untreated/non-hydrophobized wall blocks.

1. Introduction

Increased moisture has a negative effect on most building materials and acts as a degradation factor that negatively affects the properties and lifetime of building constructions [1, 2, 3].

For this reason, a number of waterproofing measures are used during construction of building constructions to protect the building construction from the adverse effects of water and moisture.

The protection of the building construction from the effect of rainfall and ground moisture is mostly covered with coating waterproofing, but there are also solutions utilizing special elements with very low water absorption, such as hydrophobized skirting masonry blocks [4]. These blocks can be used in the area of masonry establishing and serve primarily as masonry protection during construction. Hydrophobization adjustment of the blocks prevents soaking and capillary rising of the moisture from the base of the foundation and the base belts into the masonry while the construction as a whole is not protected from the effects of rainfall moisture.

Furthermore, the skirting blocks serve as additional waterproofing measure if the primary waterproofing protection - mostly asphalt or rubber belts – fails [5].

The agents used for the treatment of skirting blocks are the most commonly silicone-based polymers [6]. Even these products, like most polymers, are susceptible to degradation due to climatic effects, especially UV radiation. During construction, it is common that the period before covering of the skirting elements by surface protection ranges between 0.5 and 2 years. During this time, the skirting elements are subjected to UV radiation, which can degrade their waterproofing effects. The process of hydrophobization protection is most often carried out by dipping the blocks into a hydrophobization solution of a given concentration.

Hydrophobization preparations are currently mostly applied only to parts of the masonry elements. However, in the scope of the realized experiments, the whole hydrophobic elements were tested, loaded with the extreme case, i.e. flooding the entire wall element.

2. Methodology of testing the degradation of hydrophobization protection

In order to monitor the degradation of hydrophobization protection applied to the masonry elements due to climatic effects, ceramic masonry blocks of the Therm type for a wall thickness of 300 mm were selected. Two sets of test samples were made on which silicon-based hydrophobic agents at a concentration of 1:40 were applied by dipping for 10 s together with the pigment indicating the hydrophobized blocks (similar to commercial products). These were two sets of blocks that were hydrophobized to the full height of the blocks.

The used hydrophobization agents were based on:

- Sample A: silicon compound, nonylphenol and ethoxylate,
- Sample B: silan-siloxan.

To determine the degradation of hydrophobization protection, an accelerated aging method was chosen using load cycling due to temperature, moisture and UV radiation.

One simulation cycle consists of 170 minutes of water sprinkling using spraying jets that sprays about 103 litres of water per 1 m^2 of the effective area per hour. These 170 minutes of sprinkling is followed by a 10-minute pause. After the pause, the test sample is heated for 170 minutes using an IR lamp of about 1300 W per 1 m^2 with the addition of UV light of 250 W per 1 m^2. The temperature of the radiant area of the exposed material increases to about 60–70°C. This is again followed by a 10 minute pause. The procedure described above is repeated 17 times in succession. The samples are then left in water for 2 hours. After 2 hours of water saturation, the samples are placed in a climatic chamber at -20°C for 2 hours. This completes one simulation cycle of the durability test, which corresponds to the exposure of the samples to the outdoor environment under extreme conditions for a Central European locality for 1 year.

To create the necessary conditions, a device was used which is referred to as a „sprinkler" (see figure 1). It is equipment of Research Institute for Building Materials. The degradation simulation was carried out as part of the collaboration between research institutions.

Figure 1. The device for simulation of load cycles.

From each set of hydrophobized blocks, a sample was placed in a water bath for about 1 month (see figure 2). The water level was maintained above the upper surface of the blocks. Thus, the effectiveness of the hydrophobization protection on blocks was determined using the selected hydrophobization agents A and B. The efficacy was determined by monitoring the absorbance of the blocks over time. At the selected time intervals, the blocks were always drawn out of the water bath, for about 30 seconds the blocks were loosely placed on the pad for sufficient draining of surplus water from the surface of the block. Blocks were then weighed and put back into the water bath.

The capillary absorption (1) was calculated from the weight gain of the blocks over time (in selected periods) due to the water absorption into the pieces of block in relation to the masonry block surface.

$$C_a = \frac{m_{w,cb} - m_{d,cb}}{A} \quad (1)$$

C_a – capillary absorption [kg.m^{-2}]
$m_{w,cb}$– weight of wet ceramic block [kg]
$m_{d,cb}$– weight of dry ceramic block [kg]
A – surface ceramic block of stress of water [m^2]

Figure 2. Placing of blocks in a water bath - determination of capillary absorption.

For comparison, the absorption on the non-hydrophobized masonry block with REF marking was determined under the same conditions (see figure 3).

As can be seen from figure 3, the non-hydrophobized masonry element already has significant capillary water absorption over 8 kg.m^{-2} in the first hour of water bath placement. At 100 h, the capillary absorption rate gradually increases up to 9 kg.m^{-2}. The untreated blocks, thus absorbs more than 9 l of water over its surface.

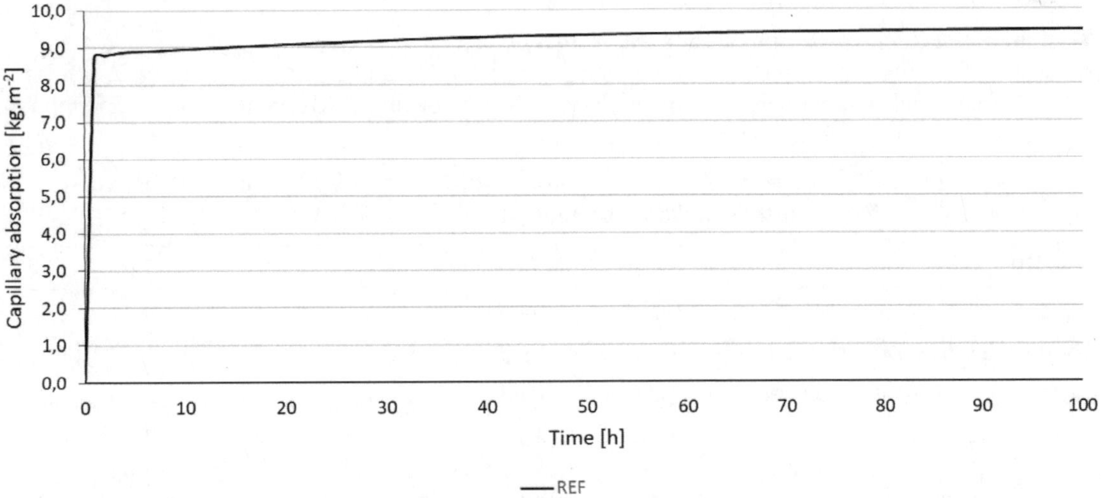

Figure 3. The capillary absorption of samples without the utilization of hydrophobization agents.

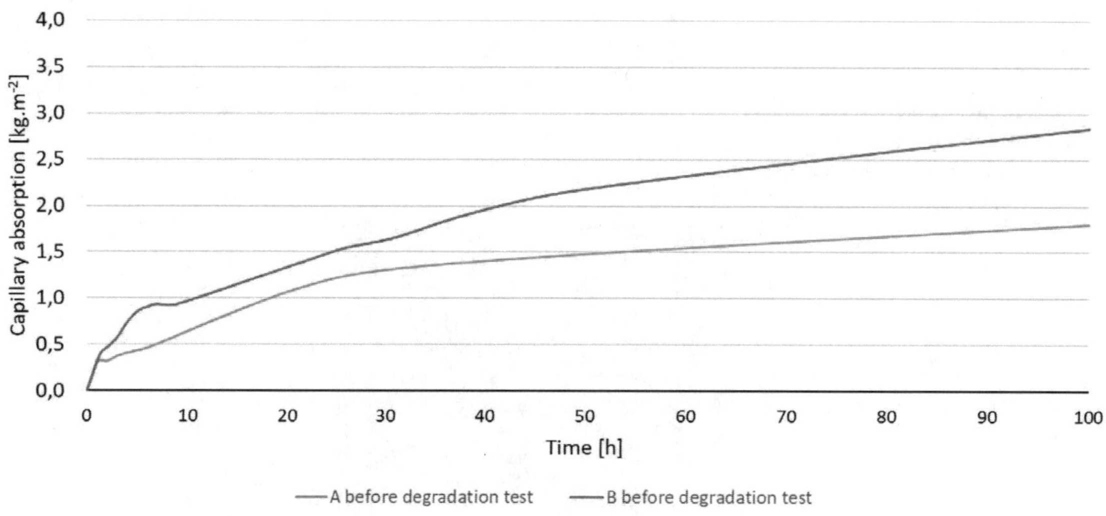

Figure 4. The absorption of test samples with hydrophobization agents.

By using hydrophobization agents, there was a slowdown of capillary absorption of the blocks in both cases (see figure 4 and table 1). The use of agent A reduced the capillary absorption rate, whereupon the treated block absorbed 70% less in the first 100 hours compared to the non-hydrophobized block, i.e. below 2 $kg.m^{-2}$.

Table 1. Change of capillary absorption after hydrophobized test samples.

Time [h]	24		48		96	
REF	9.12		9.31		9.42	
A before degradation test	1.22	- 86.6 %	1.46	- 84.3 %	1.75	- 81.4 %
B before degradation test	2.18	- 76.1 %	2.21	- 76.3 %	2.83	- 70.0 %

3. Hydrophobization protections under the influence of UV

The results of capillary absorption after exposure to accelerated aging test due to climatic effects are shown in table 2 and figure 5 below. The simulation of aging occurred within one cycle, corresponding to 1 year aging due to extreme climatic effects.

Table 2. The comparison of absorption of hydrophobized test samples before and after simulation of aging due to climatic effects.

Time [h]	0	5	24	48	72	96
REF	0.00	8.88	9.12	9.31	9.31	9.42
A before degradation test	0.00	0.43	1.22	1.46	1.57	1.75
A after degradation test simulation	0.00	0.81	1.58	2.11	2.23	2.61
B before degradation test	0.00	1.24	2.18	2.21	2.52	2.83
B after degradation test simulation	0.00	2.58	3.73	3.83	3.84	3.89

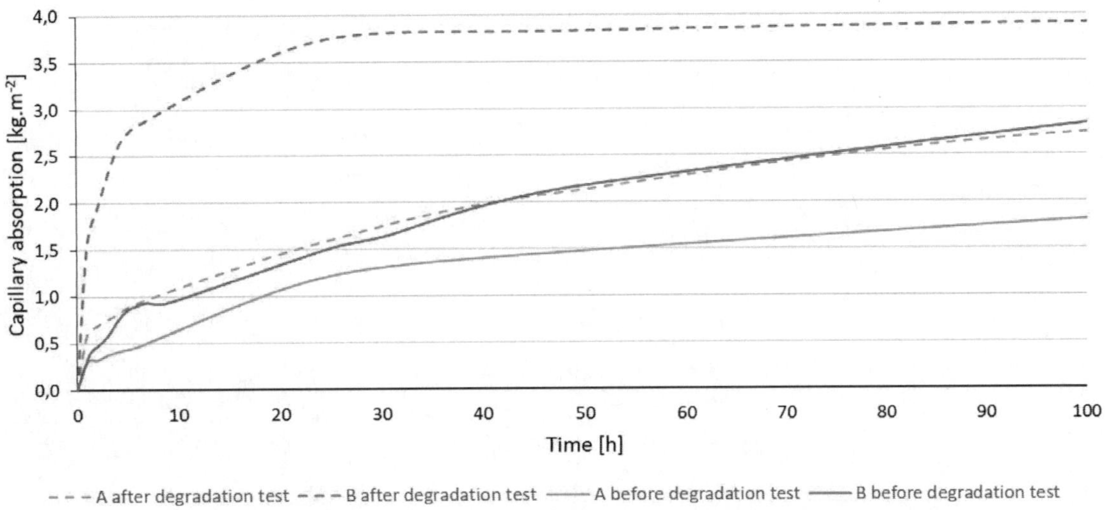

Figure 5. The comparison of absorption of hydrophobized test samples before and after simulation of aging due to climatic effects.

As expected, it is evident that the accelerated aging test has degraded the hydrophobization protection. Degradation occurred in both tested hydrophobization agents A and B in similar limits. Due to the accelerated aging test as a result of climate degradation test, the capillary absorption of hydrophobized blocks increased by about 60% over 100 hours. However, the capillary water absorption did not exceed 4 $kg.m^{-2}$ in the first 100 hours, compared to the absorbance of the reference block, which contains a capillary absorption of more than 9 $kg.m^{-2}$ at 100 hours.

Values obtained from the accelerated climate degradation test were also compared with the non-accelerated test. The non-accelerated aging test was performed on blocks on which the hydrophobization protection efficiency was determined. Based on the results of the accelerated test, it was only continued with the hydrophobization agent A testing, the application of which before and after the acceleration of the aging test resulted in a lower capillary absorption than when using hydrophobization agent B. The results of the accelerated aging due to climate degradation test are shown in table 3 and figure 6.

Table 3. Comparison of absorption of hydrophobized test samples before and after natural aging due to UV radiation.

Time [h]	0	5	24	48	72	96
REF	0.00	8.88	9.12	9.31	9.31	9.42
A before degradation test	0.00	0.43	1.22	1.46	1.57	1.75
A after degradation test	0.00	0.81	1.58	2.11	2.23	2.61
A after nature degradation test	0.00	0.71	2.04	2.48	2.84	3.23

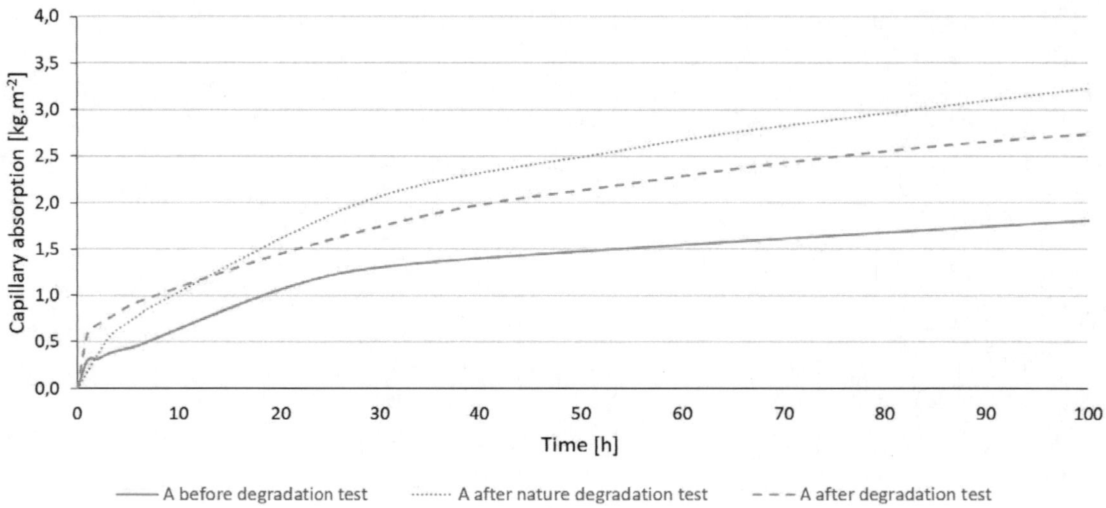

Figure 6. Comparison of absorption of hydrophobized test samples before and after natural aging due to UV radiation.

In the case of a non-accelerated/natural test of aging for 1 year, a very similar degree of degradation was demonstrated as in the case of an accelerated aging test due to a combination of climatic effects (see table 4). The capillary absorption of the samples after 1-year degradation due to climatic effects did not increase in a period of 100 hours above the value of 3.5 $kg.m^{-2}$.

Table 4. Change of capillary absorption after aging due climatic effects.

Time [h]		24		48		96
A before degradation test	1.22		1.46		1.75	
A after degradation test	1.58	29.5%	2.11	44.5%	2.61	49.1%
A after nature degradation test	2.04	67.2%	2.48	69.9%	3.23	84.6%

4. Conclusion

Tests have shown that using a suitable hydrophobization agent, ceramic masonry blocks can be effectively protected against the effects of increased moisture. If these bocks are used to construct the establishing as test samples, they effectively prevent the moisture absorption into the masonry and the subsequent capillary rising of moisture. This modification of the samples exhibits very long-term stability and is not significantly degraded by climatic effects - especially UV radiation.

It has been proven that degradation due to climatic effects does not have a significant impact on final absorption. Samples treated with a hydrophobization agent based on silicon compound, nonylphenol and ethoxylate after aging test acceleration (equivalent to 1 year in real time) due to UV radiation do not show capillary absorption higher than 4 $kg.m^{-2}$ over a period of 100 hours with a continuous load by standing in the water. Capillary absorption of untreated blocks under the same conditions is about 9.5 $kg.m^{-2}$. It is also necessary to take into account that the test carried out was an extreme case of increased moisture load, which, in real conditions, occurs exceptionally.

The results of the effect of hydrophobization coincide with the results obtained in the framework of similar scientific works [7]. The study of hydrophobic protection of ceramic elements and its degradation will continue, the aim will be to obtain comprehensive information on this issue.

The only and most significant degradation effect of the hydrophobization protection of the ceramic masonry block is therefore the mechanical damage of the hydrophobized part of the block.

Acknowledgements
This paper was written under the project No. LO1408 "AdMaS UP - Advanced Materials, Structures and Technologies", supported by the Ministry of Education, Youth and Sports under the "National Sustainability Programme I" under the project of the Technology Agency of the Czech Republic ref. No. TH04020378.

Reference

[1] Annila P J, Hellemaa M, Pakkala T A, Lahdensivu J, Suonketo J and Pentti M 2017 Extent of moisture and mould damage in structures of public buildings *Case Studies in Construction Materials* pp 103–108
[2] Zhao J and Meissener F 2017 Experimental investigation of moisture properties of historic building material with hydrophobization treatment *Energy Procedia* pp 261–266
[3] Hall Ch, Hoff W D, Viles H A and Eklund J A 2010 Moisture dynamics in walls: Response to micro-environment and climate change *Proceedings of The Royal Society* A 466 pp 194–211
[4] Barnat-Hunek D, Smarzewski P and Suchorab Z 2016 Effect of hydrophobisation on durability related properties of ceramic brick *Construction and Building Materials* pp 275–285
[5] CSN P 73 0606:2000 *Waterproofing of buildings – Continuous sheet water proofing – Basic provisions* (Prague: Czech Standards Institute)
[6] Vries J and Polder R B 1996 Hydrophobic Treatment of Concrete *Internationale Zeitschrift für Bauinstandsetzen* **2** pp 145–160
[7] Hansena T K, Bjarløva S P, Peuhkurib R H and Hansena K K 2018 Performance of hydrophobized historic solid masonry – Experimental approach *Construction and Building Materials* pp 695–708

IOP Conf. Series: Materials Science and Engineering **583** (2019) 012033 doi:10.1088/1757-899X/583/1/012033

Examples of testing methods focussed on evaluation of corrosion influence on steel structures

M Sperl [1] and L Gajdos[1]

[1] Institute of Theoretical and Applied Mechanics of the Czech Academy of Sciences, rosecká 809/76, 190 00 Prague 9, Czech Republic

Email: sperl@itam.cas.cz

Abstract. The paper presents methodologies and results of two case studies dealing with corrosion damage in steels. Specifically, it is concerned with corrosion fatigue and stress corrosion cracking. Description is given of preparing test specimens, special devices and loading frames that have been used to obtain significant experimental results. These are referred to influence of three different chemical solutions on the fatigue life of specimens taken from L485MB steel crude oil pipeline IKL. In the second part of the paper negative effects of stress corrosion cracking on the fracture toughness of structural steel CSN411353 were investigated and main results of this investigation are presented.

1. Introduction

It is well known that many civil engineering structures are subjected to degradation processes caused by various chemical substances. Degradation processes take place when protecting coats and technological anticorrosion barriers are damaged and water solutions can run into the structures. In many cases this is due to technological imperfections. These processes can also be of a natural origin and they cannot be avoided at the structure, so that it is necessary to account them in the stage of designing the structure. It is estimated that corrosion losses in industrially developed countries reach 3–5% of the gross domestic product. This means that in the Czech Republic we lose approximately 150 billions CZK per year [1].

The word corrosion is derived from the Latin word „corrodere“, which means to nibble, to gnaw away. The mechanism of rise and effectiveness of corrosion impacts are influenced by a number of factors which are dependent first of all on the aggressivity of the environment, then on the stress state of material, the quality of insulation, the state of the surface of material, etc. The main dividing of corrosion is on the basis of the principle of the corrosion process itself. There is a chemical, biological, and electrochemical corrosion [2]. According to its volumetric size and significance on the limitation of the life of a steel structure corrosion can be divided into general corrosion, pitting, stress corrosion cracking, intercrystalline, transcrystalline corrosion and selective corrosion [3]. The issue of corrosion is very important and still a timely subject which becomes a motive of many research projects.

This paper brings about two case studies which deal with corrosion problems of steels. In the first study a procedure and results of testing of the effect of corrosion on the fatigue resistance of steel are described. The motivation for this research were petrochemical industry demands for pipeline operational reliability. It is well known that separated water in crude oil can reduce steel fatigue life [4]. Our point of interest was a comparison of the fatigue properties of the concrete pipe steel in air, in crude oil, in separated water and in alkaline solution. The latter study aimed at the description of potentialities of a laboratory simulation of a very dangerous type of corrosion damage which is the stress corrosion.

It is usually associated with hydrogen embrittlement [5] which is very topical subject e.g. in the development of Tokamak fusion power plants and other specific parts of industries.

2. Corrosion fatigue

Owing to the fact that (i) operational loading of many civil engineering structures is not constant but it varies during operation, and moreover, (ii) the steel comes frequently into contact with aggressive environment, one of the factors which could reduce the life of a structure is corrosion fatigue [6]. A research of fatigue behaviour of steel in an aggressive environment appears therefore as very desirable and useful. Even though these fatigue tests should naturally be carried out at conditions close to real ones, especially as far as frequency of loading is concerned, it is often not possible to realize such tests in a real time. Therefore, in such cases the effect of the length of time of chemical action of the environment on corrosion damage is held back. However, it is advantageous to carry out at least current fatigue tests at resonant frequencies for the specimens used. These fatigue tests, of course, can not represent the real resistance of material against corrosion fatigue damage but the results obtained can be used as informative data for indication of the relative fatigue resistance of the material. We can encounter corrosion fatigue at bridges when mechanically loaded steel components come into contact with saline or at transport systems, power engineering as well as chemical and petrochemical industry. As an example of the research into the effect of corrosion environment on the fatigue life of steel an experimental work is described hereunder on a pipe segment taken from crude oil pipeline IKL (Vohburg an der Donau – Nelahozeves) after it had been in operation for 13 years.

2.1. Corrosion fatigue tests

Flat fatigue specimens with a zero gauge length (figure1) were manufactured from a segment received. Specimens were made by machining of straighten sheets taken from a pipe. The segment was taken from a crude oil pipeline DN 700/8.5 mm, the material being linepipe steel L485MB with a bainitic structure of the ultimate strength R_m = 610 MPa, the yield stress $R_{p0.2}$ = 495 MPa and ductility A_5 = 25%. These specimens are designated here by a triplet of letters IKL. Further a group of specimens were manufactured from a new pipe DN 500/10 mm of the same steel characterized by R_m = 625 MPa, $R_{p0.2}$ = 500 MPa and A_5 = 25%, manufactured at the Mittal Steel Ostrava (shortage of IKL material). These specimens are designated here by a triplet of letters MIT.

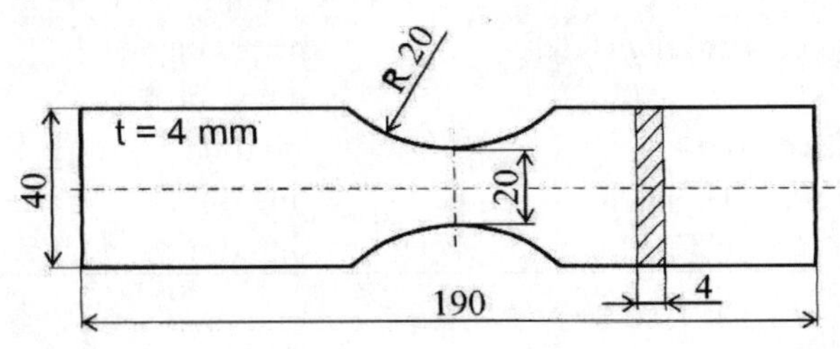

Figure 1. Fatigue specimens with zero gauge length.

Fatigue tests of both types of specimens were carried out in zero-to-tension loading in detail described in [3] using a resonant-type electromagnetic machine with the nominal force capacity 100 kN. The resonance frequency was approximately 130 Hz, F_{max} varied from 23.4 kN up to 39.9 kN. Reference tests of IKL specimens were realized in air whilst main fatigue tests were realized in liquid media: in crude oil and in water separated from the crude oil. Fatigue tests of MIT specimens were performed in 1N water alkaline solution of Na_2CO_3 with $NaHCO_3$ in the ratio 1:1 with a pH value of 9.325 [7].

A leak-proof chamber, developed in co-operation with RCP company Prague, was used for fatigue tests in liquid media. A section layout of the chamber is shown in figure 2 and a general view of the specimen with a chamber in the grips of the machine is presented in figure 3.

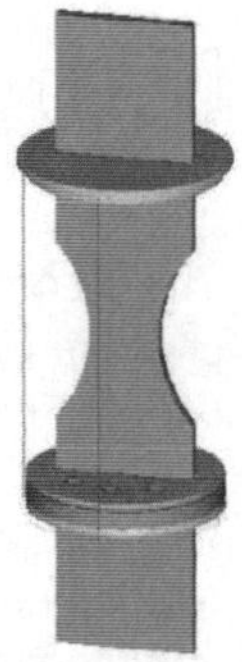

Figure 2. A section of the leak-proof chamber.

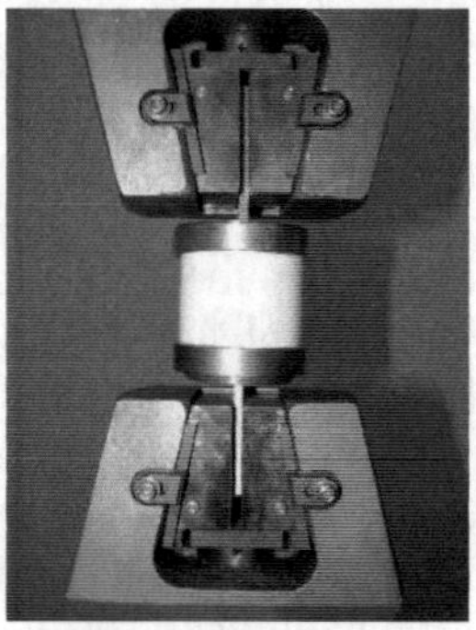

Figure 3. A specimen with the chamber in the grips of the machine.

The number of cycles to fracture, N, i.e. the life of a specimen, was determined as a number of stress cycles which the resonant type fatigue machine was able to make out without any changes in pre-set parameters. A relative crack size at the instant of automatic switching off the machine was 0.25–0.4 × the width of a specimen in its smallest cross section. Fatigue tests in the water phase separated from crude oil were, to some extent, specific, because this environment is a residue of the separated water which is present in small amounts in distributed crude oil, say either in emulsified form or as dissolved (in very small amounts). Chemical composition of the separated water is shown in table 1. Before starting the fatigue tests the separated water was in contact with air in order to be saturated with oxygen and CO_2 in the temperature range 18°C–23°C.

Table 1. The results of analysis of the water phase.

	Units	**Magnitudes**
pH	-	7,0
specific conductivity	$mS.m^{-1}$	2000
dissolved substances	$g.dm^{-3}$	14
neutralizing capacity ($KNK_{4,5}$)	$mmol.dm^{-3}$	11
∑ (Ca+Mg)	$mmol.dm^{-3}$	22
chlorides (Cl^-)	$g.dm^{-3}$	8,1
sulphates (SO_4^{2-})	$mg.dm^{-3}$	31
oxygen (15–25°C)	$mg.dm^{-3}$	9.5 – 8.5

2.2. Results of fatigue tests

All results are presented in figure 4. Besides fatigue curves relating to IKL specimens there is also shown a fatigue curve for MIT specimens. The fatigue limit in zero-to-tension loading for IKL specimens is demonstrated by a horizontal dashed line. According to [8] it is taken as $0.6 \times R_m = 366$ MPa.

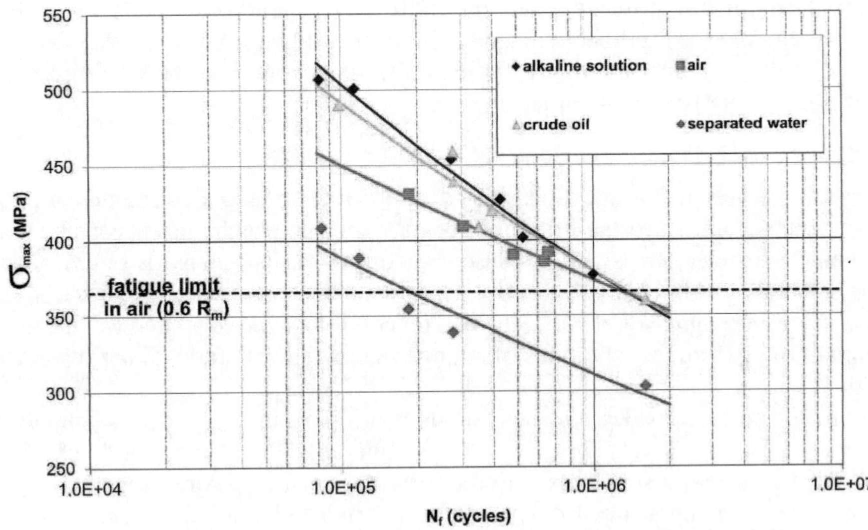

Figure 4. Summarized results of fatigue tests.

It can be seen in the figure that on the top of the set of all experimental points there are green colour triangles corresponding to the crude oil and also dark blue colour diamonds corresponding to the alkaline solution which was applied for MIT specimens. At the bottom of the set of experimental points red colour diamonds are situated with a large distance from other experimental points. These red colour points are related to the separated water. Slightly under green colour triangles there are pink colour squares which represent results obtained in air. Respective fatigue curves were obtained by smoothing experimental points with power function between stress and number of cycles to fracture. The results show that in relation to fatigue properties of IKL steel in air the effect of crude oil appears as non-aggressive and the effect of alkaline solution on MIT steel appears similarly, i.e. non-aggressive. On the other hand the separated water exhibits a very negative effect on corrosion fatigue properties of IKL steel.

If the fatigue curve for air is considered as a reference curve, then, according to this diagram, the crude oil environment will not reduce the fatigue strength of the steel investigated up to 2×10^6 cycles. It is worthy to note that the slope of the fatigue curve for crude oil is greater than for air. As it is noticeable from the position of fatigue curve for the separated water in the diagram the presence of the separated water in crude oil leads to a considerable deterioration of fatigue properties of the steel under investigation.

If we express the fatigue curve by a power function of the type (1)

$$\sigma_{max} = \frac{A}{\left(N_f\right)^b} \tag{1}$$

we shall get the following magnitudes of the constants *A, b* (these parameters were determined by least square methods from experimental data) for various environments (table 2):

Table 2. Constants *A, b* in fatigue curves.

Environment	Steel	*A*	*b*
air (5 pc.)	IKL	1147.4	0.0813
crude oil (6 pc.)	IKL	1803.1	0.1129
separated water (5 pc.)	IKL	1188.3	0.0971
alkaline solut. (6 pc.)	MIT	2098.3	0.1239

IOP Conf. Series: Materials Science and Engineering **583** (2019) 012033 doi:10.1088/1757-899X/583/1/012033

If we identify the fatigue limit with the maximum stress in a cycle for the life $N_f = 2 \times 10^6$ cycles we shall receive the following most probable magnitudes of the fatigue limit in the zero-to-tension cycle: for IKL steel it is $\sigma_f = 353$ MPa for air, $\sigma_f = 350$ MPa for trude oil and $\sigma_f = 290$ MPa for separated water; for MIT steel it is $\sigma_f = 348$ MPa for alkaline solution.

2.3. Discussion

Experimental results showed that crude oil has no adverse effect on fatigue properties of IKL steel. This finding is in an abrupt contrast to the effect of separated water which shows corrosion aggressivity towards this steel. Corrosion aggressivity of the separated water is caused by its saturation with oxygen and by a high content of chlorides [9]. On the other hand the presence of acid carbonates in high concentrations reduces its aggressivity [10]. At crude oil conditions the separated water can contain only a limited amount of oxygen but at conditions of a long time storage of crude oil the separated water can saturate with air [11].

This environment can be considered to be the most aggressive among those which are really likely [12]. In other words: as a result of a low value of pH factor and of a high ratio $[c(Cl) + c(SO_4)]/c(HCO_3$ * *c* is a substance concentration) the separated water becomes an environment with a high chemical corrosion activity. The effect of crude oil on fatigue properties of the steel investigated is comparable to the effect of an inert environment for that it expresses itself by locking the surface to air oxygen (fatigue tests in air). The composition of the alkaline solution (its alkalinity and presence of the system HCO_3^- / CO_3^{2-}) leads to the passivation of the steel which results in a lower aggressivity of the solution.

3. Stress corrosion cracking

This part of our work was aimed at the application of fracture mechanics approach to assessment of the reliability of structures containing stress corrosion (SC) cracks. Stress corrosion cracking is very dangerous type of corrosion damage. It can be found in power industry, chemical industry and other constructions in chemical and petrochemical industry (pipelines, storage tanks, etc.). The main damaging factors are (i) formation of crack networks and (ii) material embrittlement. The synergy of these factors creates a big danger for the integrity of structures.

Stress corrosion cracking may occur by several mechanisms; one of them is the hydrogen mechanism leading to hydrogen embrittlement (HE). In the HE process hydrogen atoms from the environment enter the metal and diffuse in the lattice. They tend to be attracted to regions of high triaxial tensile stress where the metal structure is dilated. Thus, the hydrogen atoms are drawn to the regions ahead of crack tips or sharp notches that are under stress. They can occupy the interstitial sites in the metal lattice, as shown by neutron diffraction in many cases [13].

The main point of this research consists in an attempt to clear up a widely discussed problem of a higher degree of dangerousness of SC cracks than fatigue (F) cracks on the integrity of structures. Seeing that character of damage is diffusive it is natural to suppose that odd atoms coming from the aggressive corrosion environment diffuse to the structure of material at the crack tip. It is also likely that due to this mechanism the fracture toughness will be reduced in comparison with a „clean" F crack. The task of the work was therefore to compare both magnitudes of the fracture toughness and to elucidate eventual differences.

The procedure of work to fulfil the main aim was divided into four steps:

- choosing a suitable aggressive environment, test material and the shape of specimens,
- design and building up of a testing equipment,
- test methods and experimental work (including specimens with F cracks),
- evaluation and analysis the results obtained.

3.1. Methods

For generating SC cracks it was necessary to develop a special equipment (figure 5). The equipment ensured the initiation and growth of SC cracks.

Figure 5. A Photo with schematic arrangement of the SC crack generator.

It is basically a horizontal cylindrical container made of linear polyethylene, inside which there is a corrosive solution with immersed compact tension (CT) specimens fixed in a loading frame. Specimens in the generator were loaded by a constant force induced by means of two pistons in an pneumatic cylinder connected in series with maximal force capacity F_{max} = 15 kN. It transmits a constant force to CT specimens with recording the deformation during the chemical process. Force transmission is via pin joints and the maximum piston displacement is 30 mm. This equipment is registered as a utility model. It was developed in our institute.

Experimental testing was carried out on CT specimens made from a commonly used structural steel CSN 411353. Three groups of specimens (A, B, C) were tested. Specimens A were provided with a F crack, whilst specimens B and C contained SC cracks which were generated in the SCC generator after the specimens were precracked by fatigue. Specimens B were tested after fortnight and specimens C immediately after taking them out of the generator. The generating of SC cracks was carried out in an acidic solution with its composition defined by the standard NACE [14] and with H_2S bubbling through. Displacement of the pistons indicated indirectly the SC crack extension on CT specimens.

The fracture resistance of the material tested (steel CSN 411353) was determined on the basis of the so called R curve with the J integral as the elasto-plastic fracture parameter. The primary step to obtain the fracture toughness was determination of the „force – force point displacement "dependence for all three groups of CT specimens, i.e. A, B, and C. The procedure for fracture toughness determination is standardized and proceeds according to the standard [15]. The J integral was introduced as a line integral [16], given for a cracked body by a general relation, equation (2):

$$J = \int_{\Gamma} W \cdot dy - \int_{\Gamma} T_i \cdot \frac{\partial u_i}{\partial x} \cdot ds \qquad (2)$$

where: *W* is the strain energy density, *T* is the traction vector, *u* is the displacement vector, *ds* is a length increment along the arbitrary contour *Γ* around the tip of a crack.

3.2 Results

A comparison of the results obtained for specimens of the group A, B, and C clearly shows that the stress corrosion process significantly influences the fracture toughness in the negative direction. This keeps not only for material in which the SC process proceeds (group C) but also for material for which the cause of SC process had been removed (group B).

All three R curves obtained for the material tested (CSN 411353) are presented on figure 6 as dependences „J integral – Δa". Altogether 7 specimens were tested for each group.

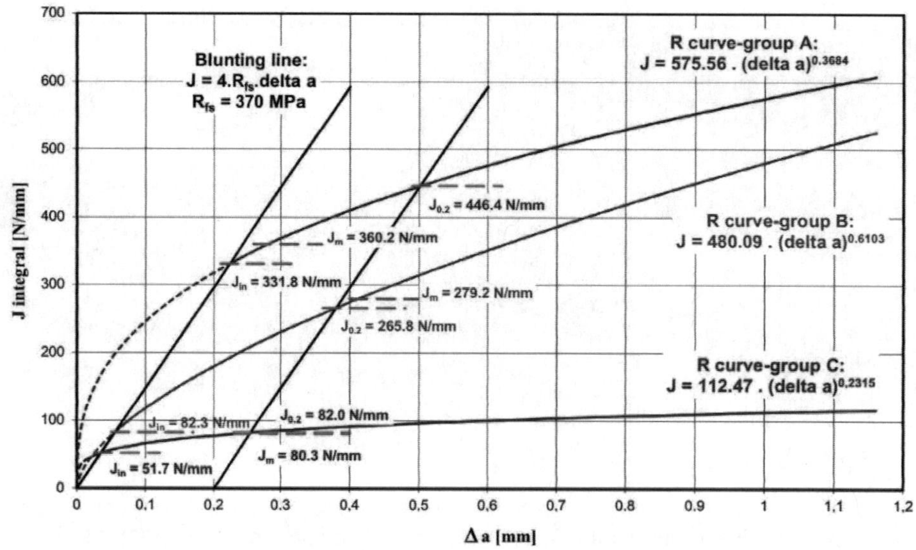

Figure 6. R curves for groups A, B and C with marked values of the J integrals J_{in}, $J_{0.2}$ a J_m.

From this figure it can readily be seen the effect of stress corrosion on the fracture toughness of material. While the R curve for F crack specimens (group A) lies in the highest position and exhibits high magnitudes of the J integral the fracture properties of B group specimens are lower. This is caused by a "residual" hydrogen occurring in a certain bulk of material ahead of the crack tip even after a 14 days pause in normal atmospheric conditions. Owing to diffusion the hydrogen atoms in the structure of material can recombine into molecules at certain suitable places inducing thus extremely high local stresses in the structure (hundreds of MPa). This causes a local or a total loss of plastic properties of the steel.

The R curve for C group specimens lies in the lowest position in figure 6 and has the flattest characteristics. The magnitudes of the J integral (J_{in}, $J_{0.2}$ a J_m) exhibit approximately 80 % decrease of the fracture toughness in comparison with the magnitudes referring to F cracks (group A). The results are summarized in table 3.

Table 3. Magnitudes of the J integral for A, B, and C group specimens from steel CSN 411353.

Group	J_{in} [N/mm]	$J_{0,2}$ [N/mm]	J_m [N/mm]
A	331.8	446.4	360.2
B	82.3	265.8	279.2
C	51.7	82.0	80.3

3.3 Discussion

The radical reduction of fracture toughness of C group specimens proves how dangerous the occurrence of SC cracks in a structure can be. The cause of the reduction of fracture toughness is the diffusion of odd atoms from aggressive medium to the structure of material ahead of the crack. These atoms induce a high local stresses and lead to a reduced plastic behavior of material. The results obtained unambiguously demonstrate an increased threat of stress-corrosion cracks in structural components. It is shown that the fracture resistance of a component depends not only on the material of the component and on the crack tip constraint but also on the origin of the crack (fatigue, stress corrosion) and thus on the corresponding crack growth mechanism.

4. Conclusions

Results of two case studies proved very detrimental effects of corrosion in either corrosion fatigue or

IOP Conf. Series: Materials Science and Engineering **583** (2019) 012033 doi:10.1088/1757-899X/583/1/012033

stress corrosion cracking in connection with pipeline steel and structural steel. From three different corrosion media, namely crude oil, alkaline solution, and separated water, it was found that the most detrimental effect on fatigue properties of the pipeline steel L485MB was that of the separated water. In contrast, the effect of crude oil as well as that of alkaline solution appear to have no detrimental effect up to the fatigue life 2 x 10^6 cycles as compared to fatigue properties of the steel in air (reference S-N curve).

As for the other case study it was found that stress corrosion of CSN 411353 steel under constant load in acidic solution, according to NACE standard, with bubbling hydrogen sulphide through the solution reduced the J - based fracture toughness of the steel by a factor of six. The methodology of these specific tests proved itself as effective and functional.

Acknowledgement
The authors acknowledge kind support from the Technological Agency of the Czech Republic (Project No. TE02000162).

References

[1] *Online:* https://cs.wikipedia.org/wiki/koroze
[2] Marcus P 2017 *Corrosion Mechanisms in Theory and Practice - Third ed.* (CRC Press) p 941
[3] Gajdoš Ľ et al. 2004 *Structural Integrity of Pressure Pipelines* (Prague: Transgas) p 234
[4] Gaverick L 2015 *Corrosion in the Petrochemical Industry (*ed. ASM International) p 263
[5] Gajdoš Ľ, Šperl M and Siegl J 2011 Apparent fracture toughness of low-carbon steel CSN 411353 as related to stress corrosion cracks *Materials & Design* vol 32 pp 4348–4353
[6] Kytyr D et al. 2013 Evaluation of impact damage effect on fatigue life of carbon fibre composites *UPB Scientific Bulletin* Series D **75**(2) pp 157–164
[7] Šperl M 2008 *The Effect of Corrosion Damage on Operation Reliability of Gas Linepipes* Ph.D. Dissertation (Prague: Czech Technical University, Faculty of Transportation Sciences) p 116
[8] Hopkins P 2001 *Defect Assessment in Pipelines* (Prague, Czech Republic: APA Course)
[9] Garverick L 1994 *Corrosion in the Petrochemical Industry* Edited by Linda Garverick (ASM International)
[10] Uhlig H H and Revie R W 1985 *Corrosion and Corrosion Control* 3rd ed. (New York, USA: Wiley – Interscience) pp 415–441
[11] DIN 50 930 *Part 1-5: Corrosion of Metallic Materials Inside Pipelines, Vessels and Apparatuses Caused by Water* (in German)
[12] Abdurahman N H, Rosli Y M et al. 2012 Pipeline Transportation of Viscous Crudes as Concentrated Oil-in-Water Emulsions *Journal of Petroleum Science and Engineering* vol 90–91 pp 139–144
[13] Čechová E and Orioni R A 1994 A brief survey of useful information about hydrogen in metals In *Proc of the international symposium on cold fusion and advanced energy sources* (Minsk, Belarus)
[14] ANSI/ASME TM0177-96 *Laboratory Testing of Metals for Resistance to Sulfide Stress Cracking and Stress Corrosion Cracking in H_2S Environments*
[15] ASTM E 1820-01 *Standard Test Method for Measurement of Fracture Toughness*
[16] Rice J R 1968 A Path Independent Integral and the Approximate Analysis of Strain Concentrations by Notches and Cracks *Journal of Applied Mechanics* **35**

IOP Conf. Series: Materials Science and Engineering **583** (2019) 012034 doi:10.1088/1757-899X/583/1/012034

Behaviour of calcined gypsum-based materials during setting

M Dolezelova[1], **M Jerman**[1] **and A Vimmrova**[1]

[1] Department of Materials Engineering and Chemistry, Faculty of Civil Engineering, Czech Technical University in Prague, Thákurova 7, 166 29 Prague, Czech Republic

Email: vimmrova@fsv.cvut.cz

Abstract. The behaviour at early age of several composed gypsum-based materials was investigated. The binary binder, composed of calcium sulphate hemihydrate and slaked lime, several ternary binders with different types of pozzolans were prepared, and their behaviour during setting was studied. Unrestrained volume changes of calcium sulphate-based pasts were measured by displacement of liquid, in which the specimens were submerged. The hydration heat of composed binders was measured by isothermal calorimetry. The behaviour of composed materials with different types of pozzolans was compared with the behaviour of calcined gypsum.

1. Introduction

Even if gypsum is one of the most environmentally friendly materials, because its low energy consumption at production and possibility to produce it from secondary materials, its utilization is restricted on the interior use only. When pure gypsum is wet, its strength decreases significantly and it starts to dissolve. To solve this problem ternary gypsum-based binders were developed. Ternary binders usually consist of calcined gypsum (calcium sulphate hemihydrate, pozzolan and activator of pozzolanic reaction (usually hydrated lime or cement). It was found, that these materials have better resistance against the water, because the insoluble phases are formed in them [1, 2]. These materials are studied intensively nowadays, but being relatively new materials, there is a lot of information missing yet. While the mechanical properties of ternary materials were studied by several authors as well as their durability in wet environment [3, 4, 5, 6], the behaviour of ternary materials at the early age was not studied very often. Singh and Garg [7] measured the setting times of secondary binder, composed of anhydrite and blast-furnace slag and they found, that the setting time shortens with increasing amount of slag. Garg, Jain and Singh [8] tested the ternary material, made from 30% of hemihydrate, 40% of fly ash and 30% of hydrated lime and they state, that the setting time of ternary binder was two times longer than setting time of pure hemihydrate. Hydration heat of composed calcium sulphate-based binders was studied very rarely. Guan et al. [9] studied the heat evolution of binary binder, prepared from calcium aluminate cement and flue gas desulfurization gypsum and found that the amount of gypsum strongly influences the hydration behaviour of the material and with increasing amount of gypsum the hydration is faster. Magallanes-Rivera and Escalante-Garcia [10] compared the results of isothermal calorimetry of commercial and fluorogypsum binders with different pozzolans and stated that the reaction of fluorogypsum is slower and less heat is liberated, but the final strength of fluorogypsum composites is higher.

The volume changes of materials during setting are also important, because they can influence significantly the properties of material in solid state. Cement based materials shrink during setting, due to the autogenous shrinkage. Immediately after mixing water and cement, a chemical shrinkage change

IOP Conf. Series: Materials Science and Engineering **583** (2019) 012034 doi:10.1088/1757-899X/583/1/012034

will occur due to the reduction in volume of the reaction products. Chemical shrinkage values for individual cement phases are given in table 1 and the total value of chemical shrinkage $V_{CS,\,cem}$ of cement can be calculated according equation (1).

Table 1. Chemical shrinkage values for individual cement phases. [11]

Phase	Chemical shrinkage [cm³/g]
C_3S	0.0532
C_2S	0.0400
C_4AF	0.1113
C_3A	0.1785

$$V_{CS,\,cem} = 0.0532\,[C_3S] + 0.0400\,[C_2S] + 0.1113\,[C_4AF] + 0.1785\,[C_3A] \qquad (1)$$

The volume changes of gypsum during setting differ from the volume changes of cement. Calcium sulphate hemihydrate starts to shrink immediately after mixing gypsum with water and then it expands slightly. The shrinkage is caused by the chemical reaction between the water and hemihydrate, when the dihydrate is formed according equation (2). If equivalent volumes of the hemihydrate, water, and the reaction product (dihydrate) are compared, the difference $\Delta V_{cs,g}$ between the volume of the dihydrate formed and the equivalent combined volumes of the hemihydrate and water will be 7.11% (equation 3) [12].

$$(CaSO4)_2 \cdot H_2O + 3H_2O \rightarrow 2\ CaSO_4 \cdot 2\ H_2O \qquad (2)$$

Table 2. Volume changes at conversion of calcium sulphate hemihydrate to dihydrate. [12]

		$(CaSO4)_2 \cdot H_2O$	$3\ H_2O$	$2\ CaSO_4 \cdot 2\ H_2O$
Molecular mass	g	290.28	54.05	344.33
Density	g/cm³	2.75	1.00	2.32
Equivalent volume	cm³	105.56	54.21	148.41

$$\Delta Vcs, g = \frac{(105.56+54.21)-148.41}{(105.56+54.21)} * 100\% = 7.11\% \qquad (3)$$

The setting expansion of gypsum is caused by the crystallization. Crystals starts to grow and intermesh and make the solid structure of gypsum. The structure after setting is therefore composed of interlocking crystals between which are micropores and pores containing the excess water required for mixing (figure 1). On drying, the excess water is lost, resulting in a porous structure.

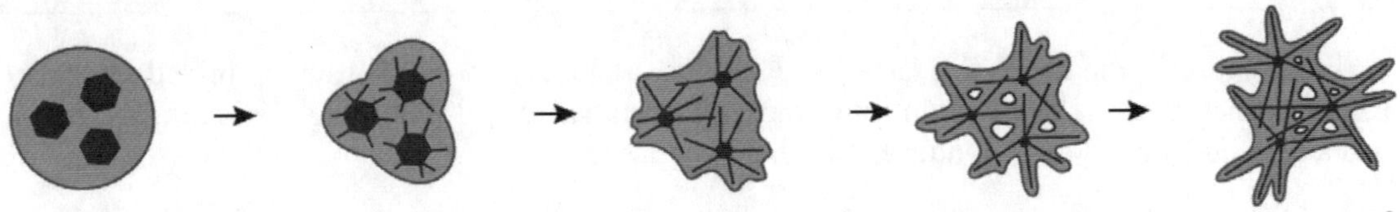

Figure1. Setting expansion of hardened gypsum due to crystallization.
a) Initial mix, b) crystal growth, c) close contact of crystals, d) expansion,
f) termination of crystal growth [12].

The volume changes of composed binders are therefore caused by several different processes. We found that behaviour of composed gypsum-based binders during setting is not described sufficiently yet, therefore we studied several composed gypsum-based binders at the early age. Volume changes during setting, hydration heat and setting times of three ternary calcium sulphate-based binders, one secondary binder and were measured and compared with behaviour of calcined gypsum.

IOP Conf. Series: Materials Science and Engineering **583** (2019) 012034 doi:10.1088/1757-899X/583/1/012034

2. Materials

Three ternary, one binary gypsum-based and one pure gypsum paste were prepared. The main component of all pastes was commercially obtained calcined gypsum (producer Rigips, s.r.o., Czech Republic). The mineral composition of calcined gypsum, determined by XRD is in table 3. As a pozzolan additives the crushed ceramic, silica fume (SF) and granulated blast-furnace slag (GGBS) were used. Crushed ceramic powder was obtained from regional producer of bricks (Heluz, Czech Republic) as a waste product from the ground brick production. The dust was sieved through the 250 μm sieve in order to exclude big particles. Silica fume and granulated blast slag were commercial products also (producers Stachema, Czech Republic and Kotouč Štramberk, Czech Republic). As an activator of pozzolanic reaction the slaked lime (producer Vápenka Čertovy Schody a.s., Czech Republic) was used. Mineral composition of slaked lime is in table 3. The amount of amorphous phase, determined by XRD analysis is given in table 4.

Table 3. Mineral composition of used calcined gypsum and slaked lime.

Mineral	Bassanite	Anhydrite	Gypsum	Calcite	Portlandite
	$CaSO_4 \cdot 1/2H_2O$	$CaSO_4$	$CaSO_4 \cdot 2H_2O$	$CaCO_3$	$Ca(OH)_2$
wt %	74.2	7.3	0.9	3.4	6.5
	-	-	-	93.6	6.4

2.1. *Composition of tested pastes*

Composition of ternary pastes was designed in such way, that the amorphous phase in each paste was the same and the amount of hydrated lime (as activator of pozzolanic reaction) was the same also. By that, the comparability of achieved results was ensured, even if the reaction of GGBS is slightly different than reaction of silica fume and crushed ceramic. The amount of amorphous phase was derived from the amount of amorphous phase in the ternary binder with 10% of metakaolin, which was found as an optimal in previous research [1]. The detailed description of used materials design is in [13]. Binary paste without pozzolan and pure gypsum paste were prepared as reference materials. The amount of water in the mixtures was determined by the flow test (flow value was established to be between 185 to 190 mm). Compositions of designed materials are in table 5.

Table 4. Amount of amorphous phase in pozzolans.

Pozzolan	Silica fume	Granulated blast slag	Ceramic powder
		% by mass	
Amorphous SiO_2	90	87.7	45.3

All pastes were prepared in the same way. All dry components were thoroughly mixed together (by hand) and then they were added into the measured amount of water during 15 seconds. The paste wasthan mixed by hand for 1 minute and immediately used.

IOP Conf. Series: Materials Science and Engineering **583** (2019) 012034 doi:10.1088/1757-899X/583/1/012034

Table 5. Compositions of tested pastes.

	Pozzolan		Gypsum	Slaked lime	Water/binder ratio
			[% by mass]		
SF	Silica fume	8.5	81.5	10	0.8
CC	Crushed ceramic	16.8	73.2	10	0.7
BS	GBS	8.7	81.3	10	0.75
GL	-	-	90	10	0.73
G	-	-	100	-	0.67

3. Measuring methods

3.1. Isothermal calorimetry
Hydration heat was measured by the isothermal heat flow calorimeter KC 01 [14]. The measurement was performed with a 1 g sample of dry mixture in a cylindrical copper vessel of inner diameter 8.25 mm and height 70 mm, with a tight rubber plug. After inserting the copper vessel with the solid sample into the stabilized calorimeter, a weighted syringe with water has been placed near the reaction vessel, and so was a plastic tubule intended for paste mixing. The calorimeter with the solid sample and liquid components was left to stabilize the temperature (25°C). After about 1 h of stabilization, the calorimeter was opened and water was inserted into the vessel. The paste was mixed for 30 s by rotation of the plastic tubule mixer and then the vessel was closed by a rubber plug.

3.2. Setting times
Setting times of all materials were tested by Vicat apparatus with needle according the EN 13454-2 [17]. The needle was indented into material with the interval 1 min.

3.3. Volume changes
Unrestrained volume changes of gypsum-based pastes and mortars were measured by displacement of liquid, in which the specimens were submerged. The fresh paste was poured into the thin rubber membrane, which was tightly tied and submerged into the vessel, filed by the silicone oil. The vessel was closed by specially adapted lid with tightly embedded glass capillary and plastic overflow pipe (figure 2). The system was filled by oil to the required level and sealed. The level of liquid was recorded by time lapse camera for one hour with the first snap at 5 min after mixing. The obtained images were manually evaluated and the level of liquid was determined for each picture (figure 3). The tested material was dried and its dry mass was recorded.

Sometimes water is used in this experiment, but some authors state that the rubber membrane is not absolutely waterproof [15] and water can penetrate into the sample during experiment. This is generally not a problem for gypsum-based materials, because water uptake starts after relatively long period of submersion (days) [16] and the hydration processes in gypsum are significantly faster (usually less than two hours). We used silicone oil in order to avoid the volume changes of water with the temperature rise, because the setting reaction of gypsum is exothermic and large amount of heat is generated.

Each material was measured three times at least. Measured values of volume changes were related to the mass of the dried sample and mean values were calculated.

IOP Conf. Series: Materials Science and Engineering **583** (2019) 012034 doi:10.1088/1757-899X/583/1/012034

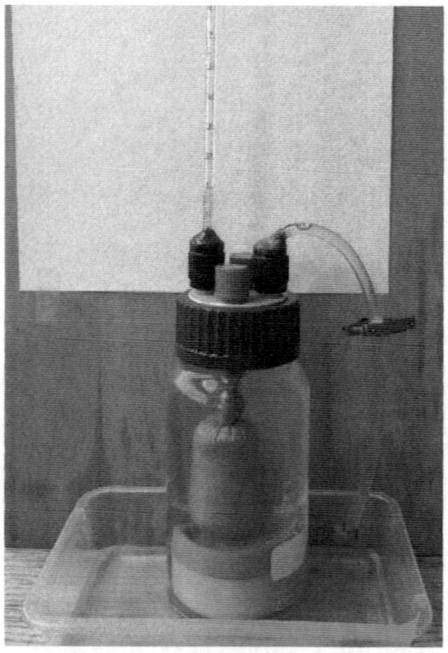

Figure 2. Arrangement of test.

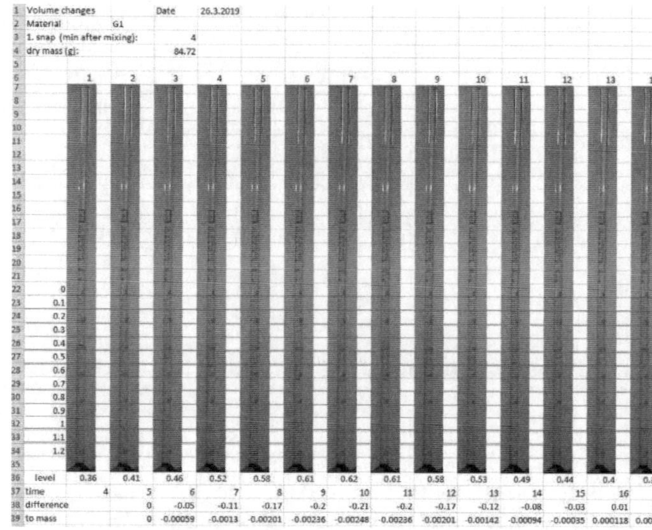

Figure 3. Part of typical data sheet for evaluation.

4. Results and discussion

4.1. Heat evolution

The rate of heat evolution during the first two hours after mixing is presented in figure 4. The shape of all curves (with the exception of calcined gypsum paste) is similar to each other. There are two significant peaks. According to Fischer [18] the first peak, developed immediately after addition of water, is due to the wetting heat together with the heat of conversion of -anhydrite III to hemihydrate. The second peak, which starts with a short delay after the first peak, represents hydration of hemihydrate to dihydrate. Similar shape of hydration curves was observed by Tydlitat et al. [14] and Singh and Middendorf [19]. In the calcined gypsum paste the first peak is missing, there can be seen only slight decrease of the line, which is in agreement with the results, obtained by Lewry [20]. This is caused by the fact, that the conversion of anhydrite II to hemihydrate does not start yet, because no activator was added and change in the heat rate evolution was caused by the wetting heat only. The second peak of pure gypsum is the earliest (20 minutes), which corresponds with the fastest setting times of calcined gypsum paste.

For the composed materials the first peak occurs between 2 to 3.5 minutes and the second peaks are achieved at about 30 minutes. After 2 hours the heat evolution is minimal and the main hydration reactions of calcium sulphate are finished. The hydration of pozzolans does not play role at this period, because it is significantly slower (days). Nevertheless, the pozzolans and lime serves as activators of anhydrite hydration and therefore the first peak in the composed binders is more pronounced than peak in the gypsum paste. The activation by slaked lime is less intense than activation by pozzolans and therefore both peaks occur later and the first peak is lower.

The amount of cumulative heat, evolved during first two hours of hydration is in direct proportion with the amount of gypsum in the mixture, as can be seen in figure 4. It confirms the fact that only reaction of gypsum occurs at the first two hours.

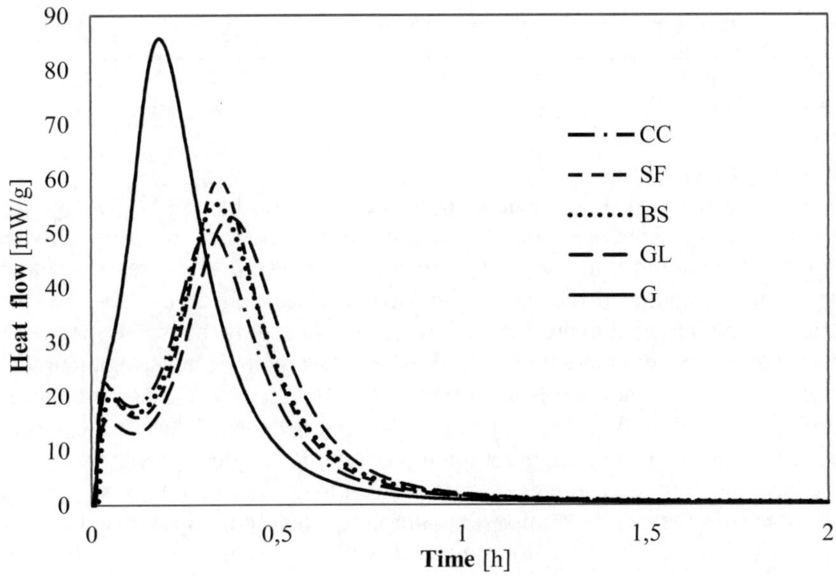

Figure 4. Heat flow evolution of tested materials.

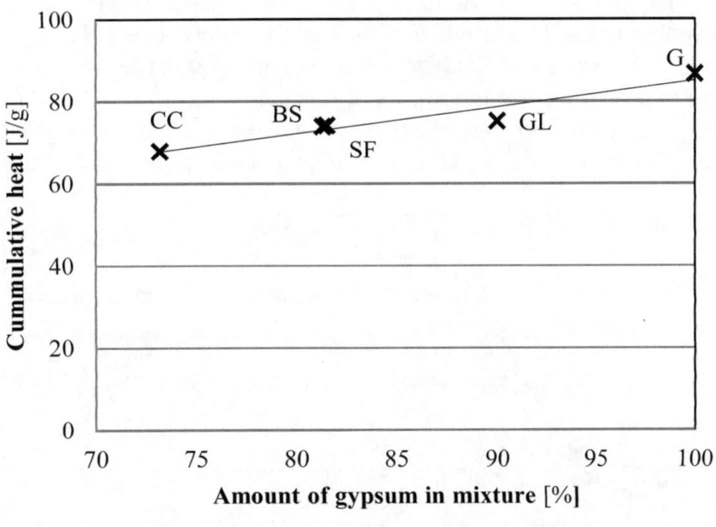

Figure 5. Relationship between the cumulative heat and amount of gypsum in material.

4.2. Setting times

The setting times are given in table 6. Generally, setting times of all materials are short. Setting of pure gypsum is fastest, as expected. The initial setting time increases with the decreasing amount of gypsum, but it does not exceed 15 minutes in all materials. The total setting time ranges from 3 minutes to 4.5 minutes and is longer in the materials with lower amount of gypsum.

The setting times are shorter than expected according the calorimetric results. It can be caused by the definition of setting times obtained by standard Vicat test, which is originally designed for cement-based materials and mechanism of gypsum setting is different from setting and hardening of the cement.

Table 6. Setting times.

Material		CC	SF	BS	GL	G
initial setting time	[min]	14	14.5	12.5	11.5	8
final setting time	[min]	18.5	17.5	16	15.5	11

4.3. Unrestrained volume changes

Volume changes of all tested materials can be seen in figure 6. The character of changes of all materials is similar. Firstly, the volume decreases, which is caused by the chemical reaction between the gypsum and water, as explained in the Introduction [12]. Then the volume starts to increase due to crystallization of gypsum. The local minimum corresponds with the initial setting times. The maximal values are achieved at the time of the maximal hydration heat evolution, found by the isothermal calorimetry. After achieving the maximal expansion, the volume of all material starts to decrease again and for all material the decrease was not finished at the end of measuring (e.g. after one hour). The final decrease of volume could be assigned to the continuing crystallization. Some experiments had to be terminated earlier, because the level of the liquid decreased under the lower end of the glass capillary.

The volume changes of pure gypsum are significantly lower than volume changes of all composed materials and occur earlier then in the composed materials, which corresponds with the shorter setting times and also with the peaks on the calorimetric curve. The biggest volume changes can be seen in the binary paste and in the paste with the silica fume, e.g. in materials with highest amount of gypsum and the smallest volume changes occur in the paste with lesser amount of gypsum. The results do not correspond with the fact, that changes of pure gypsum are smallest, but the mechanism of setting in the pure gypsum and composed materials is different. The presence of slower reacting components (either lime or pozzolans) affects the crystallization of gypsum and causes less compact structure.

There is necessary to point out, that the results of repeated measuring of the same material differ in relatively large range. The measuring was apparently affected by the amount of air in the sample and in the vessel itself, even if we tried to drive out all the air bubbles from the oil during the filling period.

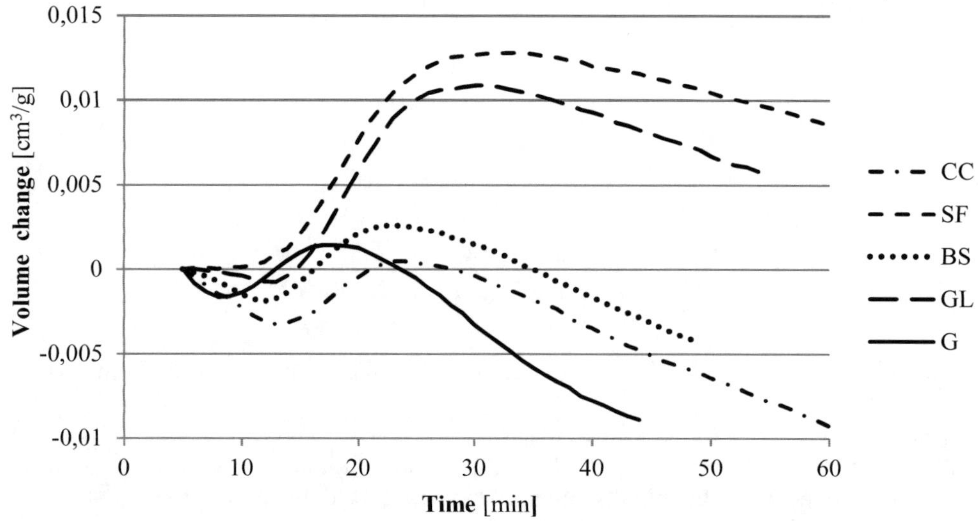

Figure 6. Volume changes during first hour.

5. Conclusions

The behaviour at early age of several composed gypsum-based material was studied by the means of isothermal calorimetry, measuring of setting times and volume changes during setting. The results were compared with the behaviour of pure gypsum paste. According the isothermal calorimetry, the most of the heat is developed during the first two hours after mixing. The heat evolution is fastest in pure gypsum and also the cumulative heat, evolved at two first two hours is biggest. The hydration of composed materials is slower and the cumulative heat corresponds with the amount of gypsum in each of them. It could be concluded, that only reaction of calcium sulphate phases occurs at the first two hours. The lime and pozzolan components do not react during that period, but they serve as activators of anhydrite conversion into hemihydrate. They also act as retarders of setting. Their presence in the materials causes the bigger volume changes during setting in comparison with volume changes of pure gypsum, because they interfere with the formation of more dense crystal structure.

Acknowledgements

This research was supported by the Czech Science Foundation, Project No. 19-08605S.

References

[1] Vimmrová A, Keppert M, Michalko O and Černý R 2014 *Cem. Concr. Compos.* **52** pp 91–6
[2] Martinez-Aguilar O, Castro-Borges P and Escalante-García J 2010 *Constr. Build. Mater.* **24** pp 631–9
[3] Fraire-Luna P, Escalante-Garcia J and Gorokhovsky A 2006 *Cem. Concr. Res.* **36** pp 1048–55
[4] Kumar S 2000 *Mater Struct* **33** pp 59–64
[5] Magallanes-Rivera R, Juarez-Alvarado C, Valdez P and Mendoza-Rangel J 2012 *Constr. Build. Mater.* **37** pp 591–6
[6] Colak A 2002 *Cem. Concr. Res.* **32** pp 109–15
[7] Singh M and Garg M 1995 *Cem. Concr. Res.* **25** pp 332–8
[8] Garg M, Jain N and Singh M 2009 *Constr. Build. Mater* **23** pp 3138–43
[9] Guan B, Lou W, Ye Q, Fu H and Wu Z 2009 *J. Therm. Anal. Calorim.* **98** pp 737–42
[10] Magallanes-Rivera R and Escalante-García J 2014 *Constr. Build. Mater.* **65** pp 20–8
[11] Paulini P 1992 *Proc. of the 9th Int. Congress on the Chemistry of Cement (New Delhi)* vol IV (New Delhi: National Council for Cement and Building Materials) pp 248–54
[12] Anusavice K, Shen Ch and Rawls HR 2016 *Phillips' Science of Dental Materials* (Singapore: Elsevier Singapore Pte Limited) p 571
[13] Doleželová M and Vimmrová A 2017 *IOP Conference Series: Materials Science and Engineering* **251** 012037
[14] Tydlitát V, Tesárek P and Černý R *J. Therm. Anal. Calorim.* **91** pp 791–796
[15] Hammer T A, Bjøntegaard Ø and Sellevold E J 2002 *Early age cracking in cementitious systems* (Cachan: RILEM) ed Bentur A pp 234–45
[16] Lura P and Jensen O 2006 *Mater. Struct.* **40** pp 431–40
[17] EN 13454-2 2003 *Binders, composite binders and factory made mixtures or floor screeds based on calcium sulfate. Test methods.*
[18] Fischer K W 1963 *Silikattechnik* **14** pp 26–33
[19] Singh N B and Middendorf B 2007 *Prog Cryst. Growth C.H.* **53** pp 57–77
[20] Lewry A J 1994 *J. Mater. Sci* **29** pp 5279–84

The assessment and prediction of concrete sustainability

T Vymazal[1], **B Teply**[1] **and K Hrabova**[1, 2]

[1] Faculty of Civil Engineering, Brno University of Technology, Brno, Czech Republic
[2] Institute of Forensic Engineering, Brno University of Technology, Brno, Czech Republic

Email: tomas.vymazal@vutbr.cz

Abstract. The sustainability of concrete and concrete structures is seeing increased attention; among other things, this is because 5% of global CO_2 emissions are produced by cement manufacturing. Responsible sustainability management requires effective tools for the quantification, i.e. measurement, or comparison of material, technological, and structural options. This is why there are a number of initiatives and approaches that aim to reduce global greenhouse gas emissions. Their project design uses techniques that assess whole structures and their usage; however, most of their input data is based on estimates. This paper presents a validation of a method of designing, assessing, and predicting sustainability for one material (concrete) with regard to its composition, relevant properties, and predicted value dispersion utilizing probabilistic approach.

1. Introduction

Sustainability is becoming one of the basic concepts in the design and operation of concrete structures throughout their life cycle, taking into account these three pillars: social, environmental and economic. In the construction sector sustainability is a rather complex issue, and all of these components must be considered from the very earliest stages of the existence of the structure in question – in other words, Life Cycle Assessment (LCA) procedures are being adopted. LCA is a method that evaluates the life cycle of a structure from the point of view of its impact on the environment. Consideration is generally given to energy and material costs, as well as to the environmental and social impact throughout the life cycle. In addition, the international document fib Mode Code 2010 [1] discusses a performance-based design of concrete structures, listing three basic performance categories: usability (i.e. also service life), safety, and sustainability. Unfortunately, the any link between these quantities is not discussed.

In June 2016, the fib Technical Council approved the start of activities on The MC2020 project under the auspices of COM10: Model Codes, which initiated Task Group 10.1: Model Code 2020 to undertake the preparation of a single general code fully integrating the provisions for the design of new concrete structures with matters relating to the through-life management and care of existing concrete structures. MC2020 will take sustainability as a fundamental requirement, based upon a holistic treatment of societal needs and impacts, life-cycle cost, and environmental impacts [2].

2. Sustainability coefficients

This multi-criteria evaluation of buildings or large structures and the assessment of their sustainability are, as a rule, accompanied by many uncertainties which can devalue the results and their applicability. It might thus prove useful to focus attention on a simpler task first – the evaluation and comparison of the sustainability of a material. To do this, the paper focuses on the quantification of sustainability related to the use of various types of concrete with regard to their durability, i.e. the resistance to degradation. Sustainability coefficients k_{SB} are determined using information about material performance, service life and eco-costs, as has been recently published [3], based on Building Material Sustainability Potential [4]. The aim of this paper is to show the possibility to approach future requirements and to simplify decision making as to the design and selection of concrete mixes in broader terms; i.e. not only with regard to load-bearing capacity or durability as is currently common. The procedure is further improved by using the probabilistic approach, i.e. by considering the input quantities as random quantities with a known probability distribution, with the output being values of statistical parameters k_{SB}, or the probability distribution of these quantities which can be also used when forming a sustainability limit state condition (as shown in [3]).

Sustainability can be quantified using normalized sustainability coefficients k_{SB} according to equation (1), where quantities L (service life), R (performance) and E (eco-cost) are divided by reference values L_{ref}, R_{ref} and E_{ref} (arbitral values hold always constant for the study in question), meaning that k_{SB} is then a dimensionless quantity whose value usually approximates 1.0.

$$k_{\mathrm{SB}} = \frac{\frac{L}{L_{ref}} \cdot \frac{R}{R_{ref}}}{\frac{E}{E_{ref}}} \tag{1}$$

This formula can be effectively utilized for to compare sustainability coefficient values between members of a group of various concretes. All cases in a given group must always be considered to be situated at the same (or similar) location and to suffer the same type of degradation and/or loading. When evaluating sustainability, a suitable type of performance is considered and service life is determined with regard to the given/chosen type of degradation/loading.

Sustainability coefficients k_{SB} can also be analysed in a probabilistic mode simply by considering quantities L, R and E to be random quantities and consequently the coefficient k_{SB} to be random as well. This might become more useful in the near future as sustainability-related issues become more frequent – see [2]. Moreover, a similar approach has been recently described in [5] with focus on a more complex task (unfortunately not fully applicable in engineering practice). Such an approach needs a definition of relevant sustainability limit state conditions; this is part of ongoing research. The present paper concentrates on statistical characteristics of k_{SB}, and presents an illustrative example, which is processed by the FReET-D software [6], supplemented by a function for k_{SB} analysis (FReET-SB option).

3. Example of the sustainability evaluation of various kinds of concrete

The Feasible Reliability Engineering Tool for Degradation effects assessment in reinforced and pre-stressed concrete structures (FReET-D) is used for probabilistic modelling of environmental degradation and can serve for design or performance-based specification. The FReET-D software is capable of predicting service life, performing statistical analysis (statistical parameters of degradation measures), sensitivity analysis (sensitivity of output on different inputs) and reliability analysis (assessment of reliability measures corresponding to target service life and different limit state functions) [6].

The composition of the mixture used in the three different concretes can be seen in table 1 (A – reference mixture, Portland cement only; B – with fly ash used as a substitute for a portion of Portland cement; C – with fine-ground slag). The strength of concrete is considered to indicate performance R in this example, and was gained by sample testing within the grant project GA ČR 103/07/0034 at Faculty of Civil Engineering, Brno University of Technology.

Table 2 shows the environmental impact values transformed into financial units. The concretes in question are supposed to suffer by carbonation, so the service life L is calculated as of the initiation period stage of reinforcement corrosion caused by concrete carbonation using an analytical model. The FReET-D tool with model Carb4b [7] are used. Basic input values were: concrete cover 30 mm; atmospheric CO_2 concentration 820 [mg/m^3]; RH = 70%; the k–value concept for concrete B k = 0.4, while k = 0.6 for concrete C [8] in addition to data listed in table 1. Statistical parameters are shown in table 3.

The decisive values to be used in equation (1) and the resulting sustainability coefficient values are listed in table 3 including their statistical parameters. Note also that the values relevant to concrete A were used as reference, i.e. $L_{ref} = L_A$, $R_{ref} = R_A$ and $E_{ref} = E_A$.

Table 1. Composition of mixtures.

Components [kg/m^3]	Concrete A	Concrete B	Concrete C
CEM I 42.5 R	389	290	290
Aggregate 0/4 mm	812	812	812
Aggregate 8/16 mm	910	910	910
Fine-ground blast furnace slag (3800 m^2/kg)	-	-	194
Fly ash from Dětmarovice	-	194	-
Water	161	182	160

Table 2. Eco-costs.

Components	Concrete A	Concrete B	Concrete C	Ref.
CEM I 42.5 R (0,109 € /t) [€]	42.4	31.6	31.6	[9]
Slag (0.069 € /t) [€]	-	-	13.4	[10]
Fly ash (0.060 € /t) [€]	-	11.6	-	[10]
Aggregate (0.007 € /t) [€]	12.1	12.1	12.1	[10]
Total eco-costs in € / m^3 of concrete	54.5	54.3	57.1	

Table 3. Concrete properties and the final values of coefficients k_{SB}.

	Concrete A			Concrete B			Concrete C		
Property	Mean	COV	Pdf	Mean	COV	Pdf	Mean	COV	Pdf
60-day cube strength [MPa]	71.9	0.06	Normal	47.2	0.06	Normal	70.5	0.06	Normal
Service life [years]	125	0.17	Normal	73	0.16	Normal	151	0.26	Normal
Eco-costs [€]	54.5	0.20	Rectangular	54.3	0.20	Rectangular	57.1	0.20	Rectangular
k_{SB}	1.04	0.28	Beta	0.40	0.27	Beta	1.18	0.33	Gamma

The resulting k_{SB} values (see table 3) show that the best choice in terms of sustainability is concrete C in spite of having higher eco-costs than variants A and B – this is clearly influenced by the fact that concrete C has the longest service life due to the content of granulated slag (considering concrete carbonation). Evidently, when the effect of other degradation types and/or the effect of mechanical load on the service life [11] are taken into account, the order of sustainability coefficient values can change. This factor is at the focus of the authors' ongoing research.

4. Conclusion

This paper can be viewed as a continuation of previous articles written by the authors, where they made attempts to utilize the sustainability coefficient to describe concrete sustainability in a deterministic approach. A full probabilistic approach is described briefly for sustainability indicator analysis together with the probabilistic modelling of degradation phenomena (due to environmental processes and used to predict the service life). To illustrate such approach, an example of three different concrete mixtures (with different additions) was studied and implications for sustainability were drawn. The presented methodology can be used for a wide range of applications, considering various degradation effects or their combinations; the authors will continue presenting the results of this as part of ongoing research. The authors believe this might become more useful in the near future as issues of sustainability enter into the centre of attention due to the current international progress being made on the new fib Model Code 2020.

Acknowledgments

This work has been supported by project No. 17-14302S “Experimental analysis of the early-age volume changes in cement-based composites”, supported by the Czech Science Foundation and project No. LO1408 “AdMaS UP – Advanced Materials, Structures and Technologies”, supported by Ministry of Education, Czech Republic.

References

[1] fib bulletin No. 65 and 66 fib Draft Model Code 2010 (2012) *International Federation for Structural Concrete (fib)* (Lausanne Switzerland)

[2] Matthews S 2017 fib Model Code 2020 - A new development in structural codes *Structural Concrete Journal of the fib* vol 18 issue 5 pp 651–652

[3] Teplý B, Rovnaníková P and Vymazal T 2018 Sustainability Quantification of Concrete Structures *Chapter in Advances in Environmental Research* vol 63 ed. Justin A. Daniels (NY, USA: NOVA Science Publishers) pp 231–248

[4] Müller H S 2013 Sustainable structural concrete – from today´s approach to future challenge *Structural Concrete* **14** (4) pp 299–300

[5] Lepech M D, Geiker M, Michel A and Stang H 2016 Probabilistic design and management of sustainable concrete infrastructure using multi-physics service life models *1st Internal Confer on Grand Chellenges in Construction Materials IGCMAT*

[6] Novák D, Vořechovský M and Teplý B 2014 FReET – Software for the statistical and reliability analysis of engineering problems and FReET-D: Degradation Module *Advances in Engineering Software* pp 179–192

[7] Papadakis V G and Tsimas S 2002 Supplementary cementing materials in concrete Part I: Efficiency and design *Cement and Concrete Research* **32** (10) pp 1525–1532

[8] EN 206 2013 *Concrete Specification, performance, production and conformity* (Brussels: European Committee for Standardization - CEN)

[9] Eco Costs Data Design-4-Sustainability: *Inspiration and knowledge by designers for designers* [cit. 2019-01-25] (Available in: http://www.design-4-sustainability.com/ecocosts)

[10] Data 2016: Emise CO2 *Svaz výrobců cementu ČR* [cit. 2019-01-25] (Available in: https://www.svcement.cz/data/data-2016)

[11] Vořechovská D, Šomodíková M, Podroužek J, Lehký D and Teplý B 2017 Concrete structures under combined mechanical and environmental actions: Modelling of durability and reliability *Computers and Concrete* **20** (1) pp 99–110